ATOMIC SPECTROSCOPY: Introduction to the Theory of Hyperfine Structure

ATOMIC SPECTROSCOPY: Introduction to the Theory of Hyperfine Structure

ANATOLI ANDREEV
M.V. Lomonosov Moscow State University
Moscow, Russia

Anatoli V. Andreev
M.V. Lomonosov Moscow State University

Atomic Spectroscopy: Introduction to the Theory of Hyperfine Structure

Consulting Editor: D. R. Vij

ISBN 10 0-387-25573-7 e-ISBN 0-387-28469-9 Printed on acid-free paper.
ISBN 13 9780387255736

Printed in the United States of America.

9 8 7 6 5 4 3 2 1 SPIN 11054344

springeronline.com

Contents

Part II Theory of Lamb Shift

Preface

There are a lot of excellent books on atomic spectroscopy today, but, hopefully, the distinctive feature of this book is its generality. We are not involved in the discussion of some specific mechanisms of formation of complex structure of atomic spectra, we are not trying to give an overview of different methods and models that are used to describe the spectra and to get a reasonable coincidence of calculated and measured data. We have tried to discuss comprehensively the general approach to the theory of atomic spectra, based on the use of the Lagrangian canonical formalism. The Lagrangian formalism enables us to easily generalize any Hamiltonian for electron motion in the external field to the Hamiltonian of many-electron problem, as a result the specific and common features of these two problems become more evident. The non-relativistic or relativistic, spin or spinless particle approximations can be used as a starting point in the general approach. All these approximations are analyzed and compared. This generality is helpful to keep the important points from technicalities of specific theories. The specific examples, that are used to illustrate the general approach, are chosen from contemporary atomic spectroscopy and light-matter interaction physics (trapped atom, mesoatom, high-precision measurements of electron anomalous magnetic moment and hydrogenic spectra, electric polarization vector of nucleons, etc.).

The book consists of two main parts. The first part deals with the hyperfine structure associated with the finite mass of nucleus, its orbital motion, and spin-spin interaction. The second part of the book deals mainly with the Lamb shift. The specific feature is that the theory of Lamb shift is based on the use of quantum mechanics. The obtained equation for hydrogenic spectrum has a very simple and compact form, as a result the physics of Lamb shift formation can be easily interpreted.

Notice, that usually the students of atomic spectroscopy theory are not deeply familiar with the methods of quantum electrodynamics theory, which is traditionally used to explain the physics of Lamb shift. Therefore the proposed approach makes the theory accessible for a wide range of specialists and students, who are familiar with the quantum mechanics and classical electrodynamics.

The basic equations and principles of quantum mechanics are briefly discussed in the book, therefore it can be used as a self-consistent textbook providing enough material for half-year or one-year course for graduate students: "Introduction into atomic spectroscopy", "Hyperfine structure of atomic spectra", etc.

Acknowledgments

I gratefully acknowledge the support of Physics Department, M.V. Lomonosov Moscow State University, in whose stimulating environment I have been working for more than twenty five years. My personal thanks to my colleagues from International Laser Center, M.V. Lomonosov Moscow State University, the interaction with them for many years and numerous discussions are a great source of inspiration in my researches. The main ideas of this book have been discussed in a number of conferences, symposiums, and scientific seminars. I am grateful to those who asked me the tricky questions. Especially, I would like to thank the participants of the scientific seminars headed by Prof. E.B. Aleksandrov (A.I. Ioffe Physical Technical Institute, St.Petersburg), Prof. L.V. Keldysh (P.N.Lebedev Physical Insitute, Moscow), Prof. V.A. Makarov (Physics Department, M.V. Lomonosov Moscow State University), Prof. M.O. Scully (TAMU, College Station, Texas). I would like to thank Prof. Olga Kocharovskaya and Prof. Vitali Kocharovskiy for hospitality during my stay at Texas A&M University, where the significant part of the book was prepared. Thanks are also due to Ilya Shutov and Eugeny Morozov, who helped me to prepare LATEX files until I learned how to do it.

Finally, very special thanks to my wife Mary and to my daughters Olga and Tatiana, who contributed with their support and patience during this long work.

The most part of the original results, which, I think, are the very important for the content of this book, was obtained during my work on the projects supported by International Science and Technology Center, Russian Foundation for Basic Research, and program "Universities of Russia".

Chapter 1

INTRODUCTION

The great number of brilliant experiments, that enable to enhance significantly the precision of the optical spectrum measurements, has been made in the last few decades. The information obtained from the spectra processing reduces significantly the uncertainties of the material constants, characterizing the material properties of the elementary particles like a charge, mass, magnitude of magnetic moment, etc. Simultaneously the tremendous successes have been achieved in development of non-optical methods of material constant measurements. The results of the precision measurements provide the powerful stimulus for researchers to verify the correctness of our description of particle interactions with electromagnetic field. Indeed, the obtained information enables to reduce significantly the uncertainty of the fundamental constants, that are not only of interest for some specific fields of research, but play the role of measure of correctness and over-all consistency of the basic theories. The speed of light c determines the ratio between the space and temporal scales. The Planck constant $\hbar$ determines the relationships between the components of the coordinate and momentum four-vectors. The elementary charge e is also the fundamental constant, because, in contrast to the other material constants, it has the same value for all elementary particles at least with the state-of-the-art accuracy. The combination of these three fundamental constants produces the fine structure constant $\alpha = e^2/(\hbar c)$, which plays the important role in the modern theory of atomic spectra.

The achieved progress in the precision measurements of atomic spectra stimulates the interest to the fundamentals of the quantum mechanic theory. Indeed it is well known that the quantum mechanics itself is originated from the problem of the explanation of the nature of spectral lines. The quantization rules proposed by Niels Bohr in 1913 and later generalized by Arnold Sommerfeld have worked well in explanation of atomic spectra. The decisive role in the formation of the particle wave mechanics plays the research of Louis de Broglie [1]. In the famous paper of Erwin Schrödinger [2] the mathematical basis of the quantum mechanics was grounded. The application of Schrödinger equation to the problem on electron motion in the Coulomb field provided the first quantum mechanical model for the hydrogen atom. The obtained formula for the hydrogenic spectra was in good agreement with the experimentally measured spectral lines of hydrogen atom and alkali atoms (the Lyman, Balmer, etc. series). The presence of the doublet lines in atomic spectra and splitting of atomic energy levels by the external magnetic field gave birth to the idea on the intrinsic angular momentum of electron. The magnitude of Zeeman splitting allowed then to estimate the magnitude of the electron magnetic moment. The apparatus of the matrix quantum mechanics for description of the intrinsic angular momentum was developed by Wolfgang Pauli [3]. The revolutionary step towards the development of the theory, giving the detailed description of the atomic spectra, was made by Paul Dirac [4] who proposed the quantum mechanical equation describing the intrinsic angular momentum of electron and its magnetic moment. The magnitude of the electron magnetic moment predicted by Dirac equation $\mu_{\mathrm{B}} = e\hbar/(2m_e c)$ was in good agreement with the experimental data. Despite its long history the theory of the hydrogenic spectra is still under development. The successes of this theory and its present-day state are discussed in the textbooks and monographes [5–8] and comprehensive review papers [9–12].

Let us mention briefly some last achievements in the spectroscopy of elementary particles and atoms.

1.1 Experiments with single particle in Penning trap

The most accurate measurements of the magnitude of electron magnetic moment were made in experiments with the single electron placed in the Penning trap at ultrahigh vacuum conditions and temperature of 4 °K [15, 16]. The trap is formed by the uniform magnetic field and weak quadrupole electric field. The electron evolves into the circular quantized motion in the plane perpendicular to the magnetic field. The quadrupole electric field forms the potential well confining the electron motion along the magnetic field direction. The configuration of the Penning trap enables to calculate the energy spectrum of the electron translational motion. The energy-level diagram includes the transversal cyclotron motion levels and longitudinal motion sublevels. The energy distance between the longitudinal sublevels is much smaller than the distance between the transversal levels. The electron cooling technique is used to shrink the radius of the orbital motion. In result the total motion occupies the very small spatial volume, where the profile of electric field is most closely coincided with the ideal model of harmonic potential well and the magnetic field is most uniform. In such conditions the electron is very weakly coupled with its environment and the electron lifetime in the trap is about ten months. Thus, following by H.G. Dehmelt, such a system may be called a "geonium atom". In addition to the translational degrees of freedom the electron possesses the spin. The spin precession around the magnetic field results in the appearance of the spin precession frequency in the spectrum of geonium atom. The accurate measurements of spin precession frequency enables to determine precisely the magnitude of the electron magnetic moment.

The Hamiltonian of electron in the Penning trap is

$$H = \frac{1}{2m_e}\left(\mathbf{p} - \frac{e}{c}\mathbf{A}\right)^2 - \mu\boldsymbol{\sigma}\mathbf{B}_0 + U(\rho, z), \tag{1.1}$$

where $\mu\boldsymbol{\sigma}$ is the electron magnetic moment, $\mathbf{B}_0$ is the magnetic field of the trap, and $U(\rho, z)$ is the potential well due to the quadrupole electric field. The vector potential of the uniform magnetic field is $\mathbf{A} = [\mathbf{B}_0\mathbf{r}]/2$, and the Hamiltonian (1.1) becomes

$$H = \frac{\mathbf{p}^2}{2m_e} + U(\rho, z) + \frac{e^2 B_0^2}{8m_e c^2}\rho^2 + (\mu_\mathrm{B} l_z + |\mu|\,\sigma_z)\, B_0, \tag{1.2}$$

where μ_B is the Bohr magneton, which is the magnitude of the electron magnetic moment in the Dirac theory,

$$\mu_\mathrm{B} = \frac{|e|\,\hbar}{2m_e c}. \tag{1.3}$$

As far as the potential well of the trap is axially symmetric then the projections of the orbital momentum l_z and spin $s_z = \sigma_z/2$ are the integrals of motion. Thus the eigenfunctions of the Hamiltonian (1.2) are simultaneously the eigenfunctions of the operators l_z and σ_z. Hence, if the electron magnetic moment coincides with the Bohr magneton, then the energy eigenvalues depend only on the sum $m + \sigma$, where m is the eigenvalue of the angular momentum projection operator l_z, and $\sigma = \pm 1$ is eigenvalue of the operator σ_z. We can see that the energy eigenvalues of the states characterized by the quantum numbers $(m = m_1,\ \sigma = +1)$ and $(m = m_2,\ \sigma = -1)$ will coincide in the case when $m_1 + 1 = m_2 - 1$. If the magnitude of the electron magnetic moment differs from $\mu_{\rm B}$, then the energy eigenvalues of the states $(m_1,\ \sigma=+1)$ and $(m_2=m_1+2,\ \sigma=-1)$ will be different. The energy difference is

$$\Delta E = 2\left(|\mu| - \mu_{\rm B}\right) B_0. \tag{1.4}$$

The measurements the energy difference (1.4) enable to determine the magnitude of the electron magnetic moment.

The values reported by Van Dyck et.al. [17] for electron μ_e and positron μ_p magnetic moments are

$$|\mu_e|/\mu_{\rm B} = 1.0011596521884(43),$$

$$|\mu_p|/\mu_{\rm B} = 1.0011596521879(43).$$

To reduce the uncertainties due to environment the special trap was constructed by Van Dyck et.al. [18]. These authors give the mean value of the 14 runs for the electron magnetic moment [18]

$$|\mu_e|/\mu_{\rm B} = 1.0011596521855(40). \tag{1.5}$$

By assuming that the CPT invariance holds for the electron-positron system the weighted mean of the data for both the electron and positron was proposed by Mohr and Taylor [9] as single experimental value

$$|\mu_{e,p}|/\mu_{\rm B} = 1.0011596521883(42).$$

A geonium atom can be also formed with the proton. The comparison of the cyclotron frequency of proton and electron enables to measure accurately the ratio of proton M_p and electron m_e masses. The value of this ratio reported by Van Dyck et al. [19] is

$$M_p/m_e = 1836.1526670(39). \tag{1.6}$$

By placing the fully ionized carbon $^{12}C^{6+}$ in the Penning trap Farnham et al. [20] have measured the ratio the ratio of carbon to electron mass

$$\frac{6m_e}{M\left(^{12}C^{6+}\right)} = 0.00027436518589(58).$$

1.2 Spectroscopy of hydrogenlike atoms

Recently there has been a drastic increase in the accuracy of measurements of transition frequencies in hydrogen and hydrogenlike ions. This progress is due to the development of the new spectroscopic methods. The interferometric methods were superseded by the absolute frequency measurement methods. The frequency of $1S-2S$ transition in hydrogen was measured with the relative uncertainty of $2\cdot 10^{-14}$

[21]. The interferometric methods of the frequency measurements are based on the comparison of the measured frequency with the frequency of interferometer modes. For example the laser cavity can play the role of the interferometer. The intermode frequency of interferometer is inversely proportional to the distance between the interferometer mirrors. However, the vibrations, thermal fluctuations, and other technical noises result in the fluctuations of the interferometer length. The various methods applied to compensate the interferometer length fluctuations enable to get the relative uncertainty up to 10^{-10}–10^{-11}. In spite of the fact that the idea of the new methods was proposed in the early works on the laser spectroscopy they were realized only when the femtosecond laser systems were developed. The spectrum of the femtosecond laser pulse, of a few optical cycles temporal width, is the frequency comb which spreads from the radio-frequency spectrum up to near ultraviolet. The intermode frequency of the comb is stabilized by the radio-frequency methods with the frequency of harmonics of the cesium atomic clock. The fluctuations of the comb frequencies are traced by heterodyne methods in the radio-frequency spectrum. As a result the measured frequency is almost directly compared with the frequency of the cesium atomic clock. The result of the most accurate measurements made by the group at the Max Plank Institute fur Quantenoptik (MPQ) in Garching, Germany [21] for the frequency of $1S - 2S$ transition in hydrogen is

$$\nu_{1S-2S} = 2\,466\,061\,413\,187\,103\,(46)\ \text{Hz}. \tag{1.7}$$

The frequency of some other transitions in hydrogen and deuterium was measured. The precision of frequency measurements for transitions including the high-lying levels is lower because the natural line-width of the high-lying states exceeds significantly the line-width of the $2S$ state. Table 1.1 shows the frequency of $\left(2S_{1/2} - 8S_J, 8D_J, 12D_J\right)$ transitions in hydrogen and deuterium made by the group at the Laboratoire Kastler-Brossel, Ecole Normale Superieure, et Universite Pierre et Marie Curie,

Table 1.1. The frequency of transitions in hydrogen and deuterium [10]

Transition	*Frequency*, MHz	
	Hydrogen	*Deuterium*
$2S_{1/2} - 8S_{1/2}$	770649350.0120(86)	770859041.2457(69)
$2S_{1/2} - 8D_{3/2}$	770649504.4500(83)	770859195.7018(63)
$2S_{1/2} - 8D_{5/2}$	770649561.5842(64)	770859252.8495(59)
$2S_{1/2} - 12D_{3/2}$	799191710.4727(93)	799409168.0380(86)
$2S_{1/2} - 12D_{5/2}$	799191727.4037(70)	799409184.9668(68)

Paris, France [10]. It is seen that the accuracy of measurements is about 10 kHz.

The detailed discussion and comparison of results of different measurements is given in [9, 10, 12, 14]. The integral results for low-lying levels of hydrogen atom are combined in schematic energy-level diagram shown in Fig. 1.1.

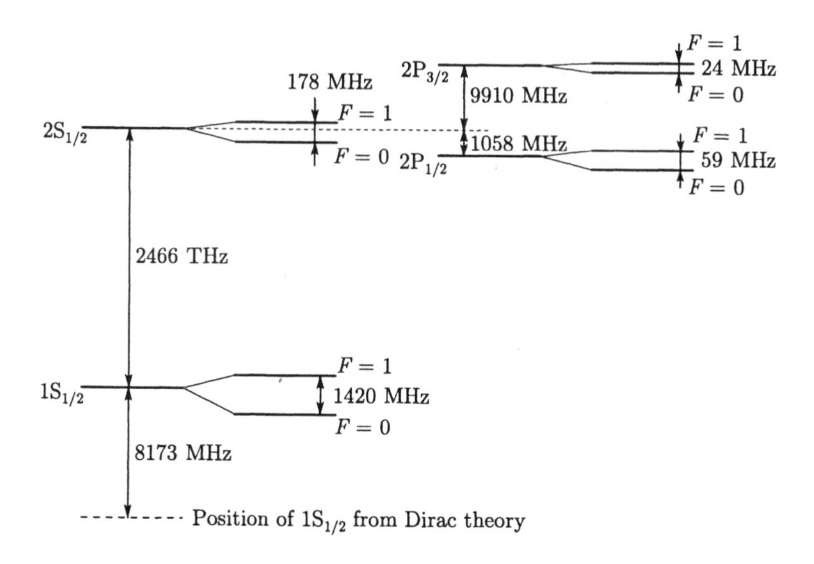

Figure 1.1. Schematic energy-level diagram for low-lying hydrogen states

In the cited above researches the Doppler-free two-photon spectroscopy method was used. This method enables to measure the frequency of the electric dipole forbidden transitions $nS \leftrightarrow n'S$ and $nS \leftrightarrow n'D$. This method was applied earlier to measure the frequency of $1S - 2S$ transition in muonium (μ^+e^- atom) [22]

$$\nu_{1S-2S}\left(\mu^+e^-\right) = 2455529002(57)\ \text{MHz}$$

and positronium [23]

$$\nu_{1S-2S}\left(e^+e^-\right) = 1233607216.4(3.2)\ \text{MHz}.$$

The adjustment of the experimental data for the transition frequencies in hydrogenlike atoms with the data obtained by other physical methods provides the self-consistent values of fundamental constants and material constants of elementary particles.

1.3 Experiments on search for electric dipole moment of elementary particles and atoms

There is the following relation between the electron magnetic moment $\mathbf{m}$ and its spin $\boldsymbol{\sigma}$

$$\mathbf{m} = \mu\boldsymbol{\sigma}, \tag{1.8}$$

where μ is magnitude of the electron magnetic moment discussed above. The same relation holds for other spin-1/2 particles. The proportionality of the particle magnetic moment to its spin is due to the fact that the spin $\boldsymbol{\sigma}$ is the intrinsic angular momentum of the particle, i.e. it is the only preferential vector in the particle rest frame. If a particle possesses the electric dipole moment (EDM), then the same arguments require that the EDM should be related with the spin operator. The simplest possible relation is

$$\mathbf{d} = d\boldsymbol{\sigma}. \tag{1.9}$$

The vectors in both sides of the equation (1.8) have the same transformation properties. Indeed, the magnetic moment and spin are invariant with respect to the space inversion, because both of them are axial vectors according to their nature. Contrary the vectors in the left-hand-side and right-hand side of the equation (1.9) differ in their transformation properties. The vector of dipole moment $\mathbf{d}$ changes sign at space inversion while the spin, being the angular momentum operator, remains invariable. Further, at the time reversal transformation the angular momentum (defined in the classical mechanics as $m\,[\mathbf{r}\,\mathbf{v}]$)) changes sign while the dipole moment remains invariable. Thus the constant d in equation (1.9) may be equal only zero, if the particle is described by an equation which is invariant with respect to the space inversion and time reversal. The equation (1.9) holds only in the case when the symmetry with respect to the space inversion (P) and time reversal (T) is violated. Indeed if we add the term $-\mathbf{dE}$ to the Hamiltonian (1.1) then the equation for the spinor wave function $\psi = (\psi_1, \psi_2)$ becomes P and T non-invariant.

The violation of the space inversion symmetry in the weak interactions [24] stimulated interest to the problem of EDM. However, the violation the P and charge conjugation (C) symmetry in weak interactions does not mean the violation of symmetry with respect to the combined CP transformation. Landau [25] pointed out that for existence of the electric dipole moment of elementary particles it is not sufficient the breakdown of P and C symmetry in separate. The violation of the combined CP symmetry is required. After the discovering of the CP violation in the decay of K^0 meson [26] the interests to the experiments on search of EDM of elementary particles and atoms has significantly enhanced [27, 28].

The leading place among the experiments on the measurements of the electric dipole moment of the elementary particles takes the experiments on neutron EDM. The main idea of the experiments proposed by Smith, Purcell, and Ramsey work [29] consists in the measurement of the neutron spin precession frequency in parallel homogeneous magnetic and electric fields. Indeed if the equation (1.9) holds then the neutron moving in the magnetic $\mathbf{B}_0$ and electric $\mathbf{E}_0$ fields will precess at frequency

$$\Omega_+ = 2\left|\mu B_0 + dE_0\right|/\hbar, \tag{1.10}$$

when the direction of electric and magnetic fields coincides. The reversion of the electric field will result in the precession frequency

$$\Omega_- = 2\left|\mu B_0 - dE_0\right|/\hbar. \tag{1.11}$$

The measurements of the difference between the frequency (1.10) and (1.11)

$$\Delta\Omega = \Omega_+ - \Omega_- = 4\,dE_0/\hbar \tag{1.12}$$

enable to determine the magnitude of the neutron EDM d.

The upper limit for the neutron EDM is estimated now [27, 28, 30] as

$$d_n/e \leq 6.3 \cdot 10^{-26}\ \text{cm}.$$

Recently the international collaboration at Paul Scherrer Institute, Switzerland has announced the program on the improved measurement of the electric dipole moment of the neutron [31]. It is planned to get a sensitivity of

$$d_n/e \approx 2 \cdot 10^{-28}\ \text{cm}.$$

It should be noted that the mechanism of the elementary particle EDM based on the violation of CP and T invariance (see (1.9)) is not the only proposed mechanism. The neutron scattering by electric field was studied in the classical paper of Schwinger [32]. He proposed the mechanism based on the interaction of the magnetic moment of moving neutron with the electric field of the atom. In this case the equation for d is $\mathbf{d}_n = (\mu/mc)\left[\boldsymbol{\sigma}\,\mathbf{p}\right]$, the scattering of neutron is due to the spin-orbital interaction [32–34]. The mechanism of the induced neutron EDM, based on the use of the interaction Hamiltonian of the type $H_{\text{int}} = -\alpha_E \mathbf{E}^2/2$, was considered in the series of papers [35–39]. The induced EDM is proportional to the strength of the electric field.

The EDM of charged particles can be measured in experiments where the neutral atom interacts with the superposition of magnetic and electric fields. However as far as the neutral atom, in contrast to charged particle, can be infinitely long in the region of space in which the electric field is non-zero, it should mean that the integral electric field at each individual charge of atom is equal to zero. It was shown by Schiff [40] that the non-zero electric field at atomic nucleus can be compensated by the forces of the non-electric nuclear interaction of nucleons, or by the interaction of the nucleus magnetic moment with the gradient of magnetic field produced by electrons of the atomic shells.

The experiments with the paramagnetic atoms give the possibility to measure, in principle, the EDM of electron. Sandars [41, 42] has demonstrated that when the relativistic effects are taken into account then the ratio of the atomic EDM d_A to the electron EDM d_e is about $d_A/d_e \approx Z^3\alpha^2$. This ratio can be quite large for suitable paramagnetic atoms. The reported results for the experimental limit on the size of electron EDM are [43, 44]

$$d_e/e \leq (6.9 \pm 7.4) \cdot 10^{-28} \text{ cm}.$$

The spin of electron shells in diamagnetic atoms is equal to zero thus the nucleus EDM can be in principle measured. In this case the nucleus spin is initially polarized by the optical methods and then the frequency of atomic spin precession in the collinear magnetic and electric fields is measured. The principle idea of the method is the same that is used to measure the EDM of neutron. The upper limits for the atomic EDM of ^{199}Hg [45] and ^{129}Xe [46] are

$$d\left(^{199}\text{Hg}\right)/e \leq 8.7 \cdot 10^{-28} \text{ cm},$$

$$d\left(^{129}\text{Xe}\right)/e \leq 0.7 \cdot 10^{-27} \text{ cm}.$$

If we assign the atomic EDM to the valent neutron in the even-odd nucleus of ^{199}Hg, then the obtained data provide the estimations for the upper limit of the neutron EDM.

Concluding the discussion we can see that the series of recent experiments bring out clearly that the behavior of elementary particles in the processes of their interaction with the electromagnetic field does not always adequately described by the basic equations of quantum mechanics. There are some specific features that require the further understanding. Indeed the magnitude of the electron magnetic moment does not coincide with the Bohr magneton, the hydrogenic spectrum differs from the spectrum calculated on the basis of non-relativistic and relativistic equations of the quantum mechanics, etc. All these discrepancies have been already explained in the modern theory, but the reasonable coincidence between the experimental and calculated data can only be obtained if we use the quantum field theory methods. The secondary quantization procedure is certainly in close connection with the methods of quantum mechanics, nevertheless it is out of the frames of the canonical quantum mechanics. This situation stimulates some essential questions of the fundamental and applied manner. Firstly, whether these discrepancies indicate on the imperfection of the basic principles of quantum mechanics, i.e. the lack of its self-consistency, or, simply, by improving the basic equations of quantum mechanics we can get the further insight into the nature of these discrepancies. Secondly, the main difference and main advantage of the quantum field theory approach is in the account for the virtual processes. As a result the number of particles involved into the process of some incident particle scattering does not fixed, while in the frame of quantum mechanics theory the number of particles is fixed by the normalization condition. Probably it gives us some indications how we should generalize the equations of quantum mechanics. Thirdly, it is evident that the problem of electron motion in the Coulomb field is not equivalent to the hydrogen atom problem, because the hydrogen atom problem is a two-body problem. The nucleus of hydrogenlike atoms has a finite mass, most of the nuclei have the non-zero spin and magnetic moment as well. The modern spectroscopy feels reliably the effects associated with the finite mass, spin, and magnetic moment of nucleus, it gives a serious motivation to develop the methods for solving of the two-body problem. This problem is two-fold: to develop the consistent procedure of deriving of the Hamiltonian of the two-body problem, and to develop the adequate methods of the mathematical analysis of the obtained equations.

I have tried in this book to present the atomic spectroscopy theory in deductive manner by starting from the simplest models to come gradually to the most general models. The book consists of the two main parts. The first part is devoted to the development of the hydrogenic spectrum theory based on the use of the Schrödinger equation, Pauli equation, Klein–Gordon–Fock equation, and Dirac equation. The comparative analysis of the spectra obtained from the solution of the above equations is given. Simultaneously, the method of deriving of the Hamiltonian for many-body problem from the equations for particle motion in the external electromagnetic field is developed. The method is based on the use of the canonical Lagrangian formalism. As an examples illustrating the main points I have tried to use the examples from the modern spectroscopy and light-matter interaction physics. The second part of the book is devoted to the development of the spin-1/2 particle theory. The base problems are here: (1) the energy spectrum of hydrogenlike atoms; (2) the spectrum of geonium atom; (3) the problem of electric dipole moment of spin-1/2 particles. The close connection between all these problems is demonstrated.

PART I

FINE AND HYPERFINE STRUCTURE OF ATOMIC SPECTRA

Chapter 2

SCHRÖDINGER EQUATION

The first quantum mechanical theory, that gave the explanation of the discrete spectra of atomic emission, was based on the equation proposed by Schrödinger [2] in 1927. In this chapter we discuss briefly the basic principles and main concepts of quantum mechanics. We start with the Schrödinger and Heisenberg equations, then we introduce the main quantum mechanical operators, and consider the properties of the wave functions and operators. The problem on the electron motion in Coulomb field for Schrödinger equation is analyzed in details. The analysis of the problem on the two oppositely charged particles interaction enables us to introduce the reduced mass. The concept of reduced mass plays the crucial role in the theory of atomic spectra. Finally we consider the problem on the energy spectra of atom placed in atomic trap and analyze the specific features of interaction of the trapped atom with electromagnetic wave.

2.1 Schrödinger equation

To remind the basic principles of the quantum mechanics we start here with the Schrödinger and Heisenberg equations and discuss briefly the boundary conditions for the states of discrete and continuous energy spectra for particle moving in attractive potential.

2.1.1 Schrödinger and Heisenberg equations

The first quantum-mechanical equation was proposed by Schrödinger [2] in 1927. The Schrödinger equation is

$$i\hbar\frac{\partial\psi\left(\mathbf{r},t\right)}{\partial t}=H_0\psi\left(\mathbf{r},t\right), \tag{2.1}$$

where H_0 is the Hamiltonian

$$H_0 = \frac{\mathbf{p}^2}{2m_0} + U(\mathbf{r}). \tag{2.2}$$

The first term in the Hamilton H_0 is the kinetic energy, which depends on the momentum operator $\mathbf{p} = -i\hbar\nabla$, and the second term is the potential energy depending on the coordinate operator $\mathbf{r}$. The particle coordinate $\mathbf{r}$ and momentum $\mathbf{p}$ operators obey the following commutation relations

$$[p_i, x_j] = -i\hbar\delta_{ij},$$

where $i = 1, 2, 3$.

The solution of equation (2.1) for the case of free particle, i.e. $U(\mathbf{r}) = 0$, is

$$\psi(\mathbf{r}, t) = \sum_{\mathbf{k}} [C_{\mathbf{k}} \exp(i\mathbf{kr}) + C_{-\mathbf{k}} \exp(-i\mathbf{kr})] \exp\left(-i\frac{E_{\mathbf{k}}t}{\hbar}\right),$$

where the energy of particle $E_{\mathbf{k}}$ in the state with the momentum $\hbar\mathbf{k}$ is

$$E_{\mathbf{k}} = \frac{\hbar^2\mathbf{k}^2}{2m_0},$$

the constants $C_{\pm\mathbf{k}}$ determine the initial state of the particle and can be determined from the initial condition

$$\psi(\mathbf{r}, 0) = \sum_{\mathbf{k}} [C_{\mathbf{k}} \exp(i\mathbf{kr}) + C_{-\mathbf{k}} \exp(-i\mathbf{kr})].$$

Thus the state of the free particle is described by the superposition of plain waves, and the particle energy depends quadratically on its momentum.

The general algorithm of obtaining equation for particle interacting with the electromagnetic field from free particle equation consists in the use of the following replacements

$$i\hbar\frac{\partial}{\partial t} \to i\hbar\frac{\partial}{\partial t} - q\varphi, \quad -i\hbar\nabla \to -i\hbar\nabla - \frac{q}{c}\mathbf{A}, \tag{2.3}$$

where q is the particle charge, $\varphi(\mathbf{r}, t)$ and $\mathbf{A}(\mathbf{r}, t)$ are the scalar and vector potentials of the electromagnetic field, respectively. By applying the replacements (2.3) to the Hamiltonian of free particle, we get the following wave equation for the particle interacting with the electromagnetic field

$$i\hbar\frac{\partial\psi(\mathbf{r}, t)}{\partial t} = \left[\frac{1}{2m_0}\left(\mathbf{p} - \frac{q}{c}\mathbf{A}\right)^2 + q\varphi\right]\psi(\mathbf{r}, t). \tag{2.4}$$

The eigenfunctions of the equation (2.1) enable us to determine the quantum mechanical average of the arbitrary functions $f(\mathbf{r}, \mathbf{p})$ of operators $\mathbf{r}$ and $\mathbf{p}$. The quantum mechanical average are determined by

$$\bar{f} = \int \psi^* (\mathbf{r}, t) f(\mathbf{r}, \mathbf{p}) \psi (\mathbf{r}, t) \, dV.$$

The quantum mechanical representation in which the operators are the function of canonically conjugated operators $\mathbf{r}$ and $\mathbf{p}$, while the wave functions are time-dependent, is called by Schrödinger representation.

Along with the Schrödinger representation the Heisenberg representation is widely used in quantum mechanics. In Heisenberg representation the operators are time-dependent. The temporal evolution of the operators is described by the Heisenberg equation

$$\frac{df}{dt} = \frac{1}{i\hbar} [f, H] + \frac{\partial f}{\partial t}. \tag{2.5}$$

If the equation (2.5) and Hamiltonian (2.2) are applied to the coordinate operator $\mathbf{r}$ then we get the following equation for the particle velocity

$$\mathbf{v} = \frac{d\mathbf{r}}{dt} = \frac{1}{i\hbar} [\mathbf{r}, H_0] = \frac{\mathbf{p}}{m_0}.$$

It is seen that the relationship between the particle velocity $\mathbf{v}$ and momentum $\mathbf{p}$ coincides with that in classical mechanics.

In the similar way, we obtain the expression for the velocity of a particle interacting with the electromagnetic field

$$\mathbf{v} = \frac{1}{i\hbar} [\mathbf{r}, H] = \frac{1}{m_0} \left(\mathbf{p} - \frac{q}{c} \mathbf{A} \right), \tag{2.6}$$

where we have used the Hamiltonian of the equation (2.4):

$$H = \frac{1}{2m_0} \left(\mathbf{p} - \frac{q}{c} \mathbf{A} \right)^2 + q\varphi. \tag{2.7}$$

The equation (2.6) shows that the operator $\mathbf{p}$ corresponds to the generalized momentum in classical electrodynamics

$$\mathbf{p} = m_0 \mathbf{v} + \frac{q}{c} \mathbf{A}.$$

The generalized momentum plays an auxiliary role in classical and quantum mechanics, but in both cases the observable value is the particle velocity.

This book is devoted to the study of energy spectra of the hydrogenlike atoms, therefore we shall use mainly the Schrödinger representation. The Heisenberg representation is convenient when we study the evolution of atom driving by some external electromagnetic wave. Nevertheless the Heisenberg representation will also widely used here, because it enables us to study the symmetry properties of different Hamiltonians and to define the integrals of motion. Indeed according to the equation (2.5) the operator $f(\mathbf{r}, \mathbf{p})$ is integral of motion if it commutes with the Hamiltonian

$$[f, H] = 0.$$

It is well known that the integrals of motion play an exceptional role in the classical and quantum mechanics.

2.1.2 Continuity equation, boundary conditions, and normalization condition

The equation for the bilinear combination of the wave function enable us to introduce the concept of the charge density and current density of the matter field. Multiplying both sides of equation (2.4) by complex conjugated wave function ψ^* and subtracting from the obtained equation its complex conjugated we get

$$\frac{\partial |\psi|^2}{\partial t} = -\frac{1}{2m_0} \nabla \left\{ \psi^* \left(-i\hbar\nabla - \frac{q}{c}\mathbf{A} \right) \psi + \left[\left(-i\hbar\nabla - \frac{q}{c}\mathbf{A} \right) \psi \right]^* \psi \right\}.$$

This equation can be written in the form

$$\frac{\partial \rho}{\partial t} + \operatorname{div} \mathbf{j} = 0, \tag{2.8}$$

where

$$\rho(\mathbf{r}, t) = q |\psi(\mathbf{r}, t)|^2,$$

$$\mathbf{j}(\mathbf{r}, t) = \frac{q}{2m_0} \left\{ \psi^* \left(-i\hbar\nabla - \frac{q}{c}\mathbf{A} \right) \psi + \left[\left(-i\hbar\nabla - \frac{q}{c}\mathbf{A} \right) \psi \right]^* \psi \right\}.$$

The equation (2.8) has the form of the classical continuity equation. Hence, the function $\rho(\mathbf{r}, t)$ can be associated with the charge density of a particle, and the function $\mathbf{j}(\mathbf{r}, t)$ plays the role of the electric current density.

Integrating the equation (2.8) over the whole space we get

$$\frac{d}{dt} \left(q \int |\psi(\mathbf{r}, t)|^2 dV \right) = - \int_{S(r\to\infty)} \mathbf{j}(\mathbf{r}, t)\, d\mathbf{S}. \tag{2.9}$$

If the particle is in the bound state of some potential well then the current density should be equal to zero at infinity. As a result we obtain the following boundary condition at infinity

$$\psi(\mathbf{r})|_{r\to\infty} \to 0. \tag{2.10}$$

The equation (2.9) together with the boundary condition (2.10) generates the following normalization condition for the wave function

$$\int |\psi(\mathbf{r},t)|^2 \, dV = 1. \tag{2.11}$$

It is seen that the condition (2.11) means that the charge associated with the particle is always equal to the elementary charge $q = \pm|e|$.

The equations (2.1) and (2.4) are the second order differential equation with respect of the space variables. Therefore to define unambiguously the radial wave function we need additionally in the second boundary condition. It is assumed usually that the wave function should be finite everywhere. For example, if we consider the particle motion in the Coulomb field it is assumed that the wave function should be finite at $r = 0$.

For the particle interacting with the attracting static electric and magnetic fields, the equation (2.4) together with the boundary conditions at $r \to \infty$ and $r = 0$ generates the eigenvalue problem

$$E_n u_n(\mathbf{r}) = \left[\frac{1}{2m_0}\left(\mathbf{p} - \frac{q}{c}\mathbf{A}(\mathbf{r})\right)^2 + U(\mathbf{r})\right] u_n(\mathbf{r}). \tag{2.12}$$

The eigenfunctions $u_n(\mathbf{r})$ corresponding to the different eigenvalues E_n are orthogonal

$$\int u_n^*(\mathbf{r})\, u_m(\mathbf{r})\, dV = \delta_{nm}.$$

Usually, the particles, producing the external (with respect to considered particle) fields, are located in the finite spatial volume, therefore the potentials of electromagnetic field, produced by them, tend to zero with the increase of distance: $\mathbf{A}(\mathbf{r})|_{r\to\infty} \to 0$, $\varphi(\mathbf{r})|_{r\to\infty} \to 0$. As a result, the potential energy of a particle, interacting with the external fields, is equal to zero at $r \to \infty$. Hence, the energy of the bound states of particle is negative, $E_n < 0$.

If $E_n > 0$ it means that the kinetic energy of a particle at $r \to \to \infty$ is non-zero, hence the particle can make an infinite motion. The spectrum of the positive energy eigenvalues is continuous. As far as $\mathbf{A}(\mathbf{r})|_{r\to\infty} \to 0$, $\varphi(\mathbf{r})|_{r\to\infty} \to 0$ the solutions of the equation (2.12) have the following asymptotic form at $r \to \infty$

$$\psi(\mathbf{r}) = C\frac{\sin(kr+\delta)}{r} Y_{lm}(\theta,\varphi),$$

where $k = \sqrt{2m_0E}/\hbar$, and Y_{lm} are the spherical harmonics.

The normalization condition for the wave functions of the continuous spectrum is also determined by the equation (2.9). The general form of solution is

$$\psi(\mathbf{r}) = R_{kl}(r)\, Y_{lm}(\theta,\varphi).$$

The spherical harmonics are normalized by the condition

$$\frac{1}{4\pi}\int Y_{lm}^{*}(\theta,\varphi)\, Y_{l'm'}(\theta,\varphi)\, d\Omega = \delta_{ll'}\delta_{mm'}.$$

In accordance with the definition (2.8), the charge of the spherical layer $(r, r+dr)$ is

$$dq = q\left(\int |\psi(\mathbf{r})|^2 r^2\, d\Omega\right) dr.$$

It is assumed that the unite charge should pass through the spherical surface of the infinite radius in the unit time. Hence

$$\frac{dq}{dt} = q\left(\int |\psi(\mathbf{r})|^2 r^2 d\Omega\right)\frac{dr}{dt} = qv$$

or

$$\int\limits_{S(r\to\infty)} \mathbf{j}(\mathbf{r},t)\, d\mathbf{S} = qv = q\frac{\hbar k}{m_0}.$$

As a result the normalization condition is

$$\int R_{kl}^{(\infty)*}(r)\, R_{k'l}^{(\infty)}(r)\, r^2\, dr = 2\pi\delta\left(k-k'\right),$$

where $R_{kl}^{(\infty)}(r)$ is the asymptotic form of the positive energy solutions of equation (2.12).

2.1.3 Gauge transformation

The equation (2.4) is gauge invariant. Indeed, if simultaneously with the gauge transformation of vector and scalar potentials

$$\mathbf{A}'(\mathbf{r},t) = \mathbf{A}(\mathbf{r},t) + \nabla\chi(\mathbf{r},t),$$

$$\varphi'(\mathbf{r},t) = \varphi(\mathbf{r},t) - \frac{1}{c}\frac{\partial\chi(\mathbf{r},t)}{\partial t},$$

we make the following transformation of the wave function

$$\psi'(\mathbf{r},t) = \psi(\mathbf{r},t)\exp\left[\frac{iq}{\hbar c}\chi(\mathbf{r},t)\right]. \tag{2.13}$$

then the Schrödinger equation (2.4)

$$i\hbar\frac{\partial\psi}{\partial t} = H(\mathbf{A},\varphi)\,\psi$$

becomes

$$i\hbar\frac{\partial\psi'}{\partial t} = H(\mathbf{A}',\varphi')\,\psi'.$$

It is seen that the Schrödinger equation does not change its form.

The gauge transformation of the wave function (2.13) does not change the quantum mechanical average of the operators $f(\mathbf{r})$ which depend on the coordinate operator only. At the same time, the quantum mechanical average of the generalized momentum operator $\mathbf{p}$ is changed

$$\int \psi'^{*}\mathbf{p}\psi'\,dV = \int \psi^{*}\left(\mathbf{p}+\frac{q}{c}\nabla\chi\right)\psi\,dV \neq \int \psi^{*}\mathbf{p}\psi\,dV.$$

This is not unexpected, because the generalized momentum operator does not correspond to the observable value. As we have mentioned above the observable value is particle velocity. For the quantum mechanical average of the particle velocity operator, $\mathbf{v} = \mathbf{p} - \frac{q}{c}\mathbf{A}$, we have

$$\int \psi'^{*}\mathbf{v}\psi'\,dV = \frac{1}{m_0}\int \psi'^{*}\left(\mathbf{p}-\frac{q}{c}\mathbf{A}'\right)\psi'\,dV =$$
$$= \frac{1}{m_0}\int \psi^{*}\left(\mathbf{p}+\frac{q}{c}\nabla\chi-\frac{q}{c}\mathbf{A}-\frac{q}{c}\nabla\chi\right)\psi\,dV = \int \psi^{*}\mathbf{v}\psi\,dV.$$

It can be easily shown also that any degrees of the velocity operator are gauge invariant too.

Hence, the quantum mechanical averages of the arbitrary functions of the coordinate and velocity operators, $f\left(\mathbf{r},\mathbf{p}-\frac{q}{c}\mathbf{A}\right)$, are the gauge invariant values.

2.2 Quantum mechanical operators

In previous section we have introduced the coordinate, momentum, and Hamiltonian operators. The exceptional role in atomic spectroscopy plays the parity operator and angular momentum operator. Here we discuss shortly the properties of these two additional quantum mechanical operators. As we have already mentioned above the Hamiltonian is the basic operator in quantum mechanical theory. Its quantum mechanical average is the energy of a system. The energy of an isolated system of particles should not vary when we make the transformations of the reference frame. The quantum mechanical operators are closely related with the operators of the orthogonal transformations of the reference frame.

2.2.1 Momentum operator

The energy of an isolated system of particles is invariant with respect to the spatial translation, i.e. when the coordinates of all particles in the system are changed in the following way: $\mathbf{r}_a \to \mathbf{r}_a + \delta\mathbf{r}$. We can consider the infinitesimally small translation $\delta\mathbf{r}$, because any finite translation is a sum of the infinitesimally small translations. If we apply this transformation to the wave function $\psi(\mathbf{r}_1, \mathbf{r}_2, \ldots)$ it becomes

$$\psi(\mathbf{r}_1 + \delta\mathbf{r}, \mathbf{r}_2 + \delta\mathbf{r}, \ldots) = \psi(\mathbf{r}_1, \mathbf{r}_2, \ldots) + \delta\mathbf{r} \sum_a \nabla_a \psi(\mathbf{r}_1, \mathbf{r}_2, \ldots)$$

or

$$\psi(\mathbf{r}_1 + \delta\mathbf{r}, \mathbf{r}_2 + \delta\mathbf{r}, \ldots) = \Big(1 + \delta\mathbf{r} \sum_a \nabla_a\Big) \psi(\mathbf{r}_1, \mathbf{r}_2, \ldots).$$

Thus the operator

$$T = 1 + \delta\mathbf{r} \sum_a \nabla_a$$

is the operator the infinitesimally small spatial translation. Since the energy of isolated system does not change under spatial translation, it means

$$TH\psi = HT\psi.$$

Hence

$$\Big(\sum_a \nabla_a\Big) H - H\Big(\sum_a \nabla_a\Big) = 0.$$

As we have already mentioned, if operator commutes with the Hamiltonian, then the physical variable corresponding to this operator is conservative. In classical mechanics the physical variable, which is conservative due to the homogeneity of space, is the momentum. Hence the operator $\sum_a \nabla_a$ is proportional to the momentum operator. The coefficient of proportionality can be found if, for example, we calculate the quantum mechanical average of operator $\mathbf{p} = -i\hbar\nabla$ for free particle describing by plane wave $\psi_{\mathbf{k}}(\mathbf{r}) = C\exp(i\mathbf{kr})$

$$\int \psi_{\mathbf{k}}^* \mathbf{p} \psi_{\mathbf{k}} \, dV = \hbar\mathbf{k} \int \psi_{\mathbf{k}}^* \psi_{\mathbf{k}} \, dV.$$

Thus, the operator

$$\mathbf{p} = -i\hbar\nabla$$

is the quantum mechanical momentum operator.

2.2.2 Space inversion and parity operator

The space inversion transformation consists in the replacement $\mathbf{r} \to \to -\mathbf{r}$. The operator P generating this transformation is called by the parity operator

$$P\psi(\mathbf{r}) = \psi(-\mathbf{r}).$$

Let us apply the parity operator to the Hamiltonian (2.7). The generalized momentum $\mathbf{p}$ and vector potential $\mathbf{A}$ are both polar vectors, therefore at the space inversion transformation we have $P\mathbf{p} = -\mathbf{p}$ and $P\mathbf{A} = -\mathbf{A}$. Hence the kinetic energy remains invariable at the space inversion. If the potential energy is invariant with respect to the space inversion $U(\mathbf{r}) = U(-\mathbf{r})$, i.e. if the external potential is centrosymmetric, then the parity operator P commutes with the Hamiltonian (2.7)

$$[P, H] = 0.$$

The commuting operators have the common set of eigenfunctions. The eigenvalues of the parity operator can be found in the following way. On the one hand

$$P^2\psi(\mathbf{r}) = Pp\psi(\mathbf{r}) = p^2\psi(\mathbf{r}).$$

On the other hand

$$P^2\psi(\mathbf{r}) = P\psi(-\mathbf{r}) = \psi(\mathbf{r}).$$

Hence

$$p_{1,2} = \pm 1.$$

As a result the wave functions of the particle, moving in the centrosymmetrical potential $\varphi(\mathbf{r}) = \varphi(-\mathbf{r})$, either remain invariable or change the sign under the space inversion. The state in which the wave function does not change its sign is called by the even state, if the wave function changes its sign under the space inversion transformation then the corresponding state is called by the odd state.

Thus the invariance of the Hamiltonian with respect to the space inversion transformation manifests the parity conservation law: if an isolated ensemble of particles has a definite parity, then the parity remains invariable in the process of ensemble evolution.

The wave functions of the even states are the scalar functions, the wave functions of the odd states are the pseudoscalar functions.

2.2.3 Three-dimensional rotations and angular momentum operator

The rotation of an isolated ensemble of particles, as a whole, around an arbitrary axis does not change the relative positions of particles, hence, the state of the whole system should remain invariable. Let us consider infinitesimally small rotation $\delta\varphi$ around the z axis. Under this rotation the particle coordinates are transformed in the following way

$$x' = x + \delta\varphi y, \quad y' = y - \delta\varphi x, \quad z' = z.$$

The transformation of wave function is

$$\psi\left(x', y', z'\right) = \psi\left(x, y, z\right) + \delta\varphi y \frac{\partial \psi}{\partial x} - \delta\varphi x \frac{\partial \psi}{\partial y} = \\ = \left[1 + \delta\varphi\left(y\frac{\partial}{\partial x} - x\frac{\partial}{\partial y}\right)\right]\psi\left(x, y, z\right).$$

Hence the operator of infinitesimally small rotation around the z axis is

$$R_z\left(\delta\varphi\right) = 1 + \delta\varphi\left(y\frac{\partial}{\partial x} - x\frac{\partial}{\partial y}\right).$$

Under the rotation around the arbitrary axis $\delta\boldsymbol{\varphi}$ the rotation operator becomes

$$R\left(\delta\boldsymbol{\varphi}\right) = 1 - i\delta\boldsymbol{\varphi}\mathbf{l},$$

where the angular momentum operator $\hbar\mathbf{l}$ is defined as

$$\hbar\mathbf{l} = [\mathbf{r}\mathbf{p}] = -i\hbar\left[\mathbf{r}\nabla\right]. \tag{2.14}$$

The components of the angular momentum operator

$$\hbar l_x = yp_z - zp_y, \quad \hbar l_y = zp_x - xp_z, \quad \hbar l_z = xp_y - yp_x$$

obey the following commutation relations

$$\left[l_\alpha, l_\beta\right] = ie_{\alpha\beta\gamma}l_\gamma, \tag{2.15}$$

where $\alpha, \beta, \gamma = x, y, z$ and $e_{\alpha\beta\gamma}$ is the completely antisymmetric tensor of the third order. The elements of this tensor are equal to zero if any two of its three indexes coincide. The non-zero elements of this tensor correspond to the three different indexes. It is usually assumed that $e_{xyz} = 1$ and any other elements obtained by permutation of these indexes are equal to minus unity, if the number of permutations is odd, and unity, if the number of permutations is even.

The commutation relations for the angular momentum operator and operators of coordinate and generalized momentum are

$$[l_\alpha, r_\beta] = ie_{\alpha\beta\gamma}r_\gamma, \quad [l_\alpha, p_\beta] = ie_{\alpha\beta\gamma}p_\gamma. \tag{2.16}$$

The operator of the angular momentum square

$$\mathbf{l}^2 = l_x^2 + l_y^2 + l_z^2$$

commutes with each of the component of operator $\mathbf{l}$. Indeed

$$\left[l^2, l_\alpha\right] = l_\beta\left[l_\beta, l_\alpha\right] + \left[l_\beta, l_\alpha\right]l_\beta = -ie_{\alpha\beta\gamma}l_\beta l_\gamma + ie_{\alpha\gamma\beta}l_\gamma l_\beta = 0. \tag{2.17}$$

In the spherical set of coordinates the angular momentum square operator is

$$\mathbf{l}^2 = -\left[\frac{1}{\sin^2\theta}\frac{\partial^2}{\partial\varphi^2} + \frac{1}{\sin\theta}\frac{\partial}{\partial\theta}\left(\sin\theta\frac{\partial}{\partial\theta}\right)\right]. \tag{2.18}$$

It is seen that the operator (2.18) coincides with the angular part of the Laplace operator, written in the spherical coordinates

$$\Delta = \frac{1}{r^2}\frac{\partial}{\partial r}\left(r^2\frac{\partial}{\partial r}\right) - \frac{1}{r^2}\mathbf{l}^2. \tag{2.19}$$

As far as operators $\mathbf{l}^2$ and l_z commute then they have the common set of eigenfunctions. With the help of commutation relations (2.15) it can be easily shown that the common eigenfunctions obey the equations

$$\mathbf{l}^2\psi_{lm}(\mathbf{r}) = l(l+1)\psi_{lm}(\mathbf{r}), \quad l_z\psi_{lm}(\mathbf{r}) = m\psi_{lm}(\mathbf{r}), \tag{2.20}$$

where l is non-negative integer, $l = 0, 1, 2, \ldots$, and the z-projection of angular momentum takes the values $m = -l, -l+1, \ldots, l$. The solutions of the equations (2.20) are the spherical harmonics

$$Y_{lm}(\theta, \varphi) = (-1)^{\frac{m+|m|}{2}} i^l \sqrt{\frac{2l+1}{4\pi}\frac{(l-|m|)!}{(l+|m|)!}} P_l^{|m|}(\cos\theta)\exp(im\varphi), \tag{2.21}$$

where $P_l^m(\cos\theta)$ is the associated Legendre polynomial.

It can be easily shown that the Hamiltonian (2.2) for the case of particle motion in the spherically symmetric potential $\varphi(\mathbf{r}) = \varphi(r)$ commutes with the angular momentum operator. Indeed the equation for the angular momentum operator in spherical coordinates is

$$\mathbf{l} = \mathbf{e}_\theta\frac{i}{\sin\theta}\frac{\partial}{\partial\varphi} - \mathbf{e}_\varphi i\frac{\partial}{\partial\theta}. \tag{2.22}$$

It is seen from (2.22) that the angular momentum operator commutes with $U(r)$. On the other hand, as it follows from the equations (2.17)–(2.19) the operator $\mathbf{l}$ commutes with the Laplace operator. Thus when we deal with the eigenvalue problem on particle motion in the spherically symmetric potential $\varphi(r)$, we can always express the eigenfunctions in terms of the spherical harmonics (2.21).

It is evident that the angular momentum operator is invariant with respect of space inversion transformation, because both coordinate and generalized momentum operators change sign under the space inversion. Hence the eigenfunctions of the problem on the particle motion in the spherically symmetric potential have to have the definite parity. The parity of the different states are determined by the parity the spherical harmonics (2.21). By applying the parity operator to the spherical harmonics we get

$$PY_{lm}(\theta,\varphi) = Y_{lm}(\pi-\theta,\varphi+\pi) = (-1)^l Y_{lm}(\theta,\varphi),$$

i.e. the parity of state is defined by

$$P_l = (-1)^l. \tag{2.23}$$

The angular momentum operator in the cylindrical set of coordinates is

$$\mathbf{l} = \mathbf{e}_\rho\left(i\frac{z}{\rho}\frac{\partial}{\partial\varphi}\right) + \mathbf{e}_\varphi\left[-i\left(z\frac{\partial}{\partial\rho} - \rho\frac{\partial}{\partial z}\right)\right] + \mathbf{e}_z\left(-i\frac{\partial}{\partial\varphi}\right). \tag{2.24}$$

It is evident from this equation, that the angular momentum projection operator l_z commutes with the Hamiltonian (2.7) when $U = U(\rho,z)$ and $\mathbf{A} = \mathbf{e}_\varphi A(\rho,z)$. In this case, as it follows from the equations (2.20) and (2.21), the angular part of the wave function is given by $\exp(im\varphi)$. Hence the parity of eigenstates for the problem of particle motion in the external fields of cylindric symmetry is defined by

$$P_m = (-1)^m. \tag{2.25}$$

2.3 Particle motion in the Coulomb field

Let us consider the problem on a particle motion in the attracting Coulomb field. In this case the potential energy of a particle is

$$U(r) = -\frac{Ze^2}{r} \tag{2.26}$$

and the equation (2.12) became

$$\left[-\frac{\hbar^2}{2m_0}\left(\frac{\partial^2}{\partial r^2} + \frac{2}{r}\frac{\partial}{\partial r} - \frac{\mathbf{l}^2}{r^2}\right) - \frac{Ze^2}{r}\right]\psi_E(\mathbf{r}) = E\psi_E(\mathbf{r}). \tag{2.27}$$

We have shown above, that the Hamiltonian of the equation (2.27) commutes with the operators of parity, angular momentum square, and projection of angular momentum. Therefore the wave function can be expressed in terms of the eigenfunctions of the parity, angular momentum, and projection of angular momentum operators. However, as we have seen, the parity of states in the spherically symmetric external field is unambiguously determined by the angular momentum, therefore to define the particle state we can use the following quantum numbers: energy E, angular momentum l, and projection of angular momentum m.

2.3.1 Discrete spectrum

As we have discussed above for the bound states of electron in the Coulomb field the boundary conditions require that the wave function should be finite at $r = 0$ and turn to zero at $r \to \infty$

$$\psi_E(\mathbf{r})|_{r\to\infty} = 0. \tag{2.28}$$

The analysis, given in the previous section, has shown that in the case of particle motion in the spherically symmetric potential the wave function can be taken in the form

$$\psi_E(\mathbf{r}) = R(r)\, Y_{lm}(\theta, \varphi).$$

By substituting the latter equation into the equation (2.27) we get the following equation for the radial part of the wave function

$$\left(\frac{d^2}{dr^2} + \frac{2}{r}\frac{d}{dr} - \frac{l(l+1)}{r^2} + \frac{2m_0 Ze^2}{\hbar^2}\frac{1}{r} + \frac{2m_0 E}{\hbar^2}\right) R(r) = 0. \tag{2.29}$$

Taking into account that the bound states are the states of the negative energy, it is convenient to introduce the following notation

$$\kappa = \sqrt{\frac{2m_0|E|}{\hbar^2}}.$$

The general solution of the equation (2.29) is

$$R(r) = \Big[C_1 F(l+1-\gamma, 2l+2, 2\kappa r) + \\ + C_2 (2\kappa r)^{-2l-1} F(-l-\gamma, -2l, 2\kappa r)\Big] r^l \exp(-\kappa r), \tag{2.30}$$

where $F(a, b, z)$ is the confluent hypergeometric function, $\gamma = Z/(\kappa a_{\mathrm{B}})$. Here a_{B} is the Bohr radius

$$a_{\mathrm{B}} = \frac{\hbar^2}{m_0 e^2}. \tag{2.31}$$

The asymptotic form of the confluent hypergeometric function $F(a, b, z)$ at $z = 0$ is

$$F(a, b, |z| \to 0) \to 1.$$

Hence we should assume that $C_2 = 0$, because the second term in the right-hand-side of the equation (2.30) does not obey the boundary condition at $r = 0$. The asymptotic form of confluent hypergeometric function $F(a, b, z)$ at $z \to \infty$ is

$$F(a, b, z) = \frac{\Gamma(b)\exp(i\pi a)}{\Gamma(b-a)} z^{-a} + \frac{\Gamma(b)}{\Gamma(a)} z^{a-b} \exp(z) + \ldots.$$

It is seen that the second term in this equation infinitely increases when $r \to \infty$. However, this term vanishes when the argument a of the confluent hypergeometric function $F(a, b, z)$ is a non-positive integer. Thus the solution (2.30) satisfies the boundary conditions when $C_2 = 0$ and

$$l + 1 - \gamma = -n_r, \tag{2.32}$$

where n_r is the non-negative integer.

The latter equation yields the following equation for the energy spectrum of bound states

$$E_n = -\frac{Z^2\hbar^2}{2m_0 a_{\mathrm{B}}^2 n^2} = -\frac{Z^2 e^4 m_0}{2\hbar^2 n^2}, \tag{2.33}$$

where

$$n = n_r + l + 1 = 1, 2, 3, \ldots$$

Notice, that the energy spectrum (2.33), resulted from the solution of the quantum mechanical problem on the electron motion in the Coulomb field, coincides with the spectrum that was obtained with the help application of the Bohr–Sommerfeld quantization rules to the classical equations. The quantum number n_r is called by the radial quantum number. We shall see later that the radial quantum number determines the number of nodes of the radial wave function $R(r)$. The quantum number l is usually called by the azimuthal quantum number. The quantum number n is called by the principle quantum number.

Before the quantum mechanics was completely worked out, the spectroscopic notations were developed to describe the different hydrogen-like energy levels in an atom. Basically, the notation consisted of a number (representing the value of n) followed by a letter (representing the value of l). The letters originally described the characteristics of the spectral lines, like "sharp", "principal", etc. The correspondence between the values of l and letters is given in Tab. 2.1.

Table 2.1. Spectroscopic notation

l	$=$	0	1	2	3	4	5	...
letter	$\longrightarrow$	s	p	d	f	g	h	...

It can be easily shown that the ground state of the hydrogenlike atom is always the s state. Indeed, the substitution $R(r) = f(r)/r$ transforms the equation (2.29) to the form

$$\left(\frac{d^2}{dr^2} + \frac{2m_0E}{\hbar^2} - \frac{l(l+1)}{r^2} - U(r)\right) f(r) = 0.$$

It is seen that the last equation coincides with the Schrödinger equation for particle moving in the one-dimensional potential well of the form

$$U_{\text{eff}}(r) = U(r) + \frac{\hbar^2}{2m_0}\frac{l(l+1)}{r^2}.$$

The second term in the right-hand-side of this equation is the energy of centrifugal motion. This energy is definitely positive at $l > 0$. Hence, the energy of fundamental states at $l > 0$ is always higher than the energy of the s state. It can also be stated that the energy of the fundamental state for a given l increases with the increase of l.

In the case when a is a non-positive integer, the confluent hypergeometric functions $F(a, b, z)$ can be expressed in terms of the Laguerre polynomials: $F(-n, b+1, z) = (\Gamma(b+1)n!/\Gamma(b+1+n))L_n^{(b)}(z)$. Hence the radial wave function $R(r)$ can be rewritten in the following equivalent form

$$R_{nl}(r) = Cr^l \exp(-\kappa_n r) L_{n-l-1}^{(2l+1)}(2\kappa_n r), \tag{2.34}$$

where $\kappa_n = 1/na_{\text{B}}$. By using the normalization condition

$$\int_0^\infty R_{nl}^2(r)\, r^2\, dr = 1,$$

we get the following equation for the normalized wave function

$$R_{nl}(r) = \sqrt{\frac{(2\kappa_n)^3(n-l-1)!}{2n(n+l)!}}\,(2\kappa_n r)^l \exp(-\kappa_n r) L_{n-l-1}^{(2l+1)}(2\kappa_n r) =$$

$$= \frac{2}{n^2}\sqrt{\frac{Z^3(n-l-1)!}{a_{\text{B}}^3(n+l)!}} \left(\frac{2Zr}{na_{\text{B}}}\right)^l \exp\left(-\frac{Zr}{na_{\text{B}}}\right) L_{n-l-1}^{(2l+1)}\left(\frac{2Zr}{na_{\text{B}}}\right) \tag{2.35}$$

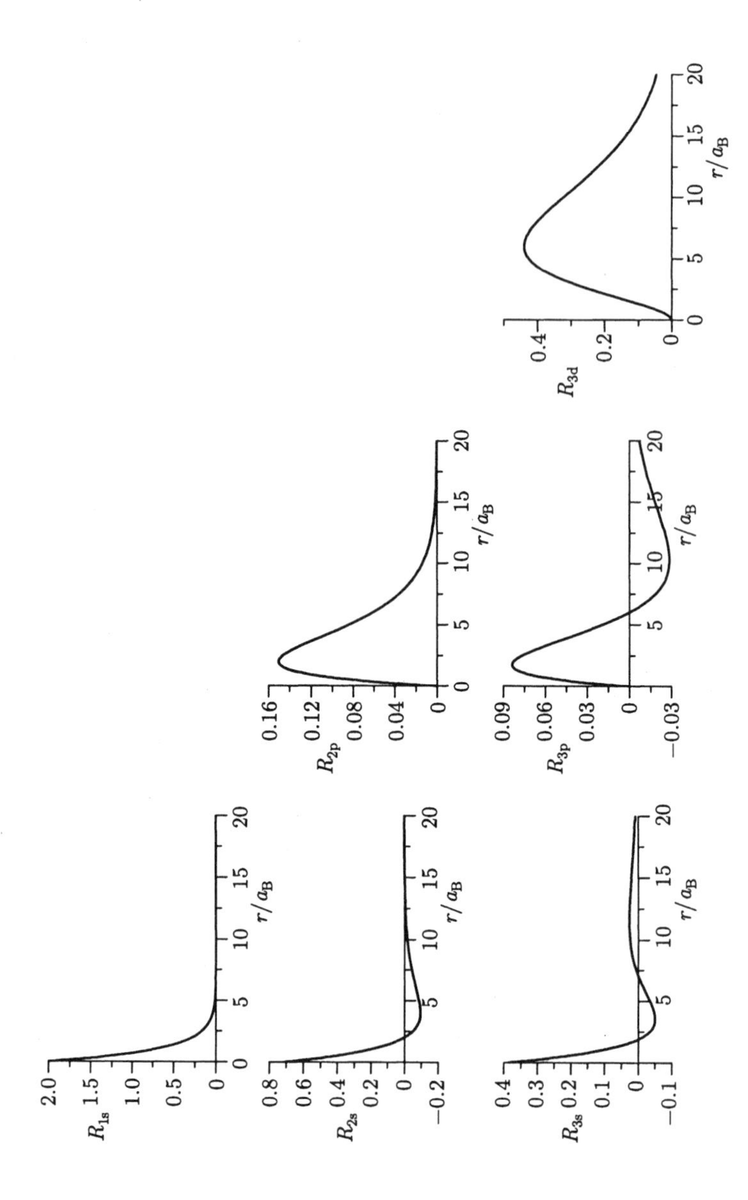

Figure 2.1. Radial wave functions of hydrogen atom

The explicit form of the wave functions for a number of states in hydrogen atom is given below:

1 $1s$ – state ($n = 1$, $l = 0$)

$$R_{1s}(r) = \frac{2}{\sqrt{a_{\mathrm{B}}^3}} \exp\left(-\frac{r}{a_{\mathrm{B}}}\right);$$

2 $2s$ – state ($n = 2$, $l = 0$)

$$R_{2s}(r) = \sqrt{\frac{1}{2a_{\mathrm{B}}^3}} \exp\left(-\frac{r}{2a_{\mathrm{B}}}\right)\left(1 - \frac{r}{2a_{\mathrm{B}}}\right);$$

3 $2p$ – state ($n = 2$, $l = 1$)

$$R_{2p}(r) = \sqrt{\frac{1}{6a_{\mathrm{B}}^3}} \exp\left(-\frac{r}{2a_{\mathrm{B}}}\right)\frac{r}{2a_{\mathrm{B}}};$$

4 $3s$ – state ($n = 3$, $l = 0$)

$$R_{3s}(r) = \frac{2}{3\sqrt{3a_{\mathrm{B}}^3}} \exp\left(-\frac{r}{3a_{\mathrm{B}}}\right)\left(1 - 2\frac{r}{3a_{\mathrm{B}}} + \frac{2}{3}\left(\frac{r}{3a_{\mathrm{B}}}\right)^2\right);$$

5 $3p$ – state ($n = 3$, $l = 1$)

$$R_{3p}(r) = \frac{2}{9}\sqrt{\frac{2}{3a_{\mathrm{B}}^3}} \exp\left(-\frac{r}{3a_{\mathrm{B}}}\right)\left(2 - \frac{r}{3a_{\mathrm{B}}}\right)\frac{r}{3a_{\mathrm{B}}};$$

6 $3d$ – state ($n = 3$, $l = 2$)

$$R_{3d}(r) = \frac{2}{9}\sqrt{\frac{2}{15a_{\mathrm{B}}^3}} \exp\left(-\frac{r}{3a_{\mathrm{B}}}\right)\left(\frac{r}{3a_{\mathrm{B}}}\right)^2.$$

The graphs of the corresponding functions is shown in Fig. 2.1. By taking into account the definition of the principle quantum number n:

$$n = n_r + l + 1, \tag{2.36}$$

we can see that the number of zeros of the wave function is really determined by the radial quantum number n_r. The wave functions of the s states are maxima at $r = 0$, the wave functions of states with $l > 0$ turn into zero at this point.

We have mentioned above that the product $R^2(r)\,r^2$, proportional to the probability for particle to be inside the spherical layer $(r, r + dr)$, is called by the charge density. The charge density distribution for the above states is shown in Fig. 2.2. It is seen that the maximum of charge density moves away from the center with the increase of the principle quantum number n.

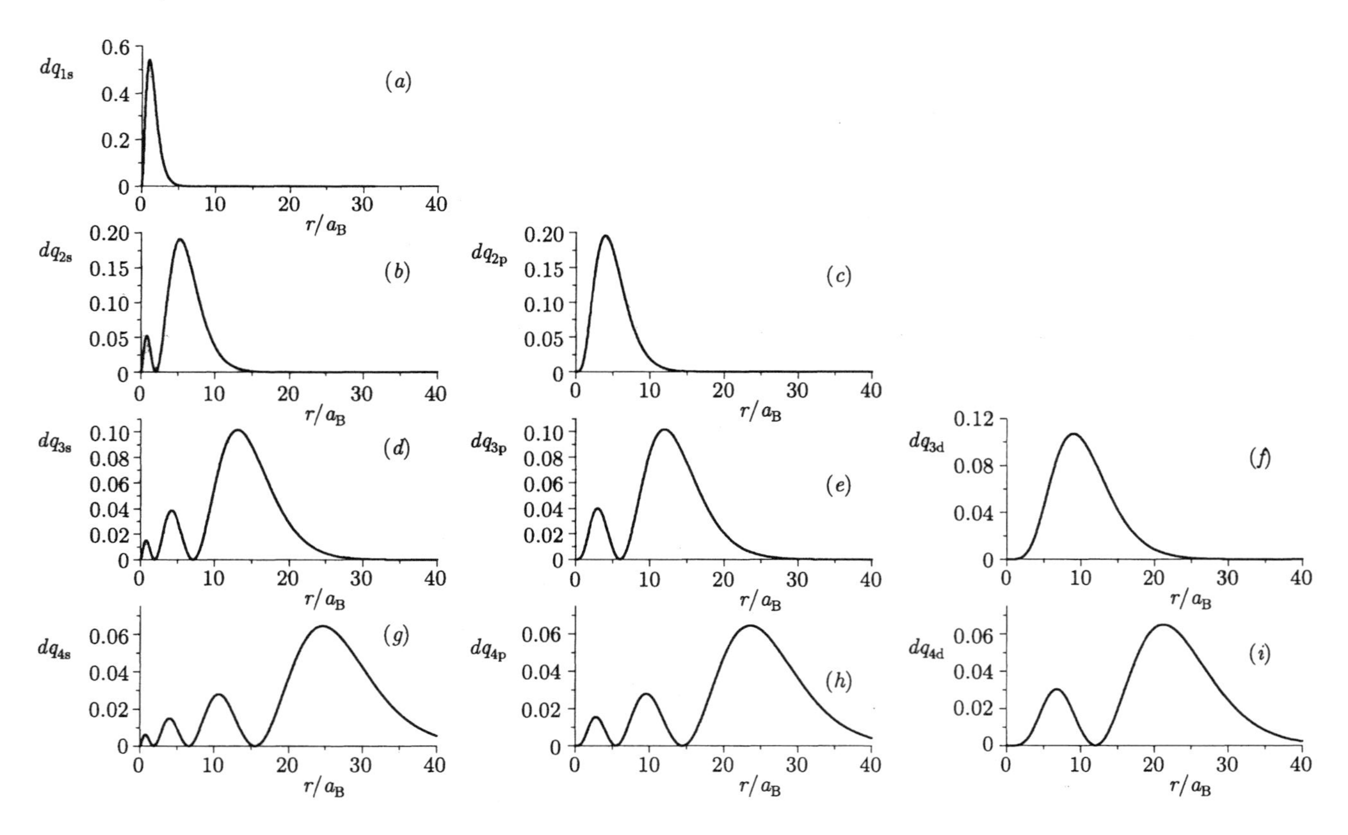

Figure 2.2. Density of charge distribution in hydrogen atom

2.3.2 Continuous spectrum

The continuous spectrum of the positive energy eigenvalues is stretched from zero up to infinity. Each energy eigenvalue is infinitely degenerated with respect to the angular momentum l, which runs all integers from zero up to infinity, and its projection m, taking all possible values, $|m| \leq l$, at given l.

The general solution is given by the equation (2.30), where we should again assume the coefficient C_2 equal to zero in order to satisfy the boundary condition at $r = 0$:

$$R_{kl}(r) = C_{kl}(2kr)^l \exp(-ikr) F\left(l + 1 + \frac{iZ}{ka_{\mathrm{B}}}, 2l + 2, i2kr\right), \quad (2.37)$$

where

$$k = \frac{\sqrt{2m_0 E}}{\hbar}.$$

The asymptotical form of the solution (2.37) at infinity is

$$R_{kl}(r) = C_{kl} \frac{\Gamma(2l+2)\exp\left(-\dfrac{\pi Z}{2ka_{\mathrm{B}}}\right)}{\left|\Gamma\left(l+1+\dfrac{iZ}{ka_{\mathrm{B}}}\right)\right|} \frac{\sin\left(kr + \left(\dfrac{Z}{ka_{\mathrm{B}}}\right)\ln 2kr - \dfrac{\pi l}{2} + \delta_l\right)}{kr}, \quad (2.38)$$

where $\delta_l = \arg\Gamma(l + 1 - iZ/(ka_{\mathrm{B}}))$. By normalizing the wave function in accordance with the procedure discussed in subsection 2.1.2, we get the following equation for the normalization coefficient C_{kl}:

$$C_{kl} = \sqrt{\frac{8\pi kZ/a_{\mathrm{B}}}{1 - \exp(-2\pi Z/(ka_{\mathrm{B}}))} \frac{1}{\Gamma(2l+2)}} \prod_{s=1}^{l} \sqrt{s^2 + \left(\frac{Z}{ka_{\mathrm{B}}}\right)^2}.$$

2.3.3 Matrix elements of transitions

The rate of the radiative transitions between the atomic states depends on the magnitude of the matrix elements of transitions

$$\langle n_1 l_1 m_1 | \mathbf{r} | n_2 l_2 m_2 \rangle = \int \psi^*_{n_1 l_1 m_1}(\mathbf{r})\, \mathbf{r} \psi_{n_2 l_2 m_2}\, dV. \quad (2.39)$$

It is convenient to make the following transformation of the radius vector $\mathbf{r}$:

$$\mathbf{r} = \mathbf{e}_+ r \sin\theta \exp(i\varphi) + \mathbf{e}_- r \sin\theta \exp(-i\varphi) + \mathbf{e}_z r \cos\theta,$$

where $\mathbf{e}_\pm = (\mathbf{e}_x \mp i\mathbf{e}_y)/2$. In this case the right-hand-side of the equation (2.39) transforms into the product of integrals over the radial and angular variables. The integrals over the angular variables give us the selection rules for the dipole allowed transitions. They are

$$\langle l_1 m_1 | \cos\theta \, | l_2 m_2 \rangle = i \sqrt{\frac{(l_1 - m_1 + 1)(l_1 + m_1 + 1)}{(2l_1 + 1)(2l_1 + 3)}} \delta_{l_2, l_1+1} \delta_{m_2, m_1},$$

$$\langle l_1 m_1 | \cos\theta \, | l_2 m_2 \rangle = -i \sqrt{\frac{(l_2 - m_2 + 1)(l_2 + m_2 + 1)}{(2l_2 + 1)(2l_2 + 3)}} \delta_{l_1, l_2+1} \delta_{m_1, m_2},$$

$$\langle l_1 m_1 | \sin\theta \exp(i\varphi) \, | l_2 m_2 \rangle = \tag{2.40}$$

$$= i \sqrt{\frac{(l_1 - m_1 + 1)(l_1 - m_1 + 2)}{(2l_1 + 1)(2l_1 + 3)}} \delta_{l_2, l_1+1} \delta_{m_2, m_1-1},$$

$$\langle l_1 m_1 | \sin\theta \exp(i\varphi) \, | l_2 m_2 \rangle =$$

$$= i \sqrt{\frac{(l_2 + m_2 + 1)(l_2 + m_2 + 2)}{(2l_2 + 1)(2l_2 + 3)}} \delta_{l_1, l_2+1} \delta_{m_1, m_2+1}.$$

The matrix elements for the component $(\mathbf{r}\mathbf{e}_-)$ can be easily obtained from the last two equations in (2.40) with the help of the equality

$$\langle l_1 m_1 | \sin\theta \exp(-i\varphi) \, | l_2 m_2 \rangle = \langle l_2 m_2 | \sin\theta \exp(i\varphi) \, | l_1 m_1 \rangle^* .$$

Thus the selection rules for the dipole allowed transitions are:

a) linear polarized wave

$$\Delta l = l_1 - l_2 = \pm 1, \quad \Delta m = m_1 - m_2 = 0;$$

b) circular polarized wave

$$\Delta l = l_1 - l_2 = \pm 1, \quad \Delta m = m_1 - m_2 = \pm 1.$$

In the last case the signs plus and minus correspond to the right and left circular polarized waves, respectively.

The radial matrix elements are

$$\langle n_1 l_1 | \, r \, | n_2 l_2 \rangle = \frac{a_B}{Z} \frac{2^{l_1 + l_2 + 2}}{n_1^{l_1+2} n_2^{l_2+2}} \sqrt{\frac{(n_1 - l_1 - 1)! \, (n_2 - l_2 - 1)!}{(n_1 + l_1)! \, (n_2 + l_2)!}} \times$$

$$\times \int_0^\infty x^{l_1 + l_2 + 3} \exp\left[-x\left(\frac{1}{n_1} + \frac{1}{n_2}\right)\right] L_{n_1 - l_1 - 1}^{(2l_1+1)}\left(\frac{2x}{n_1}\right) L_{n_2 - l_2 - 1}^{(2l_2+1)}\left(\frac{2x}{n_2}\right) dx \tag{2.41}$$

Particularly, for the transition $1S \to nP$ we get

$$\langle nP | \, r \, | 1S \rangle = \frac{a_B}{Z} \sqrt{\frac{(n+1)!}{(n-2)!} \frac{16 n^3 (n-1)^{n-3}}{(n+1)^{n+3}}}. \tag{2.42}$$

Table 2.2. Radial matrix elements of transitions $1S \to nP$

n	**2**	**3**	**4**	**5**	**6**	**7**	**8**
$\frac{Z}{a_B}\langle nP\|r\|1S\rangle$	1.29027	0.51669	0.30458	0.2087	0.15514	0.12142	0.0985
n	**9**	**10**	**11**	**12**	**13**	**14**	**15**
$\frac{Z}{a_B}\langle nP\|r\|1S\rangle$	0.08205	0.06975	0.06026	0.05276	0.0467	0.04173	0.03758
n	**16**	**17**	**18**	**19**	**20**	**21**	**22**
$\frac{Z}{a_B}\langle nP\|r\|1S\rangle$	0.03408	0.03109	0.02852	0.02628	0.02432	0.02259	0.02106

Table 2.2 shows the numerical values of the matrix elements $1S \to nP$ when n is varied in the range $2 \le n \le 22$.

The equation (2.42) enables us to find the asymptotic form of the matrix elements from the ground state to the high-lying quasiclassical nP states. In the case when $n \gg 1$ we get from the equation (2.42) the following result

$$\langle nP|\,r\,|1S\rangle = \frac{a_B}{Z}\sqrt{\frac{(n+1)!}{(n-2)!}\,\frac{16n^3(n-1)^{n-3}}{(n+1)^{n+3}}} \xrightarrow[n\gg 1]{} \frac{C}{n^{3/2}}. \tag{2.43}$$

Thus, for these transitions, the oscillator force decreases with the increase of the principle quantum number as $1/n^3$. In Fig. 2.3 the matrix

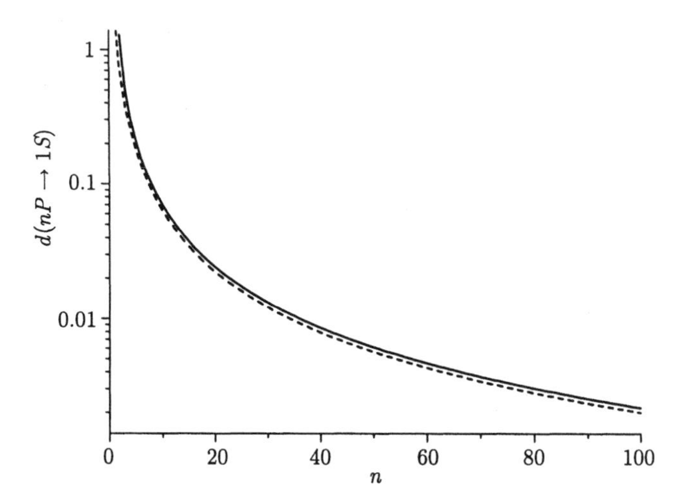

Figure 2.3. The magnitude of the matrix elements of $nP \to 1S$ transitions in hydrogen as a function of the principle quantum number n. The solid curve is the exact function (2.42), the dashed curve is approximation (2.43). The dashed curve is shifted down for convenience

elements calculated on the basis of equations (2.42) and (2.43) are shown in comparison. For the illustration purposes we assume the coefficient C equal to $C = 2$, but at $C = 2.2$ the two curves in Fig. 2.3 coincide almost completely in the region $n \geqslant 7$. There is some discrepancy in the region $n < 7$. But by comparing the equations (2.42) and (2.43) we can see that according to equation (2.42) in the region $n < 7$ the decrease is more fast, therefore in this region the higher powers of $1/n$ should be included in asymptotic equation (2.43).

2.4 Hydrogen atom

The hydrogenlike atom consists of the electron and nucleus. The atomic nucleus has the finite mass, therefore the nucleus of the hydrogen atom is also involved into the motion. Hence, the energy spectrum must depend on the nucleus mass. If the ratio of electron mass to nucleus mass (in the hydrogen atom this is the ratio of electron mass to proton mass $m_e/m_n \approx 5 \cdot 10^{-4}$) is taken as a smallness parameter, then the energy spectrum of electron in the Coulomb field gives us only the zero order approximation for the hydrogen atom spectrum. The total energy of atom is the sum of the electron energy and nucleus energy. Similar the total momentum, and total angular momentum of atom are the sums of them for electron and nucleus. In the processes of absorption or emission of photons by an atom, the conservation laws hold for whole isolated system, therefore the motion of electron in the process of photon absorption or emission is always accompanied by the motion of nucleus. Therefore if we would like to increase the accuracy of calculated energy spectra for hydrogenlike atom we should take into account the motion of the atomic nucleus.

The account for the finite nucleus mass provides the simplest hydrogen atom model. The further development of this model will be given in the next chapters. Here, we start with the study of the influence of the finiteness of the nucleus mass on the energy spectra of hydrogenlike atoms.

2.4.1 Hamiltonian of two-particle problem

The Hamiltonian of system consisting of two charged particles with the Coulomb interaction is

$$H_0 = \frac{\mathbf{p}_a^2}{2m_a} + \frac{\mathbf{p}_b^2}{2m_b} + \frac{q_a q_b}{|\mathbf{r}_a - \mathbf{r}_b|}, \tag{2.44}$$

where q_a and q_b are the charges of particles. It is seen that the Hamiltonian (2.44) does not commute with momentum operators for each individual particle, but it commutes with the total momentum operator

$$\mathbf{P} = \mathbf{p}_a + \mathbf{p}_b. \tag{2.45}$$

Indeed

$$[\mathbf{P}, H_0] = -i\hbar q_a q_b \left(\frac{\partial}{\partial \mathbf{r}_a} + \frac{\partial}{\partial \mathbf{r}_b} \right) \frac{1}{|\mathbf{r}_a - \mathbf{r}_b|} = 0.$$

It means that the variation in the electron momentum of free atom is always accompanied by the variation in the nucleus momentum. However, the coordinate of the atomic center of mass does not vary, because the total momentum is an integral of motion. Therefore it is convenient to introduce the center-of-mass coordinate, $\mathbf{R}$, and relative position coordinate, $\mathbf{r}$:

$$\mathbf{r} = \mathbf{r}_a - \mathbf{r}_b, \quad \mathbf{R} = \frac{m_a \mathbf{r}_a + m_b \mathbf{r}_b}{m_a + m_b}. \tag{2.46}$$

Similar to the momentum operators, the operators of angular momentum of each individual particle do not commute with the Hamiltonian (2.44), but the total angular momentum operator

$$\hbar \mathbf{L} = \hbar \mathbf{l}_a + \hbar \mathbf{l}_b = [\mathbf{r}_a \mathbf{p}_a] + [\mathbf{r}_b \mathbf{p}_b] \tag{2.47}$$

commutes with the Hamiltonian (2.44). By taking into account that the angular momentum operators $\mathbf{l}_a$ and $\mathbf{l}_b$ commute with each other and both of them commute with the Laplace operators $\Delta_{a,b}$ we get for the total angular momentum operator

$$[\mathbf{L}, H_0] = -i q_a q_b \left(\left[\mathbf{r}_a \frac{\partial}{\partial \mathbf{r}_a} \right] + \left[\mathbf{r}_b \frac{\partial}{\partial \mathbf{r}_b} \right] \right) \frac{1}{|\mathbf{r}_a - \mathbf{r}_b|} = 0.$$

Thus the total angular momentum operator $\mathbf{L}$ is the integral of motion, while the angular momenta of the individual particles are not conserved.

If the transformations (2.46) are applied to the equation (2.47) we get

$$\hbar \mathbf{L} = [\mathbf{r}\mathbf{p}] + [\mathbf{R}\mathbf{P}], \tag{2.48}$$

where

$$\mathbf{p} = -i\hbar \frac{\partial}{\partial \mathbf{r}} = \frac{m_b}{m_a + m_b} \mathbf{p}_a - \frac{m_a}{m_a + m_b} \mathbf{p}_b, \quad \mathbf{P} = -i\hbar \frac{\partial}{\partial \mathbf{R}} = \mathbf{p}_a + \mathbf{p}_b. \tag{2.49}$$

Thus the total angular momentum $\mathbf{L}$ is the sum of the angular momentum of center of mass and angular momentum of the relative motion of particles.

2.4.2 Reduced electron mass

If the transformations (2.46) and (2.49) are applied to the Hamiltonian (2.44), we get

$$H_0 = -\frac{\hbar^2}{2m_r}\Delta_r + \frac{q_a q_b}{r} - \frac{\hbar^2}{2(m_a+m_b)}\Delta_R, \tag{2.50}$$

where m_r is the reduced electron mass, defined as

$$m_r = \frac{m_a m_b}{m_a + m_b}. \tag{2.51}$$

If an atom is placed into the atomic trap, the potential of which possesses the central symmetry, then the total angular momentum of atom $\mathbf{L}$ is still the integral of motion. Indeed, if the potential energy of a trap depends only on the magnitude of the radius vector $\mathbf{R}$, i.e. $U_{\text{trap}} = U(R)$, we get

$$[\mathbf{L}, U(R)] = -i\left([\mathbf{r}_a \nabla_a] + [\mathbf{r}_b \nabla_b]\right) U(R) =$$
$$= -i\left[\frac{m_a \mathbf{r}_a + m_b \mathbf{r}_b}{m_a + m_b}\mathbf{R}\right]\frac{1}{R}\frac{\partial U}{\partial R} = -i\,[\mathbf{R}\mathbf{R}]\,\frac{1}{R}\frac{\partial U}{\partial R} = 0.$$

The Hamiltonian for an atom, placed in the atomic trap, is

$$H = H_0 + U(R) = -\frac{\hbar^2}{2m_r}\Delta_r + \frac{q_a q_b}{r} - \frac{\hbar^2}{2(m_a+m_b)}\Delta_R + U(R). \tag{2.52}$$

Thus we can see that the Hamiltonian (2.52) is the sum of two terms. One of them depends on the relative position coordinate, $\mathbf{r}$, another term depends on the center-of-mass coordinate, $\mathbf{R}$. It is seen that the angular momentum operator of the relative motion of particles $[\mathbf{r}\mathbf{p}]$ and angular momentum operator of center of mass $[\mathbf{R}\mathbf{P}]$ commutes separately with the Hamiltonian (2.52). Hence, both angular momenta are the integrals of motion. In this case the two-particle wave function $\psi(\mathbf{r}_a, \mathbf{r}_b)$ is factorized, i.e. it becomes a product of the wave functions depending on the coordinates $\mathbf{r}$ and $\mathbf{R}$ in separate

$$\psi(\mathbf{r}_a, \mathbf{r}_b) = f(\mathbf{r})\, g(\mathbf{R}).$$

The wave functions $f(\mathbf{r})$ and $g(\mathbf{R})$ obey the following equations

$$\left(-\frac{\hbar^2}{2m_r}\Delta_r + \frac{q_a q_b}{r}\right) f(\mathbf{r}) = E^{(1)} f(\mathbf{r}), \tag{2.53a}$$

$$\left(-\frac{\hbar^2}{2(m_a+m_b)}\Delta_R + U(R)\right) g(\mathbf{R}) = E^{(2)} g(\mathbf{R}). \tag{2.53b}$$

It is seen that the Hamiltonian of the equation (2.53a) will completely coincide with the Hamiltonian for the problem of electron motion in Coulomb field, if we substitute the electron mass in equation (2.29) by the reduced mass defined by the equation (2.51). Hence without any additional analysis, we can easily write the equation for the energy spectrum of the hydrogenlike atom, when $q_a = -|e|$ and $q_b = Z|e|$

$$E_n^{(1)} = -\frac{Z^2 e^4}{2\hbar^2 n^2} \frac{m_e m_n}{m_n + m_e}, \tag{2.54}$$

where m_e is the electron mass, and m_n is the nucleus mass.

2.4.3 Atom in trap

In the case of the free atom ($U(R) = 0$) the solutions of the equation (2.53b) are the plain waves $\psi(\mathbf{R}) = C_{\mathbf{K}} \exp(i\mathbf{KR})$. To find the wave function $\psi(\mathbf{R})$ in the case of trapped atom, we need in the profile of the potential well of atomic trap. In the vicinity of its bottom, the potential energy of the atomic trap can be approximated by the parabolic potential well:

$$U(\mathbf{R}) = U(0) + \alpha R^2.$$

By accounting the spherical symmetry of the problem, the wave function of the equation (2.53b) can be taken in the following form

$$g(\mathbf{R}) = g_l(R) Y_{lm}(\theta_R, \varphi_R),$$

where the radial wave function $g_l(R)$ obeys the following equation

$$\frac{d^2 g_l}{dR^2} + \frac{2}{R}\frac{dg_l}{dR} + \left(\frac{2ME}{\hbar^2} - \frac{l(l+1)}{R^2} - \left(\frac{M\Omega}{\hbar}\right)^2 R^2\right) g_l = 0, \tag{2.55}$$

where

$$M = m_a + m_b, \quad \Omega^2 = 2\alpha/M.$$

The solutions of the equation (2.55) are again the confluent hypergeometric functions. By taking into account the boundary conditions that were discussed in the previous section, for eigenfunctions of the equation (2.53b) we get

$$g_{n_R lm}(\mathbf{R}) = \sqrt{\frac{2\beta^{l+3/2} n_R!}{(n_R + l + 1/2)!}} Y_{lm}(\theta_R, \varphi_R) \times \\ \times R^l \exp\left(-\frac{\beta R^2}{2}\right) L_{n_R}^{(l+1/2)}(\beta R^2), \tag{2.56}$$

where $\beta = M\Omega/\hbar$, and n_R is the non-negative integer. The spectrum of the energy eigenvalues is defined by

$$E^{(2)}_{n_R l} = \hbar\Omega\left(2n_R + l + \frac{3}{2}\right). \tag{2.57}$$

Similarly to the electron motion in the Coulomb field, the spectrum (2.57) is degenerated with respect to the combination of the quantum numbers n_R and l. In this case, the combination is the sum of doubled radial quantum number and orbital quantum number, $p = 2n_R + l$.

2.4.4 Interaction of trapped atom with electromagnetic field

Let us consider the interaction of the trapped atom with the electromagnetic field. For the hydrogenlike atom, we have $q_a = -|e|$ and $q_b = Z|e|$, and the Hamiltonian of the problem can be written in the form

$$\begin{aligned} H = {} & \frac{1}{2m_a}\left(\mathbf{p}_a - \frac{q_a}{c}\mathbf{A}(\mathbf{r}_a)\right)^2 + \frac{1}{2m_b}\left(\mathbf{p}_b - \frac{q_b}{c}\mathbf{A}(\mathbf{r}_b)\right)^2 + U(r) + U(R) = \\ & = \frac{\mathbf{p}^2}{2m_r} + \frac{|e|}{c}\left(\frac{1}{m_a}\mathbf{A}(\mathbf{r}_a) + \frac{Z}{m_b}\mathbf{A}(\mathbf{r}_b)\right)\mathbf{p} + U(r) + \\ & + \frac{\mathbf{P}^2}{2(m_a + m_b)} - \frac{|e|}{Mc}\left(\mathbf{A}(\mathbf{r}_a) - Z\mathbf{A}(\mathbf{r}_b)\right)\mathbf{P} + U(R) + \\ & + \frac{1}{2c^2}\left(\frac{q_a^2}{m_a}\mathbf{A}^2(\mathbf{r}_a) + \frac{q_b^2}{m_b}\mathbf{A}^2(\mathbf{r}_b)\right). \end{aligned} \tag{2.58}$$

The characteristic spatial width of the potential well of atomic trap is significantly greater than the Bohr radius, i.e. $\sqrt{\beta}a_{\mathrm{B}} \ll 1$, and we can use the following expansions

$$\begin{aligned} \mathbf{A}(\mathbf{r}_a) &= \mathbf{A}(\mathbf{R}) + \frac{m_b}{M}(\mathbf{r}\nabla)\mathbf{A}(\mathbf{R}) + \ldots, \\ \mathbf{A}(\mathbf{r}_b) &= \mathbf{A}(\mathbf{R}) - \frac{m_a}{M}(\mathbf{r}\nabla)\mathbf{A}(\mathbf{R}) + \ldots \end{aligned} \tag{2.59}$$

The leading term of both expansions is

$$\mathbf{A}(\mathbf{r}_{a,b}) \approx \mathbf{A}(\mathbf{R}),$$

Hence, the Hamiltonian (2.58) takes the form

$$\begin{aligned} H = {} & \frac{\mathbf{p}^2}{2m_r} + U(r) + \frac{|e|}{c}\left(\frac{1}{m_a} + \frac{Z}{m_b}\right)\mathbf{A}(\mathbf{R})\mathbf{p} + \\ & + \frac{\mathbf{P}^2}{2M} + U(R) + \frac{|e|}{Mc}(1 - Z)\mathbf{A}(\mathbf{R})\mathbf{P} + \frac{e^2}{2c^2}\left(\frac{1}{m_a} + \frac{Z^2}{m_b}\right)\mathbf{A}^2(\mathbf{R}). \end{aligned} \tag{2.60}$$

The last term in (2.60) is usually omitted, because its mean value over the period of optical oscillations does not depend on the coordinate. Therefore, for the hydrogen atom (when $Z = 1$), we get

$$H = -\frac{\hbar^2}{2m_r}\Delta_r + U(r) + \frac{|e|}{m_r c}\mathbf{A}(\mathbf{R})\mathbf{p} - \frac{\hbar^2}{2M}\Delta_R + U(R). \qquad (2.61)$$

By comparing the equations (2.60) and (2.61) we can see that in hydrogen atom, in contrast to the hydrogenlike ions, the transitions between the trap levels without change in the intra-atomic electron state are prohibited. However it should be noted that the energy distance between the states of atom in the atomic trap $\hbar\Omega$ is usually much smaller than the energy distance between the different electron states in atom $\hbar\Omega \ll \ll E_n - E_m$. Therefore, if the frequency of electromagnetic wave is close to the frequency of the intra-atomic electron transitions $\omega_{nm} = = (E_n - E_m)/\hbar \approx \omega$ and at the same time $\Omega \ll \omega$, then the probability of the above mentioned transitions is very low for ions too.

As already mentioned, the angular momenta $\hbar\mathbf{l}_1 = [\mathbf{rp}]$ and $\hbar\mathbf{l}_2 = = [\mathbf{RP}]$ are both the integrals of motion. As a result the wave function of the trapped atom can be written as the following product

$$\psi(\mathbf{r}_a, \mathbf{r}_b, t) = f_{n_1 l_1 m_1}(\mathbf{r})\, g_{n_2 l_2 m_2}(\mathbf{R}) \exp\left[-i\frac{E_1 + E_2}{\hbar}t\right],$$

where the values of $E_{1,2}$ are defined by the equations (2.54) and (2.57), respectively:

$$E_1 = -\frac{Z^2 e^4 m_r}{2\hbar^2 (n_1 + l_1 + 1)^2}, \quad E_2 = \hbar\Omega\left(2n_2 + l_2 + \frac{3}{2}\right).$$

The probability amplitude for the transition between the trapped hydrogen atom states of energy $E = E_1 + E_2$ and $E' = E_1' + E_2'$ is defined, in the frame of the first order approximation, by the following equation

$$a_{E'E}(t) = \frac{|e|}{m_r c}\langle n_1' l_1' m_1' | \mathbf{p} | n_1 l_1 m_1 \rangle \times$$
$$\times \int_0^t \langle n_2' l_2' m_2' | \mathbf{A}(\mathbf{R}, t') | n_2 l_2 m_2 \rangle \exp\left[i\frac{E' - E}{\hbar}t'\right] dt'. \qquad (2.62)$$

It is convenient to express the matrix elements of the momentum operator in terms of the matrix elements of the coordinate operator. The commutator of the Hamiltonian (2.53a) and operator $\mathbf{r}$ is

$$[H_{01}, \mathbf{r}] = -\frac{i\hbar}{m_r}\mathbf{p}.$$

Hence

$$\left\langle n_1'l_1'm_1'\right| H_{01}\mathbf{r} - \mathbf{r}H_{01} \left|n_1l_1m_1\right\rangle = \left(E_1' - E_1\right) \left\langle n_1'l_1'm_1'\right| \mathbf{r} \left|n_1l_1m_1\right\rangle = \\ = -\frac{i\hbar}{m_r} \left\langle n_1'l_1'm_1'\right| \mathbf{p} \left|n_1l_1m_1\right\rangle .$$

Now, we can use the matrix elements of the coordinate operator, that were calculated in the previous section.

Let atom interact with the plain wave

$$\mathbf{A}\left(\mathbf{R}, t\right) = \mathbf{A}_0 \sin\left(\omega t - \mathbf{kR}\right).$$

To calculate the matrix elements $\langle n_2'l_2'm_2'| \mathbf{A}\left(\mathbf{R}\right) |n_2l_2m_2\rangle$ we shall use the following expansion of plain wave onto the spherical harmonics

$$\exp(ikZ) = \sum_{l=0}^{\infty} i^l \left(2l+1\right) P_l\left(\cos\theta\right) j_l(kr),$$

where $j_l\left(x\right)$ is the spherical Bessel functions. We have assumed that the wave vector $\mathbf{k}$ of incident wave is directed along the z axis of the given reference frame. Thus, the required matrix elements are

$$\left\langle n_2'l_2'm_2'\right| P_l\left(\cos\theta\right) j_l(kr) \left|n_2l_2m_2\right\rangle = \\ = \int Y^*_{l_2'm_2'}\left(\theta,\varphi\right) P_l\left(\cos\theta\right) Y_{l_2m_2}\left(\theta,\varphi\right) \sin\theta\, d\theta\, d\varphi \times \\ \times \int R^{l_2'+l_2} \exp\left(-\beta R^2\right) L_{n_2'}^{\left(l_2'+1/2\right)}\left(\beta R^2\right) L_{n_2}^{\left(l_2+1/2\right)}\left(\beta R^2\right) j_l(kR) R^2\, dR \tag{2.63}$$

The first integral in (2.63) results in the following selection rules

$$l_2' = l_2 + l, \quad m_2' = m_2.$$

To calculate the second integral in (2.63) we can use the following formula

$$\int_0^{\infty} x^{\alpha-1} \exp\left(-px^2\right) J_\nu(cx)\, dx = \\ = \frac{c^\nu}{p^{(\alpha+\nu)/2}} \frac{\Gamma\left(\left(\alpha+\nu\right)/2\right)}{2^{\nu+1}\Gamma\left(\nu+1\right)} F\left(\frac{\alpha+\nu}{2}, \nu+1, -\frac{c^2}{4p}\right),$$

where $F\left(a,b,z\right)$ is the confluent hypergeometric function.

As an example for the matrix elements of the transition between the initial state $n_2 = 0$, $l_2 = 0$, $m_2 = 0$ and final state $n_2' = n$, $l_2' = l$, $m_2' = 0$ we get

$$\langle nl0|\exp(ikZ)|000\rangle =$$
$$= \sqrt{\frac{\pi}{2}\left(l+\frac{1}{2}\right)\frac{n!\,(n+l+1/2)!}{(1/2)!}}\exp\left(-\frac{k^2}{4\beta}\right)\cdot\left(\frac{k^2}{4\beta}\right)^{l/2}\times$$
$$\times\sum_{m=0}^{n}\frac{(-1)^m}{(n-m)!\,(l+m+1/2)!}L_m^{(l+1/2)}\left(\frac{k^2}{4\beta}\right). \quad (2.64)$$

In particular

$$\begin{aligned}
|\langle 000|\exp(ikZ)|000\rangle|^2 &= \exp\left(-\frac{k^2}{2\beta}\right),\\
|\langle 010|\exp(ikZ)|000\rangle|^2 &= \frac{k^2}{2\beta}\exp\left(-\frac{k^2}{2\beta}\right),\\
|\langle 100|\exp(ikZ)|000\rangle|^2 &= \frac{1}{6}\left(\frac{k^2}{2\beta}\right)^2\exp\left(-\frac{k^2}{2\beta}\right),\\
|\langle 110|\exp(ikZ)|000\rangle|^2 &= \frac{1}{10}\left(\frac{k^2}{2\beta}\right)^3\exp\left(-\frac{k^2}{2\beta}\right).
\end{aligned} \quad (2.65)$$

The interpretation of the obtained equations becomes more obvious, if we substitute the parameter β in the last equations by its explicit expression:

$$\frac{k^2}{2\beta} = \frac{(\hbar k)^2}{2M}\frac{1}{\hbar\Omega} = \frac{E_k}{\hbar\Omega}. \quad (2.66)$$

Let atom be initially in the ground trap state. In the process of photon absorption the atom should accept the recoil momentum $\hbar k$ and, hence, the recoil energy $E_k = \hbar^2k^2/(2M)$. Thus we can see from the equations (2.65) that the probability of atom transition from the ground to excited trap state, in the process of photon absorption, depends on the ratio of the recoil energy to the energy difference between the trap states.

Notice that the process of the emission or absorption of photons by the trapped atom is similar to the process of emission or absorption of gamma photons by nuclei in crystals (Mossbauer effect). In the latter case the probability of the recoilless emission depends on the ratio of the recoil energy to the phonon energy.

The time integration in the equation (2.62) results in the energy conservation law

$$E_1' + E_2' = E_1 + E_2 + \hbar\omega.$$

Thus the equations (2.62)–(2.64) determine completely the selection rules and the probabilities of the radiative transitions for the trapped hydrogen atom. In contrast to hydrogen atom, the hydrogenlike ions ($Z > 1$) can make the transitions between the atomic trap levels without change in the intra-atomic electron state. This type of transitions is described by the two last terms of the Hamiltonian (2.60).

Chapter 3

VARIATIONAL PRINCIPLE FOR SCHRÖDINGER EQUATION: ORBITAL INTERACTION IN HYDROGENLIKE ATOMS

Here we start with the application of the canonical Lagrangian formalism to the problem of the many-particle system, in which the particles are coupled by the electromagnetic field. The Lagrangian L is the space integral of the local functions of the electromagnetic and material fields and their space and time derivatives. The field equations are determined from the principle that the action $\int L\,dt$ should be stationary when the fields are varied. The variational derivative of the Lagrangian is the 'momentum' conjugate to that field. We explore the various invariance and conservation laws. The Hamiltonian function is the sum of all canonical momenta times the time derivatives of the corresponding fields, minus the Lagrangian. The Hamiltonian function provides us with the energy functional for the hydrogen atom, the variation of which yields the equation for the hydrogen atom wave function. The hydrogen atom spectrum is determined by the solution of this equation.

3.1 Particle wave fields

In the frame of quantum mechanics it is assumed that the microparticles are the point particles, but the probability amplitudes for them to be in definite point in space at definite moment of time obey the wave equations. The wave equations reflect the presence of the wave properties in the behavior of microparticles. The matter fields, as well as the electromagnetic field, are characterized by their amplitudes at each spatial point at any moment of time, therefore they are equivalent to the mechanical systems that have the infinite number of degrees of freedom. In the frame of the field theory formalism the wave function $\psi(\mathbf{r}, t)$ plays the role of the coordinate of the matter field.

3.1.1 Lagrange function

The Lagrange function of the point particle depends in classical mechanics on the coordinate, velocity, and time. Similarly, the Lagrange function of the matter field depends on the wave function, its space and time derivatives, and time .

$$L\left(\psi, \nabla\psi, \dot{\psi}, t\right),$$

where $\dot{\psi} = \partial\psi/\partial t$. The appearance of the space derivative, in addition to time derivative, is quite understandable, because the wave function is the continuous function of coordinates. It is this feature that reflects the infinite number of degrees of freedom of the matter fields.

By integrating the Lagrange function over the space and time, we get the action

$$S = \iint L\left(\psi, \nabla\psi, \dot{\psi}, t\right) dV\, dt.$$

The matter field equations are generated by the principle of the least action

$$\delta S = \iint \left[\frac{\partial L}{\partial \psi} - \frac{\partial}{\partial x_\alpha}\left(\frac{\partial L}{\partial \psi_\alpha}\right) - \frac{\partial}{\partial t}\left(\frac{\partial L}{\partial \dot{\psi}}\right)\right] \delta\psi\, dV\, dt = 0, \qquad (3.1)$$

where $\psi_\alpha = \partial\psi/\partial x_\alpha$. As well as the variation $\delta\psi$ is arbitrary, then the equation (3.1) is equivalent to the following differential equation

$$\frac{\partial L}{\partial \psi} - \frac{\partial}{\partial x_\alpha}\left(\frac{\partial L}{\partial\left(\partial\psi/\partial x_\alpha\right)}\right) - \frac{\partial}{\partial t}\left(\frac{\partial L}{\partial\left(\partial\psi/\partial t\right)}\right) = 0. \qquad (3.2)$$

The last equation is called by the Euler–Lagrange equation. In general case, the wave function is a complex function, therefore the matter field is really the two-component field

$$\psi\left(\mathbf{r}, t\right) = \psi_1(\mathbf{r}, t) + i\psi_2(\mathbf{r}, t).$$

Hence, the Lagrange function is the function of the two real wave fields in general case. Of course, instead of real and imaginary parts of the wave function, we can use the functions $\psi\left(\mathbf{r}, t\right)$ and $\psi^*\left(\mathbf{r}, t\right)$ as the field coordinates:

$$L\left(\psi, \nabla\psi, \dot{\psi}, \psi^*, \nabla\psi^*, \dot{\psi}^*, t\right).$$

The action should be independently varied over the functions ψ and ψ^*.

3.1.2 Hamiltonian function

The generalized momentum, canonically conjugated to the field coordinate ψ, is defined by

$$\pi = \frac{\partial L}{\partial \dot{\psi}}.$$

With the help of this definition we can introduce the Hamiltonian function

$$H = \pi\dot{\psi} - L, \tag{3.3}$$

which enables us to introduce the Hamilton equations for field

$$\begin{aligned} \frac{\partial \psi}{\partial t} &= \frac{\partial H}{\partial \pi} - \frac{\partial}{\partial x_\alpha}\left(\frac{\partial H}{\partial\left(\partial \pi/\partial x_\alpha\right)}\right), \\ \frac{\partial \pi}{\partial t} &= -\left[\frac{\partial H}{\partial \psi} - \frac{\partial}{\partial x_\alpha}\left(\frac{\partial H}{\partial\left(\partial \psi/\partial x_\alpha\right)}\right)\right]. \end{aligned} \tag{3.4}$$

The Hamilton equations are convenient when we make the secondary quantization. In this case we should substitute the classic Poisson brackets by the quantum Poisson brackets.

Let the Lagrange function be

$$L = i\hbar\psi^*\dot{\psi} - \frac{\hbar^2}{2m_0}\nabla\psi^*\nabla\psi - \psi^* U\left(\mathbf{r}, t\right)\psi. \tag{3.5}$$

The variation of action with respect to the wave function ψ^* results

$$i\hbar\frac{\partial \psi}{\partial t} + \frac{\hbar^2}{2m_0}\Delta\psi - U\psi = 0. \tag{3.6}$$

The obtained equation coincides with the Schrödinger equation (2.1) with the Hamiltonian (2.2).

The variation of action with respect to the wave function ψ results in the following equation

$$-i\hbar\frac{\partial \psi^*}{\partial t} = -\frac{\hbar^2}{2m_0}\Delta\psi^* + U\left(\mathbf{r}, t\right)\psi^*.$$

It is seen that the obtained equation is complex conjugate of the equation (3.6).

For the Lagrange function (3.5) the generalized momentum canonically conjugated to ψ is

$$\pi = \frac{\partial L}{\partial \dot{\psi}} = i\hbar\psi^*.$$

The Lagrange function (3.5) does not include the derivative $\dot{\psi}^*$, therefore the generalized momentum conjugated to the field coordinate ψ^* is identically equal to zero.

Thus we can see that the principle of the least action enables us to derive the Schrödinger equation from the Lagrange function (3.5). Therefore the equation (3.5) is equivalent, in certain sense, to the equation (3.6).

It should be noted that there is a qualitative difference in transformation properties of the Lagrange function given by equation (3.5) and the Lagrange function for a classical particle. In classical mechanics, the Lagrange function depends only on the square of the particle momentum both in relativistic and non-relativistic cases. The motivations for this are twofold. Firstly, the Lagrange function of free particle can not depend on the coordinate and time, this is due to homogeneity of space and time. Secondly, it can not depend on the direction of particle propagation, this is due to the isotropy of the space. In contrast to this the Lagrange function (3.5) depends on the product of field coordinate ψ^* and generalized momentum $\dot{\psi}^*$. As a result the Lagrange function (3.5) is not invariant with respect to transformation $t \to -t$. If we substitute the product $\psi^*\dot{\psi}$ by the term proportional to $\dot{\psi}^*\dot{\psi}$, which is invariant with respect to the time reversal, then the resultant equation for particle will be the differential equation of the second order with respect to the time derivative. It means that the Lagrange function in the form of (3.5) selects one of the two possible solutions of the second order differential equation. Let us try to answer the question which of the two solutions should be selected. Notice that the Lagrange function (3.5), and hence the action, is not self-conjugate. As a result the selected solution could not be the real function of time. The wave function should definitely be the complex function of time. Only in this case the Lagrange function is invariant with respect to combined transformation including the time reversal and complex conjugation. To satisfy this requirement the Lagrange function (3.5) is taken in such a form that the generalized momentum conjugated to the field coordinate ψ^* is equal to zero. As a result the Lagrange function (3.5) selects from the two possible solutions those which corresponds to the desired relationship between the particle energy and momentum: the particle energy should increase with the increase of its momentum, in complete analogy with the classical mechanics.

It should be noted, that the Lagrange function does not correspond to some observable, therefore it needs not to be a hermitian function. It should only provide us the correct equations of motion. Contrary, the Hamiltonian function corresponds to the observable, because the volume integral of the Hamiltonian function is the energy. By substituting the equation (3.5) into the equation (3.3) we get the following equation for the Hamiltonian function

$$H = \frac{\hbar^2}{2m_0}\nabla\psi^*\nabla\psi + \psi^* U\psi.$$

It is seen that the Hamiltonian function is hermitian. The energy of a particle is

$$E = \int \psi^*(\mathbf{r},t)\left[-\frac{\hbar^2}{2m_0}\Delta + U\right]\psi(\mathbf{r},t)\,dV = \int \psi^*(\mathbf{r},t)\,\hat{H}_0\psi(\mathbf{r},t)\,dV.$$

We can see that, in accordance with the general definitions of the quantum mechanics, the energy is the quantum mechanical average of the Hamiltonian operator $\hat{H}_0$. We have used the hat symbol in the last equation to distinguish the Hamiltonian function from the Hamiltonian operator.

3.1.3 Action for particle interacting with electromagnetic field

The action for a particle interacting with the electromagnetic field depends on the particle variables and potentials of the electromagnetic field. The Lagrange function of the free electromagnetic field is

$$L_f = \frac{1}{8\pi}\left(\mathbf{E}^2 - \mathbf{B}^2\right), \tag{3.7}$$

where the strength of the electric $\mathbf{E}$ and magnetic $\mathbf{B}$ fields is defined by the well known equations

$$\mathbf{E} = -\frac{1}{c}\frac{\partial\mathbf{A}}{\partial t} - \nabla\varphi, \quad \mathbf{B} = \text{curl}\,\mathbf{A}. \tag{3.8}$$

The action for the particle interacting with the electromagnetic field can be obtained from the action of free particle with the help of standard replacement of the four-momentum operator by the generalized four-momentum operator

$$-i\hbar\nabla \to -i\hbar\nabla - \frac{q}{c}\mathbf{A}, \quad i\hbar\frac{\partial}{\partial t} \to i\hbar\frac{\partial}{\partial t} - q\varphi.$$

In result we get the following equation for action of the particle interacting with the electromagnetic field

$$S = \frac{1}{8\pi}\iint\left[\left(\frac{1}{c}\frac{\partial\mathbf{A}}{\partial t} + \nabla\varphi\right)^2 - (\text{curl}\,\mathbf{A})^2\right]dV\,dt +$$

$$+\iint\left[\psi^*\left(i\hbar\frac{\partial\psi}{\partial t} - q\varphi\psi\right) - \frac{1}{2m_0}\left(i\hbar\nabla\psi^* - \frac{q}{c}\mathbf{A}\psi^*\right)\left(-i\hbar\nabla\psi - \frac{q}{c}\mathbf{A}\psi\right)\right]dV\,dt \tag{3.9}$$

The Euler–Lagrange equation, when S is varied with respect to ψ^*, is

$$i\hbar\frac{\partial\psi}{\partial t} = -\frac{\hbar^2}{2m_0}\Delta\psi - \frac{q}{2m_0c}\left(\mathbf{pA} + \mathbf{Ap}\right)\psi + \frac{q^2}{2m_0c^2}\mathbf{A}^2\psi + q\varphi\psi. \quad (3.10)$$

It is seen that the obtained equation coincides with the equation (2.4). The variation of S with respect to ψ results in the equation complex conjugated to (3.10).

Variation of action (3.9) with respect to $\mathbf{A}$ and φ results in the following equations for vector and scalar potentials of the electromagnetic field

$$\frac{1}{c^2}\frac{\partial^2\mathbf{A}}{\partial t^2} + \operatorname{curl}\operatorname{curl}\mathbf{A} + \frac{1}{c}\nabla\frac{\partial\varphi}{\partial t} = 4\pi\mathbf{j},$$

$$\Delta\varphi + \frac{1}{c}\frac{\partial}{\partial t}\operatorname{div}\mathbf{A} = -4\pi\rho,$$

where the charge ρ and current $\mathbf{j}$ density are defined by the equation (2.8).

If we shall use the Lorentz gauge

$$\frac{1}{c}\frac{\partial\varphi}{\partial t} + \operatorname{div}\mathbf{A} = 0$$

then the electromagnetic field equations become

$$\begin{aligned}\Delta\mathbf{A} - \frac{1}{c^2}\frac{\partial^2\mathbf{A}}{\partial t^2} &= -\frac{4\pi}{c}\mathbf{j},\\ \Delta\varphi - \frac{1}{c^2}\frac{\partial^2\varphi}{\partial t^2} &= -4\pi\rho.\end{aligned} \quad (3.11)$$

It should be noted that, in the frames of the probability interpretation of the wave function, the particle is the point particle and the particle wave field is the field of amplitude of probability for particle to be at given point in space at given moment of time. Therefore the vector and scalar potentials in the equation (3.10) are the potentials that are produced by the external particles. The potentials of the electromagnetic field produced by particle itself are determined by the solution of equations (3.11). The electromagnetic field produced by a particle exists in the area outside of the particle localization point. It is seen from the equations (3.11), that the velocity of electromagnetic wave propagation is equal to the light velocity. The velocity of particle propagation is always smaller that the light velocity, therefore one can say that the particle has no opportunity to interact with its own field.

3.2 Symmetry properties with respect to orthogonal transformations

The symmetry properties of solutions of the eigenvalue problem for equation (2.12) are completely determined by the symmetry properties of the physical system under consideration. For example, if we study the motion of a particle in the external field of the axial symmetry, then the admitted solutions should reflect this symmetry, i.e. the rotation of the coordinate frame around the symmetry axes should not affect on the quantum mechanical averages of the observable variables. Another example is an isolated ensemble of the interacting particles. It is well known that an isolated ensemble of interacting particles possesses symmetry with respect to rotation of ensemble, as a whole, around any axis; the particle permutations should not change the energy of ensemble if the particles in ensemble are identical, and so on. Hence, it is very usefull to examine the symmetry and invariance properties of the equations. This analysis gives us information on the properties of the admitted solutions.

3.2.1 Orthogonal transformations

The basic quantum mechanical operators are the generators of orthogonal transformations such as the spatial translation, three-dimensional rotation, space inversion, etc. The energy of an isolated ensemble of particles is invariant with respect to these transformations, therefore these operators commute with the Hamiltonian and, hence, the observable corresponding to these operators are conservative.

The transformations of translation, rotation, and inversion can be considered in the alternative way. For example, the rotation can be considered as rotation of some object at the fixed position of the reference frame, or as rotation of the reference frame at the fixed position of the object. These two transformations are close connected with each other, because in both cases we transform the particle coordinate. The orthogonal transformations can be represented in the following general form

$$x_i' = a_{ij}x_j + a_i, \tag{3.12}$$

where the matrices a_{ij} satisfy the condition $a_{ij}a_{jk} = \delta_{ik}$. The transformations (3.12) include the spatial translations, three-dimensional rotations, and space inversion.

The transformations of the particle coordinates with respect to the given reference frame and transformations of the reference frame at a given position of particle are mutually reciprocal one to another. Indeed the spatial translation of the particle results in the following transformation of its coordinates $\mathbf{r}' = \mathbf{r} + \mathbf{a}$. On the other hand, the translation of the reference frame at the same vector $\mathbf{a}$ results in the following transformation of the particle coordinates $\mathbf{r}' = \mathbf{r} - \mathbf{a}$. Nevertheless, these two transformations are really not absolutely equivalent.

In general case, the equations for a particle moving in some external fields are not possessed by the translational invariance. By moving the particle alone we change the magnitude of forces that act on the particle from the particles producing these external fields. The transformations of the reference frame mean always that the coordinates of the considered particle and coordinates of the particles, producing the external fields, are simultaneously changed. The translational invariance is the property appropriate to an isolated ensemble of particles only. The presence of the external particles, having the fixed positions, means that the whole system is not isolated.

The second difference is in the fact that the equations for interacting particles, along with the particle equations, should include the equation for the fields realizing the interaction. The symmetry properties of the equations for the fields with respect to transformations (3.12) could not coincide with the symmetry properties of the particle wave equations. Therefore the symmetry properties of the equation for single particle can be significantly different from the symmetry properties of equations for an ensemble of interacting particles.

Let us consider the spatial transformations. By applying the transformation (3.12) to the Schrödinger equation (3.10) for particle a we get

$$i\hbar \frac{\partial \psi'_a(\mathbf{r}'_a)}{\partial t} = \left[\frac{1}{2m_0} \left(\mathbf{p}'_a - \frac{q_a}{c} \mathbf{A}'_b(\mathbf{r}'_a) \right)^2 + q_a \varphi'_b(\mathbf{r}'_a) \right] \psi'_a(\mathbf{r}'_a), \qquad (3.13)$$

where $\mathbf{A}_b(\mathbf{r}_a)$ and $\varphi_b(\mathbf{r}_a)$ are the potentials of the electromagnetic field produced by the particle b in the spatial position of the particle a. In accordance with the equations (3.11) these potentials obey the following equations

$$\left(\Delta' - \frac{1}{c^2} \frac{\partial^2}{\partial t^2} \right) \mathbf{A}'_b(\mathbf{r}', t) = -\frac{4\pi}{c} \left\{ -\frac{i\hbar q_b}{2m_b} \left[\psi'^*_b(\mathbf{r}', t) \cdot \nabla' \psi'_b(\mathbf{r}', t) - \right. \right.$$
$$\left. \left. - \nabla' \psi'^*_b(\mathbf{r}', t) \cdot \psi'_b(\mathbf{r}', t) \right] - \frac{q_b^2}{m_b c} \psi'^*_b(\mathbf{r}', t) \mathbf{A}'_a(\mathbf{r}', t) \psi'_b(\mathbf{r}', t) \right\}, \qquad (3.14a)$$

$$\left(\Delta' - \frac{1}{c^2} \frac{\partial^2}{\partial t^2} \right) \varphi'_b(\mathbf{r}', t) = -4\pi q_b \psi'^*_b(\mathbf{r}', t) \psi'_b(\mathbf{r}', t). \qquad (3.14b)$$

Notice that the d'Alamber operator $\Delta - \dfrac{1}{c^2} \dfrac{\partial^2}{\partial t^2}$ is invariant with respect to any transformation prescribed by equation (3.12). Hence the variations of the field potentials can only be due to the variations in the right-hand-sides of the equations (3.14).

The invariance of the equation with respect to the orthogonal transformations (3.12) means, that there are such operators T, defined by

$$\psi'(\mathbf{r}') = T^{(1)}\psi(\mathbf{r}), \quad \varphi'(\mathbf{r}') = T^{(1)}\varphi(\mathbf{r}), \quad A_i'(\mathbf{r}') = T_{ij}^{(2)}A_j(\mathbf{r}),$$

which transform the equations in the primed reference frame into the equation in the initial unprimed reference frame. The operators T, realizing the above transformations, are the generators of the corresponding transformations.

Under the spatial transformations (3.12), the momentum operator is transformed in the following way

$$p_i' = a_{ij}p_j.$$

Let us initially consider the infinitesimal transformations when the matrix a_{ij} is infinitesimally close to the identity matrix:

$$a_{ij} = \delta_{ij} + \varepsilon_{ij} \tag{3.15}$$

where ε_{ij} is the matrix of the infinitesimal transformation. As far as matrix a_{ij} is unitary matrix, then the matrix ε_{ij} is antisymmetric. Indeed

$$a_{ij}a_{ik} = \delta_{jk} + \varepsilon_{ij} + \varepsilon_{ji} + \ldots = \delta_{jk}.$$

At the infinitesimal transformation (3.15), the momentum operator is transformed in the following way

$$p_i' = (\delta_{ij} + \varepsilon_{ij})\, p_j,$$

therefore it is convenient to write the primed vector potential $\mathbf{A}'(\mathbf{r}')$ in the following form

$$A_i'(\mathbf{r}') = (\delta_{ij} + \varepsilon_{ij})\, A_j(\mathbf{r}') + \delta A_i(\mathbf{r}). \tag{3.16}$$

The first term in the last equation is due to the vectorial nature of $\mathbf{A}$, the second term is due to the variation of $\mathbf{A}$ in result of the infinitesimal transformation of the radius vector $\mathbf{r}$.

By substituting the equation (3.16) into the equation (4.13a), we get

$$\left(\Delta - \frac{1}{c^2}\frac{\partial^2}{\partial t^2}\right)\left[a_{ij}A_{bj}(\mathbf{r}) + \delta A_{bi}(\mathbf{r})\right] =$$
$$= -\frac{4\pi}{c}\left[a_{ij}j_{bj}(\mathbf{r}) + \delta j_{bi}(\mathbf{r}) - \frac{q_b^2}{m_b c^2}\psi_b^*(\mathbf{r})\,\delta A_{ai}\psi_b(\mathbf{r})\right], \tag{3.17}$$

where $\delta \mathbf{j}_b(\mathbf{r})$ is the variation of the current density $\mathbf{j}_b(\mathbf{r})$ associated with the transformation of the wave function due to the infinitesimal transformation of the radius vector $\mathbf{r}$

$$\psi'(\mathbf{r}') = \psi(\mathbf{r}) + \delta\psi(\mathbf{r}).$$

The first terms in the both sides of the equation (3.17) are mutually canceled, because they coincide with the equation in the initial non-primed reference frame. The residual terms establish the connection between the variation of vector potential and variation of the wave function, where both of them are due to the infinitesimal transformation of the radius vector $\mathbf{r}$.

The similar relation can be easily obtained for scalar potential $\varphi(\mathbf{r}, t)$, but this relation is evident because both $\varphi(\mathbf{r}, t)$ and wave function are scalar functions. Notice that under the infinitesimal transformation the charge density is transformed in the following way

$$\rho'(\mathbf{r}', t) = q(\psi^*(\mathbf{r}, t) + \delta\psi^*(\mathbf{r}, t))(\psi(\mathbf{r}, t) + \delta\psi(\mathbf{r}, t)) = \rho(\mathbf{r}, t) + \delta\rho(\mathbf{r}, t). \tag{3.18}$$

By substituting the transformation (3.16) into the equation (3.13), we get

$$\hat{H}' = \frac{1}{2m_a}\left[\left(\mathbf{p} - \frac{q_a}{c}\mathbf{A}_b(\mathbf{r}_a)\right)^2 - 2\frac{q_a}{c}\left(\mathbf{p} - \frac{q_a}{c}\mathbf{A}_b(\mathbf{r}_a)\right)\delta\mathbf{A}_b(\mathbf{r}_a)\right] + \\ + q_a(\varphi_b(\mathbf{r}_a) + \delta\varphi_b(\mathbf{r}_a)) = \hat{H} + \delta\hat{H}. \tag{3.19}$$

Hence, the variations of Hamiltonian are completely due to the infinitesimal transformation of the radius vector $\mathbf{r}$ and they are not associated with the vectorial manner of the generalized momentum operator.

Thus the relativistic invariant form of the electromagnetic field equations and the quadratic dependency of the Hamiltonian on the generalized momentum operator exclude the variations of Hamiltonian associated with the vectorial manner of the appropriate variables and remain only the variations that are due to the infinitesimal transformation of the radius vector $\mathbf{r}$.

3.2.2 Space inversion

The matrix a_{ij} of the space inversion transformation is

$$a_{ij}^{(P)} = \begin{pmatrix} -1 & 0 & 0 \\ 0 & -1 & 0 \\ 0 & 0 & -1 \end{pmatrix}.$$

The space inversion transformation is the discrete transformation. The space inversion transformation of the wave function is given by

$$\psi'(\mathbf{r}') = \psi(-\mathbf{r}) = P\psi(\mathbf{r}). \tag{3.20}$$

Hence, the operator, P, of the space inversion transformation is defined by

$$P\psi(\mathbf{r}) = \psi(-\mathbf{r}).$$

The eigenvalues of the operator P are determined by the solutions of the following equation

$$P\psi(\mathbf{r}) = \lambda\psi(\mathbf{r}). \tag{3.21}$$

As we have shown in the previous chapter, the eigenvalues are

$$\lambda = \pm 1.$$

Thus, the space inversion transformation shows us that the eigenfunctions of the Schrödinger equation can be scalar or pseudoscalar functions.

As well as $P\mathbf{p} = -\mathbf{p}$ and $P\psi'^{*}\psi' = \psi^{*}\psi$, then it follows from the equations (3.14a), (3.14b) that under space inversion transformation the electromagnetic field potentials are transformed in the following way

$$\mathbf{A}'(\mathbf{r}') = \mathbf{A}'(-\mathbf{r}) = -\mathbf{A}(\mathbf{r}), \quad \varphi'(\mathbf{r}') = \varphi'(-\mathbf{r}) = \varphi(\mathbf{r}).$$

Hence, by applying operator P^{-1} to the equations (3.13), (3.14) we transform these equation to their initial unprimed form.

Thus, the coupled set of equations (3.13), (3.14) is invariant with respect to the space inversion transformation.

3.2.3 Spatial translation

Let us consider now the spatial translation transformation. At the infinitesimal translation of reference frame, $\delta\mathbf{a}$, the particle coordinates are transformed in the following way

$$\mathbf{r}' = \mathbf{r} - \delta\mathbf{a}.$$

By applying this transformation to the wave function we get

$$\psi'(\mathbf{r}') = \psi(\mathbf{r} - \delta\mathbf{a}) = \psi(\mathbf{r}) - \delta\mathbf{a}\nabla\psi(\mathbf{r}) = (1 - \delta\mathbf{a}\nabla)\psi(\mathbf{r}). \tag{3.22}$$

Therefore, the operator of the infinitesimal spatial translation is defined by

$$T_{\delta\mathbf{a}} = 1 - \delta\mathbf{a}\nabla. \tag{3.23}$$

In the case of the spatial translation transformation, the matrix ε_{ij} is $\varepsilon_{ij} = 0$, hence, in accordance with the equations (3.17)–(3.19), we get

$$\hat{H}' = T_{\delta\mathbf{a}}\hat{H}, \quad \mathbf{A}'(\mathbf{r}') = T_{\delta\mathbf{a}}\mathbf{A}(\mathbf{r}), \quad \varphi'(\mathbf{r}') = T_{\delta\mathbf{a}}\varphi(\mathbf{r}).$$

According to the definition of the operator of the infinitesimal spatial translation (see (3.23)), we have

$$T_{\delta\mathbf{a}}^{-1} f'(\mathbf{r}') \, g'(\mathbf{r}') = T_{\delta\mathbf{a}}^{-1} f'(\mathbf{r}') \, T_{\delta\mathbf{a}}^{-1} g'(\mathbf{r}') \, .$$

Notice, that the equation (3.18) is the particular case of this general relationship. It can be easily seen, that if we apply the operator $T_{\delta\mathbf{a}}^{-1}$ to both sides of the equations (3.13), (3.14) we transform them into their initial unprimed form. Thus, the coupled set of equations (3.13), (3.14) is invariant with respect to spatial translation transformation and the operator of this transformation is defined by the equation (3.23).

3.2.4 Three-dimensional rotations

The matrix a_{ij} describing the rotation by the angle φ around the z-axis is

$$a_{ij}^{(R)} = \begin{pmatrix} \cos\varphi & \sin\varphi & 0 \\ -\sin\varphi & \cos\varphi & 0 \\ 0 & 0 & 1 \end{pmatrix} .$$

The matrix $a^{(R)}$ is the matrix of continuous transformation because it depends on the rotation angle φ, which can be varied continuously.

At the infinitesimal rotation of reference frame around the z-axis, the particle coordinates are transformed in the following way

$$x' = x + \delta\varphi y, \quad y' = y - \delta\varphi x, \quad z' = z.$$

Hence, the matrix ε_{ij} of the infinitesimal three dimensional rotation is

$$\varepsilon_{ij} = \begin{pmatrix} 0 & \delta\varphi & 0 \\ -\delta\varphi & 0 & 0 \\ 0 & 0 & 0 \end{pmatrix} .$$

At the infinitesimal rotation of reference frame around an arbitrary axis the particle coordinates are transformed as

$$\mathbf{r}' = \mathbf{r} - [\delta\boldsymbol{\varphi}\,\mathbf{r}] \, . \tag{3.24}$$

Hence, the transformation of the wave function at the infinitesimal three-dimensional rotations is defined by

$$\psi'(\mathbf{r}') = \psi(\mathbf{r}) - \delta\boldsymbol{\varphi}\,[\mathbf{r}\,\nabla]\,\psi(\mathbf{r}) = (1 - \delta\boldsymbol{\varphi}\,[\mathbf{r}\,\nabla])\,\psi(\mathbf{r})$$

Thus, the operator of the infinitesimal three dimensional rotation is

$$T_R = 1 - \delta\boldsymbol{\varphi}\,[\mathbf{r}\,\nabla] = 1 - i\delta\boldsymbol{\varphi}\mathbf{l}. \tag{3.25}$$

Hence, we get

$$\mathbf{p}' = \mathbf{p} - [\delta\boldsymbol{\varphi}\,\mathbf{p}], \quad \mathbf{A}'(\mathbf{r}') = T_R\mathbf{A}(\mathbf{r}) - [\delta\boldsymbol{\varphi}\,\mathbf{A}(\mathbf{r})], \quad \varphi'(\mathbf{r}') = T_R\varphi(\mathbf{r}).$$

By substituting the last equations into the equations (3.13), (3.14) and taking into account that the terms proportional to ε_{ij} are mutually canceled (see eq. (3.17)), we obtain finally

$$i\hbar\frac{\partial(T_R\psi)}{\partial t} = T_R\hat{H}(T_R\psi),$$

$$\left(\Delta - \frac{1}{c^2}\frac{\partial^2}{\partial t^2}\right)T_R\mathbf{A} = -\frac{4\pi}{c}T_R\mathbf{j},$$

$$\left(\Delta - \frac{1}{c^2}\frac{\partial^2}{\partial t^2}\right)T_R\varphi = -4\pi T_R\rho.$$

Thus, by applying operator T_R^{-1} to both sides of the last equations we transform them into the initial unprimed form.

3.2.5 Transformations including time axis

The Schrödinger equation is non-relativistic equation, therefore the orthogonal transformations (3.12) do not concern the time axis in this case. Nevertheless it is useful to discuss some transformations including the time axis.

Time shift

The time shift transformation

$$t' = t + \delta t$$

is similar to the spatial translation transformation, and, as a result, it can be analyzed in completely similar way. At the infinitesimal time shift, the wave function is transformed in the following way

$$\psi'(t') = \psi(t + \delta t) = \psi(t) + \delta t\frac{\partial\psi(t)}{\partial t} = \left(1 + \delta t\frac{\partial}{\partial t}\right)\psi(t).$$

Hence, the operator of the infinitesimal time shift is defined by

$$T_{\delta t} = 1 + \delta t\frac{\partial}{\partial t}.$$

Hence, if we apply the operator $T_{\delta t}^{-1}$ to both sides of the equations (3.13), (3.14) we transform them to the initial unprimed form.

Time reversal

There is a significant difference between the time reversal transformation and other transformations (the spatial translation, three dimensional rotations, etc.). Indeed, the orthogonal transformations discussed above are the transformations of the reference frame, which mean that, if the observer O sees a system in a state ψ, then the equivalent observer O', who looks at the same system, will observe it in a different state ψ', but the two observers must find the same quantum averages. When we change sign of time in classical equations of motion, then the particle velocity change sign. The symmetry of classical equations with respect to time reversal means that we are interested in the conditions, at which a particle will be involved into the motion reversed in time. Here, we analyze the time reversal invariance of the Schrödinger equation.

So, at the time reversal transformation we change the sign of time, $t \to -t$. It is seen that at the time reversal transformation the left-hand-side of the equation (3.10) changes sign. However if we make the time reversal and complex conjugation simultaneously, then the equation (3.10) becomes

$$i\hbar\frac{\partial\psi^*\left(\mathbf{r},-t\right)}{\partial t}=\left[\frac{1}{2m_0}\left(i\hbar\nabla-\frac{q}{c}\mathbf{A}\left(\mathbf{r},-t\right)\right)^2+q\varphi\left(\mathbf{r},-t\right)\right]\psi^*\left(\mathbf{r},-t\right). \tag{3.26}$$

It is seen that with the help of transformations

$$\begin{aligned}\psi'\left(\mathbf{r},t'\right)&=\psi^*\left(\mathbf{r},-t\right)\to\psi\left(\mathbf{r},t\right),\\ \mathbf{A}'\left(\mathbf{r},t'\right)&=\mathbf{A}\left(\mathbf{r},-t\right)\to-\mathbf{A}\left(\mathbf{r},t\right),\\ \varphi'\left(\mathbf{r},t'\right)&=\varphi\left(\mathbf{r},-t\right)\to\varphi\left(\mathbf{r},t\right)\end{aligned} \tag{3.27}$$

we return the equations (3.13), (3.14) to their initial unprimed form.

The transformations (3.27) can be easily interpreted. Indeed in the classical electrodynamics the particle will make motion reversed in time, only in the case when the electric field remains invariable, $\mathbf{E}(-t) = \mathbf{E}(t)$, while the magnetic field changes sign, $\mathbf{B}(-t) = -\mathbf{B}(t)$. By taking into account the definition of the electric and magnetic filed vectors (3.8) we can see that at transformations (3.27) the electric field is the even function of time and the magnetic field is the odd function of time. Thus the time-reversed motion of particle is described by the wave function $\psi^*(\mathbf{r},-t)$.

Charge conjugation

We have seen that under the transformations (3.27) the particle makes the time-reversed motion. But in the previous subsection we have assumed that the charges of the particles remain invariable. If we assume now that charges of the particles change their signs $q_{a,b} \to -q_{a,b}$, then the equations (3.13), (3.14) can again be transformed to the initial unprimed form. In this case we should make the following transformation

$$\begin{aligned} \psi'(\mathbf{r},t') &= \psi^*(\mathbf{r},-t) \to \psi(\mathbf{r},t), \\ \mathbf{A}'(\mathbf{r},t') &= \mathbf{A}(\mathbf{r},-t) \to \mathbf{A}(\mathbf{r},t), \\ \varphi'(\mathbf{r},t') &= \varphi(\mathbf{r},-t) \to -\varphi(\mathbf{r},t). \end{aligned} \tag{3.28}$$

It is seen that these transformations are again in agreement with the properties of the classical equations of motion.

The opposite parity of the vector and scalar potentials with respect to the time reversal transformation follows from the relativistic invariant Lorentz gauge condition

$$\frac{1}{c}\frac{\partial\varphi}{\partial t} + \operatorname{div}\mathbf{A} = 0,$$

therefore there is no necessity to consider the transformations differing from (3.27), (3.28), because any other transformations will result in the violation of the Lorentz gauge that we have used in deriving of equations (3.11).

3.3 Many-electron atom

3.3.1 Action principle for many-electron atom

The discussion given in the previous section has shown us that the equations for an ensemble of particles interacting via the electromagnetic field are invariant with respect of spatial translation, time shift and three dimensional rotations. It means that for the whole isolated system including both particles and electromagnetic field the conservation laws associated with the above mentioned transformations hold. Here, the detailed analysis of the equations for an ensemble of particles coupled by the electromagnetic field is presented.

The action for the ensemble of particles coupled by the electromagnetic field is

$$\begin{aligned} S = \frac{1}{8\pi}\int\left[\left(\frac{1}{c}\frac{\partial\mathbf{A}}{\partial t} + \nabla\varphi\right)^2 - (\operatorname{curl}\mathbf{A})^2\right]dV\,dt + \\ + \sum_i \int\left[\psi_i^*\left(i\hbar\frac{\partial\psi_i}{\partial t} - q_i\sum_{j(\neq i)}\varphi_j(\mathbf{r}_i)\psi_i\right) - \right. \\ \left. - \frac{1}{2m_i}\left(i\hbar\nabla\psi_i^* - \frac{q_i}{c}\sum_{j(\neq i)}\mathbf{A}_j(\mathbf{r}_i)\psi_i^*\right)\left(-i\hbar\nabla\psi_i - \frac{q_i}{c}\sum_{j(\neq i)}\mathbf{A}_j(\mathbf{r}_i)\psi_i\right)\right]dV_i\,dt. \end{aligned} \tag{3.29}$$

It is seen that the action is additive.

The variation of action (3.29) with respect to ψ_i^* and electromagnetic field potentials results in the following equations

$$i\hbar\frac{\partial\psi_i}{\partial t} = \left[\frac{1}{2m_i}\left(\mathbf{p}_i - \frac{q_i}{c}\sum_{j(\neq i)}\mathbf{A}_j\left(\mathbf{r}_i\right)\right)^2 + q_i\varphi_j\left(\mathbf{r}_i\right)\right]\psi_i, \tag{3.30a}$$

$$\Delta\mathbf{A} - \frac{1}{c^2}\frac{\partial^2\mathbf{A}}{\partial t^2} = -\frac{4\pi}{c}\sum_i \mathbf{j}_i\left(\mathbf{r},t\right), \tag{3.30b}$$

$$\Delta\varphi - \frac{1}{c^2}\frac{\partial^2\varphi}{\partial t^2} = -4\pi\sum_i \rho_i\left(\mathbf{r},t\right), \tag{3.30c}$$

where

$$\mathbf{j}_i\left(\mathbf{r},t\right) = \frac{q_i}{m_i}\left\{\frac{i\hbar}{2}\left(\nabla_i\psi_i^*\left(\mathbf{r},t\right)\cdot\psi_i\left(\mathbf{r},t\right) - \psi_i^*\left(\mathbf{r},t\right)\nabla_i\psi_i\left(\mathbf{r},t\right)\right) - \right.$$
$$\left. - \frac{q_i}{c}\sum_{j(\neq i)}\psi_i^*\left(\mathbf{r},t\right)\mathbf{A}_j\left(\mathbf{r},t\right)\psi_i\left(\mathbf{r},t\right)\right\}, \tag{3.31a}$$

$$\rho_i\left(\mathbf{r},t\right) = q_i\left|\psi_i\left(\mathbf{r},t\right)\right|^2. \tag{3.31b}$$

As well as the electromagnetic field equations are linear, then the field potentials are the sums of potentials associated with the individual particles of ensemble

$$\mathbf{A}\left(\mathbf{r},t\right) = \sum_i \mathbf{A}_i\left(\mathbf{r},t\right), \quad \varphi\left(\mathbf{r},t\right) = \sum_i \varphi_i\left(\mathbf{r},t\right).$$

In particular, if the ensemble of particles interacts with the electromagnetic fields produced by some external particles (usually they are at infinitely large distance from the considered ensemble), then the integral field is the sum of the external field and fields produced by the particles of the considered ensemble

$$\mathbf{A}\left(\mathbf{r},t\right) = \mathbf{A}_{\text{ext}}\left(\mathbf{r},t\right) + \sum_i \mathbf{A}_i\left(\mathbf{r},t\right), \quad \varphi\left(\mathbf{r},t\right) = \varphi_{\text{ext}}\left(\mathbf{r},t\right) + \sum_i \varphi_i\left(\mathbf{r},t\right).$$

Following the general formalism, discussed above, we introduce the generalized momenta of electromagnetic and matter fields

$$\mathbf{\Pi} = \frac{\partial L}{\partial\dot{\mathbf{A}}} = \frac{1}{4\pi c}\left(\frac{1}{c}\frac{\partial\mathbf{A}}{\partial t} + \nabla\varphi\right), \quad \pi_i = \frac{\partial L}{\partial\dot{\psi}_i} = i\hbar\psi_i^*$$

and the Hamiltonian function

$$H = \mathbf{\Pi}\dot{\mathbf{A}} + \sum_i \pi_i\dot{\psi}_i - L.$$

The space-integral of the Hamiltonian function is the energy of the system

$$E = \frac{1}{8\pi}\int\left[\frac{1}{c^2}\left(\frac{\partial \mathbf{A}}{\partial t}\right)^2 + (\mathrm{curl}\ \mathbf{A})^2 - (\nabla\varphi)^2\right] dV + \\ + \sum_{\substack{i,j \\ (i\neq j)}} \int \psi_i^* \left[\frac{1}{2m_i}\left(\mathbf{p} - \frac{q_i}{c}\mathbf{A}_j(\mathbf{r}_i)\right)^2 + q_i\varphi_j(\mathbf{r}_i)\right]\psi_i\, dV_i. \quad (3.32)$$

Let us use the following transformations

$$(\nabla\varphi)^2 = -\varphi\Delta\varphi + \mathrm{div}\,(\varphi\nabla\varphi),$$

$$\frac{1}{c}\,\mathrm{div}\left(\varphi\frac{\partial \mathbf{A}}{\partial t}\right) = \frac{1}{c}\frac{\partial \mathbf{A}}{\partial t}\nabla\varphi + \frac{1}{c}\varphi\frac{\partial}{\partial t}\,\mathrm{div}\ \mathbf{A} = \frac{1}{c}\frac{\partial \mathbf{A}}{\partial t}\nabla\varphi - \frac{1}{c^2}\varphi\frac{\partial^2\varphi}{\partial t^2},$$

where the Lorentz gauge was used. By applying these transformations to the equation (3.32) we obtain

$$E = \frac{1}{8\pi}\int\left(\mathbf{E}^2 + \mathbf{H}^2\right) dV + \sum_{\substack{i,j \\ (i\neq j)}} \int \psi_i^* \left[\frac{1}{2m_i}\left(\mathbf{p}_i - \frac{q_i}{c}\mathbf{A}_j(\mathbf{r}_i)\right)^2\right]\psi_i\, dV_i. \quad (3.33)$$

Thus, the energy of an ensemble of particles coupled by the electromagnetic field is the sum of the electromagnetic field energy and kinetic energy of particles. It should be reminded here, that the operator $\left(\mathbf{p}_i - (q_i/c)\,\mathbf{A}_j(\mathbf{r}_i)\right)/m_i = \mathbf{v}_i$ is the operator of i-th particle velocity.

In the absence of the external fields the first term in the right-hand-side of equation (3.33) is the energy of the electromagnetic field produced by the particle of the ensemble. This energy depends on the relative positions of the particles of ensemble. For example if the ensemble consists of two oppositely charged particles the energy of electromagnetic field is equal to zero when the positions of the particles coincide. If the particles are far away from each other then the energy of the integral electromagnetic field is not equal to zero. The energy of the integral field increases with the increase of distance between the particles, because the integral field produced by the two oppositely charged particles is non-zero in a volume with the radius comparable with the distance between the particles. The steady states of an ensemble of particles correspond to the local or global minima of the energy functional (3.33) in the space of the particle wave functions ψ_i. When the interacting particles approach to each other then the potential energy of their interaction is transformed into the kinetic energy of their motion. Hence the kinetic energy increases. In the steady states the optimal ratio between the potential and kinetic energy is realized. Thus the wave functions of the steady states of an ensemble of particles can be determined from the minima of the energy functional given by the equation (3.33).

3.3.2 Hydrogen atom

Let the considered ensemble of particles be the ensemble of the two oppositely charged particles. This ensemble is equivalent to the hydrogenlike atom. In the hydrogenlike atom one of the particles is electron, another is nucleus. As we have mentioned above we can find the wave functions of the steady states of this system by calculating the minima of the energy functional given by the equation (3.33). The global minimum of this functional is realized in the ground state of the system. The subsequent steady states realize the local minima, therefore we can find them by varying the energy functional under the additional condition of the orthonormalization of any new wave function with the wave functions of all previous steady states.

The energy given by (3.33) is functional of the electromagnetic field potentials and particle wave functions. On the other hand the electromagnetic field potentials are in their turn the functionals of the particle wave functions. Therefore it is convenient to exclude the electromagnetic field potentials from the equation (3.33) and vary the functional with respect to the particle wave functions only. The strength of the static electric and magnetic fields is given by

$$\mathbf{E} = -\nabla\varphi, \quad \mathbf{B} = \mathrm{curl}\,\mathbf{A}.$$

Therefore with the help of the vectorial transformations

$$(\nabla\varphi)^2 = -\varphi\Delta\varphi + \mathrm{div}\,(\varphi\nabla\varphi)\,,$$

$$(\mathrm{curl}\,\mathbf{A})^2 = \mathrm{div}\,[\mathbf{A}\;\mathrm{curl}\,\mathbf{A}] + \mathbf{A}\,\mathrm{curl}\;\mathrm{curl}\,\mathbf{A},$$

we get the following equation for the electromagnetic field energy

$$E_f = \frac{1}{8\pi}\int \left(\mathbf{E}^2 + \mathbf{H}^2\right) dV = -\frac{1}{8\pi}\int \left(\varphi\Delta\varphi + \mathbf{A}\Delta\mathbf{A}\right) dV = \\ = \frac{1}{2}\int \varphi\rho\, dV + \frac{1}{2c}\int \mathbf{A}\mathbf{j}\, dV, \quad (3.34)$$

where we have used the Lorentz gauge, which, in the static case, is

$$\mathrm{div}\,\mathbf{A} = 0.$$

We have already mentioned that, in the frames of the probability interpretation of the wave function, the elementary particles are the point particles and the wave function $\psi(\mathbf{r}, t)$ determines the amplitude for particle to be at the moment of time t in the spatial point $\mathbf{r}$, therefore the particle could not interact with its own field. Hence, in the case of the hydrogen atom, the equation (3.34) reads

$$E_f = \frac{1}{2} \int \varphi_n(\mathbf{r}_e) \rho_e(\mathbf{r}_e) \, dV_e + \frac{1}{2} \int \varphi_e(\mathbf{r}_n) \rho_n(\mathbf{r}_n) \, dV_n + \\ + \frac{1}{2c} \int \mathbf{A}_n(\mathbf{r}_e) \mathbf{j}_e(\mathbf{r}_e) \, dV_e + \frac{1}{2} \int \mathbf{A}_e(\mathbf{r}_n) \mathbf{j}_n(\mathbf{r}_n) \, dV_n,$$

where ρ_a and $\mathbf{j}_a$ are defined by the equations (3.31a) and (3.31b) respectively; $\varphi_b(\mathbf{r}_a)$ and $\mathbf{A}_b(\mathbf{r}_a)$ are the potentials produced by the particle $b = (n, e)$ at the position of the particle $a = (e, n)$.

By substituting the last equation into the equation (3.33) we get

$$E = \int \psi_e^* \left(\frac{\mathbf{p}_e^2}{2m_e} + \frac{1}{2} q_e \varphi_n(\mathbf{r}_e) \right) \psi_e \, dV_e + \\ + \int \psi_n^* \left(\frac{\mathbf{p}_n^2}{2m_n} + \frac{1}{2} q_n \varphi_e(\mathbf{r}_n) \right) \psi_n \, dV_n - \frac{q_e}{2m_e c} \int \psi_e^* \mathbf{A}_n(\mathbf{r}_e) \mathbf{p}_e \psi_e \, dV_e - \\ - \frac{q_n}{2m_n c} \int \psi_n^* \mathbf{A}_e(\mathbf{r}_n) \mathbf{p}_n \psi_n \, dV_n \quad (3.35)$$

where $\psi_e = \psi_e(\mathbf{r}_e)$ is the electron wave function in the hydrogenlike atom, and $\psi_n = \psi_n(\mathbf{r}_n)$ is the nucleus wave function. In derivation of the equation (3.35) we have used the Lorentz gauge condition. It is helpful to rewrite its again here

$$\mathrm{div}_a \, \mathbf{A}_b(\mathbf{r}_a) = 0. \quad (3.36)$$

The solutions of the static field equations (3.30b) and (3.30c) are

$$\varphi_b(\mathbf{r}_a) = \int \frac{\rho_b(\mathbf{r}_b)}{|\mathbf{r}_a - \mathbf{r}_b|} \, dV_b, \quad \mathbf{A}_b(\mathbf{r}_a) = \frac{1}{c} \int \frac{\mathbf{j}_b(\mathbf{r}_b)}{|\mathbf{r}_a - \mathbf{r}_b|} \, dV_b.$$

These equations enable us to exclude the field potentials from the equation (3.35). By substituting in the latter equations the equations (3.31a) and (3.31b) we get

$$\varphi_b(\mathbf{r}_a) = \int \psi_b^*(\mathbf{r}_b) \frac{q_b}{|\mathbf{r}_a - \mathbf{r}_b|} \psi_b(\mathbf{r}_b) \, dV_b, \quad (3.37a)$$

$$\mathbf{A}_b(\mathbf{r}_a) = \frac{q_b}{c} \int \frac{\psi_b^* \mathbf{v}_b \psi_b}{|\mathbf{r}_a - \mathbf{r}_b|} \, dV_b, \quad (3.37b)$$

where

$$\mathbf{v}_b = \frac{1}{m_b}\left(\mathbf{p}_b - \frac{q_b}{c}\mathbf{A}_a(\mathbf{r}_b) - \frac{i\hbar\,|\mathbf{r}_a - \mathbf{r}_b|}{2}\nabla_b\frac{1}{|\mathbf{r}_a - \mathbf{r}_b|}\right). \tag{3.38}$$

The last term in the equation (3.38) arises from the transformation of the expression $(\mathbf{p}\psi)^*\psi - \psi^*\mathbf{p}\psi$ to the form $\psi^*\mathbf{p}\psi$.

By substituting the equations (3.37a), (3.37b) into the equation (3.35) and using the wave function normalization condition we get

$$E = \int \psi_e^*\psi_n^*\left\{ -\frac{\hbar^2}{2m_e}\Delta_e - \frac{\hbar^2}{2m_n}\Delta_n + \frac{q_e q_n}{r_{en}} - \right.$$
$$\left. - \frac{q_e q_n}{2c^2 r_{en}}\left(\mathbf{v}_n\frac{\mathbf{p}_e}{m_e} + \mathbf{v}_n\frac{\mathbf{p}_n}{m_n}\right)\right\}\psi_e\psi_n\, dV_e\, dV_n. \tag{3.39}$$

The terms in the braces of the equation (3.39) have the following physical meaning.

The first three terms in equation (3.39) are exactly coincide with the hydrogen atom Hamiltonian (2.44), that we have used in the previous chapter:

$$H_0 = -\frac{\hbar^2}{2m_e}\Delta_e - \frac{\hbar^2}{2m_n}\Delta_n + \frac{q_e q_n}{r_{en}}. \tag{3.40}$$

To clarify the nature of the last term in the equation (3.39) it is convenient to use the following transformations. Firstly, the products $\mathbf{v}_a\mathbf{p}_b$ can be identically transformed to the following form

$$\mathbf{v}_n\mathbf{p}_e = \frac{1}{r_{en}^2}\left\{[\mathbf{r}_{en}\mathbf{v}_n][\mathbf{r}_{en}\mathbf{p}_e] + (\mathbf{r}_{en}\mathbf{v}_n)(\mathbf{r}_{en}\mathbf{p}_e) + \frac{i\hbar}{m_n}(\mathbf{r}_{en}\mathbf{v}_e) + \frac{\hbar^2}{2m_n}\right\},$$
$$\mathbf{v}_e\mathbf{p}_n = \frac{1}{r_{en}^2}\left\{[\mathbf{r}_{en}\mathbf{v}_e][\mathbf{r}_{en}\mathbf{p}_n] + (\mathbf{r}_{en}\mathbf{v}_e)(\mathbf{r}_{en}\mathbf{p}_n) - \frac{i\hbar}{m_e}(\mathbf{r}_{en}\mathbf{v}_n) + \frac{\hbar^2}{2m_e}\right\}. \tag{3.41}$$

Secondly, by substituting the equation (3.37b) into the Lorentz gauge condition (3.36) we get

$$\operatorname{div}_a \mathbf{A}_b(\mathbf{r}_a) = -\frac{q_b}{c}\int \frac{\psi_b^*\mathbf{r}_{ab}\mathbf{v}_b\psi_b}{r_{ab}^3}\, dV_b = 0. \tag{3.42}$$

Notice here that in the previous chapter we have shown that the eigenfunctions of the Hamiltonian (3.40) have the form $\psi(\mathbf{r}) = f_{lm}(r,\theta) \times \exp(im\varphi)$, where the functions $f_{lm}(r,\theta)$ can always be chosen as the real functions. Hence, we have

$$\mathbf{j}_{lm} = \frac{iq\hbar}{2m_0}\left[(\nabla\psi)^*\psi - \psi^*\nabla\psi\right] = \mathbf{e}_\varphi\frac{q\hbar m}{m_0}\left|f_{lm}(r,\theta)\right|^2,$$

and the condition (3.42) certainly holds.

Thus if the transformations (3.41) and (3.42) are applied to the last term in the equation (3.39) it finally becomes

$$H_{ll} = \frac{q_e q_n \hbar^2}{2m_e m_n c^2} \frac{\mathbf{l}_e \mathbf{l}_n + \mathbf{l}_n \mathbf{l}_e - 1}{r_{en}^3}, \tag{3.43}$$

where

$$\hbar \mathbf{l}_e = [\mathbf{r}_{en} \mathbf{p}_e], \quad \hbar \mathbf{l}_n = [\mathbf{r}_{ne} \mathbf{p}_n]. \tag{3.44}$$

It should be noted that the last term in the equation (3.43) is due to the fact that the operators $\mathbf{l}_e$ and $\mathbf{l}_n$ are the noncommuting operators.

Thus we can see that the last term in the right-hand-side of the equation (3.39) is responsible for the orbital interaction in the hydrogen atom or the interaction of the electron and nucleus currents. This interaction can be more precisely defined in the following way: the magnetic field, resulted from the orbital motion of nucleus, acts on the electron, moving in its orbit in hydrogen atom, and vise versa.

3.3.3 Integrals of motion

As we have mentioned above the wave functions of the steady states can be determined with the help of variational principle

$$\delta \int \psi_e^* \psi_n^* (H - E) \psi_e \psi_n \, dV_e \, dV_n = 0,$$

where

$$H = H_0 + H_{ll}. \tag{3.45}$$

By varying the energy functional with respect to the function $\Psi(\mathbf{r}_e, \mathbf{r}_n) = \psi_e(\mathbf{r}_e) \psi_n(\mathbf{r}_n)$, we get the following wave equation

$$H\Psi = E\Psi. \tag{3.46}$$

It is convenient to introduce the center-of-mass reference frame

$$\mathbf{r} = \mathbf{r}_e - \mathbf{r}_n, \quad \mathbf{R} = \frac{m_e \mathbf{r}_e + m_n \mathbf{r}_n}{m_e + m_n}.$$

The momentum operators, associated with the radius vectors $\mathbf{r}$ and $\mathbf{R}$, are the relative motion momentum operator $\mathbf{p}$ and center-of-mass motion momentum operator $\mathbf{P}$, respectively. They are

$$\mathbf{p} = -i\hbar \frac{\partial}{\partial \mathbf{r}} = \frac{m_n}{M} \mathbf{p}_e - \frac{m_e}{M} \mathbf{p}_n,$$

$$\mathbf{P} = -i\hbar \frac{\partial}{\partial \mathbf{R}} = \mathbf{p}_e + \mathbf{p}_n,$$

where $M = m_e + m_n$. The equation for the Hamiltonian H_0 in the center-of-mass reference frame becomes

$$H_0 = -\frac{\hbar^2}{2m_r}\Delta_r - \frac{\hbar^2}{2M}\Delta_R + \frac{q_e q_n}{r}, \tag{3.47}$$

where $m_r = m_e m_n / M$ is the reduced mass.

It is seen from the equations (3.47) and (3.43) that the Hamiltonian (3.45) depends only on the radius vector $\mathbf{r}_{en}$. Hence, the Hamiltonian (3.45) commutes with the operator of the total momentum

$$[\mathbf{P}, H_0 + H_{ll}] = 0. \tag{3.48}$$

The total angular momentum operator is defined as

$$\hbar\mathbf{L} = [\mathbf{r}_e\mathbf{p}_e] + [\mathbf{r}_n\mathbf{p}_n] = [\mathbf{r}\mathbf{p}] + [\mathbf{R}\mathbf{P}]. \tag{3.49}$$

The orbital interaction Hamiltonian H_{ll}, defined by the equation (3.43), depends on the angular momentum operators $\hbar\mathbf{l}_e$ and $\hbar\mathbf{l}_n$ (see (3.44)). Let us introduce the new auxiliary operators

$$\hbar\mathbf{l}_1 = [\mathbf{r}\mathbf{p}], \quad \hbar\mathbf{l}_2 = [\mathbf{R}\mathbf{P}], \quad \hbar\mathbf{l}_3 = [\mathbf{r}\mathbf{P}]. \tag{3.50}$$

These operators enable us to rewrite the angular momentum operators for electron and nucleus in the following form

$$\mathbf{l}_e = \mathbf{l}_1 + \frac{m_e}{M}\mathbf{l}_3, \quad \mathbf{l}_n = \mathbf{l}_1 - \frac{m_n}{M}\mathbf{l}_3$$

The commutation relations for operators $\mathbf{L}$ and $\mathbf{l}_i$ are

$$[L_\alpha, l_{i\beta}] = ie_{\alpha\beta\gamma}l_{i\gamma}. \tag{3.51}$$

Hence

$$[L_\alpha, \mathbf{l}_e\mathbf{l}_n] = [L_\alpha, l_{e\beta}]\, l_{n\beta} + l_{e\beta}\, [L_\alpha, l_{n\beta}] = \\ = ie_{\alpha\beta\gamma}l_{e\beta}l_{n\gamma} - ie_{\alpha\gamma\beta}l_{e\gamma}l_{n\beta} = 0. \tag{3.52}$$

The Laplace operators Δ_r and Δ_R of the Hamiltonian H_0 can be written in the following form

$$\Delta_r = \frac{1}{r^2}\frac{\partial}{\partial r}\left(r^2\frac{\partial}{\partial r}\right) - \frac{1}{r^2}\mathbf{l}_1^2, \quad \Delta_R = \frac{1}{R^2}\frac{\partial}{\partial R}\left(R^2\frac{\partial}{\partial R}\right) - \frac{1}{R^2}\mathbf{l}_2^2.$$

It can be easily shown with the help of calculations similarly to (3.52) that the total angular momentum operator $\mathbf{L}$ commutes with the operators $\mathbf{l}_1^2$ and $\mathbf{l}_2^2$.

Thus, we can see, the total angular momentum operator $\mathbf{L}$ commutes with the Hamiltonian (3.45)

$$[\mathbf{L}, H_0 + H_{ll}] = 0. \qquad (3.53)$$

Hence, the total angular momentum of the hydrogen atom is the integral of motion. This is in complete agreement with the general consideration given in previous section, where we have shown that the equations for an isolated ensemble of particles are invariant with respect to the three dimensional rotations of the reference frame.

3.3.4 Energy level shift due to orbital interaction in hydrogenlike atoms

We have shown that the Hamiltonian (3.45)commutes with the total momentum and total angular momentum operators, therefore we can take the eigenvalues of momentum operator $\mathbf{P}$, angular momentum l and its projection m as the quantum numbers characterizing the state of the hydrogen atom. If the atom is motionless then the total momentum is equal to zero, $\mathbf{P} = 0$. In this case the quantum mechanical average of the operator $\mathbf{l}_3$ is equal to zero and we can use the following replacement

$$\mathbf{l}_e\mathbf{l}_n + \mathbf{l}_n\mathbf{l}_e - 1 \to 2\mathbf{l}_1^2.$$

Notice here again, that minus unity in the left-hand-side is due to the fact that $\mathbf{l}_e$ and $\mathbf{l}_n$ are non-commuting operators. The total angular momentum is the sum of the electron and nucleus angular momenta, or the sum of the relative motion and center-of-mass motion angular momenta. In motionless atom the total angular momentum is the angular momentum of the relative motion only. Of course, this operator commutes with itself. Thus for the case of motionless atom the equation (3.46) becomes

$$\left(-\frac{\hbar^2}{2m_r}\Delta - \frac{Ze^2}{r} - \frac{Ze^2\hbar^2}{m_e m_n c^2}\frac{\mathbf{l}^2}{r^3}\right)\psi(\mathbf{r}) = E\psi(\mathbf{r}). \qquad (3.54)$$

It is seen that the spherical harmonics are the solutions of the angular part of this equation. Hence, we have

$$\psi(\mathbf{r}) = f(r)\, Y_{lm}(\theta, \varphi).$$

and for radial part of the wave function we get the following equation

$$\frac{d^2 f}{dr^2} + \frac{2}{r}\frac{df}{dr} + \left(\frac{2m_r E}{\hbar^2} + \frac{2m_r Ze^2}{\hbar^2}\frac{1}{r} - \frac{l(l+1)}{r^2} + \frac{2Ze^2}{Mc^2}\frac{l(l+1)}{r^3}\right) f = 0. \qquad (3.55)$$

It is convenient to introduce the dimensionless coordinate x and energy E':

$$x = \frac{r}{a_{\mathrm{B}}}, \quad E = \frac{\hbar^2 E'}{2m_r a_{\mathrm{B}}^2},$$

where $a_{\mathrm{B}} = \hbar^2/(m_r e^2)$. By substituting these transformations into the equation (3.55) we get

$$\frac{d^2 f}{dx^2} + \frac{2}{x}\frac{df}{dx} + \left(E' + \frac{2Z}{x} - \frac{l\,(l+1)}{x^2} + 2Z\alpha^2 \frac{m_r}{M}\frac{l\,(l+1)}{x^3}\right) f = 0, \quad (3.56)$$

where α is the fine structure constant

$$\alpha = \frac{e^2}{\hbar c}. \quad (3.57)$$

As far as the relative energy of the orbital interaction is proportional to $\alpha^2 m_e/m_n$ we can consider it as a correction to the energy of Coulomb interaction of particles. According to perturbation theory methods the first order correction to the energy of steady states is determined by

$$\Delta E_{nlm}^{(1)} = \int \psi_{nlm}^{*} H_{ll} \psi_{nlm} dV. \quad (3.58)$$

To make the numerical estimations we need in the following integrals

$$\int_0^\infty \frac{1}{r^3} R_{nl}^2(r)\, r^2\, dr$$

where R_{nl} are the radial wave functions for the problem of electron motion in Coulomb field. These functions were calculated in the previous chapter. By substituting R_{nl} given by (2.30) into the above integrals we get

$$\int_0^\infty \frac{1}{r^3} R_{nl}^2(r)\, r^2\, dr = \left(\frac{2Z}{n a_{\mathrm{B}}}\right)^3 \frac{1}{4l\,(l+1)\,(2l+1)}. \quad (3.59)$$

By substituting the equation (3.59) into the equation (3.58) we obtain the following equation for the magnitude of the energy shift

$$\Delta E_{nl}^{(1)} = -\frac{4Z^4\alpha^2}{n^3(2l+1)}\frac{m_r}{M} Ry, \quad (3.60)$$

where the Rydberg constant Ry is defined by

$$Ry = \frac{m_r e^4}{2\hbar^2}. \quad (3.61)$$

It is useful to make the following remark concerning the formula (3.60). We can see from the equation (3.59) that the diagonal matrix elements (3.59) are divergent at $l = 0$. On the other hand the energy shift (3.60) is not. This is due to the fact that the energy of orbital interaction is proportional to $l\,(l+1)$, and this product, being in the numerator of the equation (3.60), cancels the same product in the denominator of this equation. On the first glans, it is looks like that including of the case of $l = 0$ into the general formula (3.59) is an artificial trick, but we should remember that the quantum number l determines the angular momentum of the relative motion of particles. If the angular momentum of relative motion of particles is equal to zero it does not necessary mean that the particles are immovable. As we have mentioned above, the origin of the orbital interaction is in the interaction of the currents due to the motion of particles. Hence if the particles are not motionless then the currents are non-zero.

Thus the energy level shift due to the orbital interaction (i.e. the interaction of particle currents via the magnetic field) is about $\Delta E \sim 10^{-8}\,\mathrm{Ry}$ for the ground state and decreases with the increase of the principle quantum number n. According to equation (3.60) the scaling law for this correction is

$$\Delta E_{nl}^{(ll)} \sim \frac{Z^4\alpha^2}{n^3(2l+1)}\frac{m_e}{m_n}.$$

The Fig. 3.1 shows the diagram of $2S$, $2P$ and $3S$, $3P$, $3D$ energy levels in hydrogen atom. We can see that the account for the orbital interaction

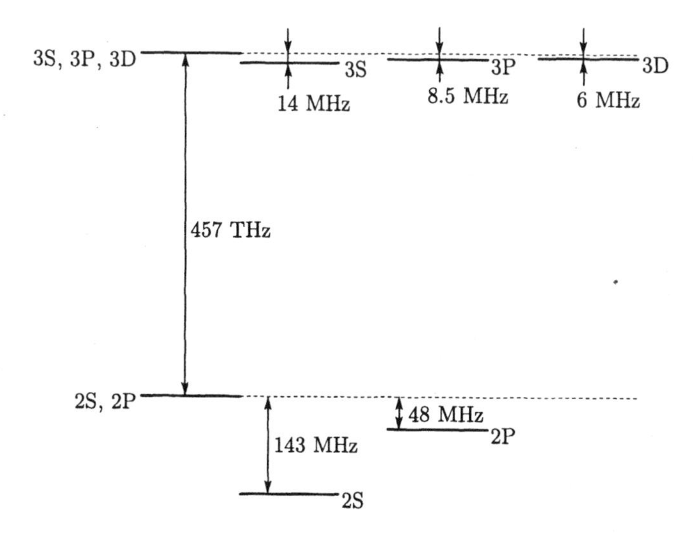

Figure 3.1. The hydrogen level shifts due to the orbital interaction

removes the degeneracy of the nL energy states and as a result the frequencies of $nL \leftrightarrow n'L'$ transitions are changed. But the comparison of the Fig. 3.1 and Fig. 1.1 shows that the calculated spectrum is not yet coincide with the results of the experimental measurements, the experimentally measured spectra show that there is a splitting of the energy levels in addition to their shift with respect to the Bohr formula. The further progress in the description of the hydrogen spectrum was achieved when the idea on the electron spin was implemented into the theory.

Chapter 4

PAULI EQUATION

The series of experiments made in 1921–1925 years gave basis to assume [47–49] that the electron possesses the inner angular momentum of $\hbar/2$ and magnetic moment of $\mu = e\hbar/(2m_0c)$. The operator of the electron inner angular momentum, or spin, can be introduced in the way similar to the angular momentum associated with the translational motion of particle.

4.1 Spin

It has been shown in the previous chapters that the angular momentum operator $[\mathbf{rp}]$ is the generator of the group of three dimensional rotations. The angular momentum l and its projection m determine the angular dependency of the particle wave function and the transformation properties of wave function with respect to the three dimensional rotations. The angular momentum of an isolated system of particles is the integral of motion, and it does not depend on the choice of the reference frame. However, the projection of the angular momentum m is conservative, only in the case, when we rotate the reference frame around the axis of the system rotation. If we decline the z-axis of the reference frame from the axis of particle rotation then the projection of the angular momentum on the new axis z' ceases to be the conservative value. Hence, in this case, the wave function becomes a superposition of the wave functions with all possible projections of the angular momentum onto the z'-axis of the new reference frame. As far as the angular momentum conserves, then the wave function of particle with the definite angular momentum and its projection, ψ_{lm}, becomes a superposition of $2l + 1$

components with the different projections m' in the new reference frame

$$\psi_{lm} = \sum_{m'} \langle lm' | \, U | lm \rangle \psi_{lm'}.$$

Thus, under the three-dimensional rotations, the $2l+1$ wave functions ψ_{lm} are expressed in terms of $2l+1$ wave functions $\psi_{lm'}$, therefore these functions form the irreducible representation of the rotation group. Hence, the angular momentum l defines unambiguously the classification of the particle states with respect to the three-dimensional rotation transformation.

4.1.1 Spin operator

It is clear from the preceding discussion, that if the particle possesses the inner angular momentum s then to describe the inner degrees of freedom we can introduce the multi-component wave function $\psi(\mathbf{r}, \sigma)$ which depends on the quantum number σ having $(2s+1)$ possible values. The angular momentum operator acts on the space coordinates of the particle wave function, therefore, in this case, the rotations of the particle and the rotations of the reference frame are equivalent transformations. The spin operator $\mathbf{s} = (s_x, s_y, s_z)$ acts on the spin variable of the wave function σ therefore the form of this operator are exclusively determined by transformation the reference frame rotation. Indeed, we can always assume that the particle is at the origin of the reference frame, hence its coordinates remain invariable. However, the equivalence of the rotations of particle and reference frame for the translational degrees of freedom results in the coincidence of the commutation relations for the spin operator with commutations relations for the angular momentum operator

$$[s_x, s_y] = is_z, \quad [s_y, s_z] = is_x, \quad [s_z, s_x] = is_y. \tag{4.1}$$

These commutations relations can be obtained directly. Indeed it can be easily shown that the successive infinitesimal rotations of the reference frame, initially around the x and y axes and then around the same axes but in the inverted sequence, are equivalent to the rotation around the z-axis by the angle equal to the product of the rotation angles around the x and y axes.

The similarity in the commutation relations results in the similar properties of the spin and angular momentum operators. The spin square operator

$$\mathbf{s}^2 = s_x^2 + s_y^2 + s_z^2 \tag{4.2}$$

commutes with any spin projection operators. Hence the spin square operator can have the common set of eigenfunctions with the spin projection operator. The joint eigenfunctions of the operators $\mathbf{s}^2$ and s_z obey the following equations

$$\mathbf{s}^2 u_{s\sigma} = s\,(s+1)\,u_{s\sigma}, \quad s_z u_{s\sigma} = \sigma u_{s\sigma}, \tag{4.3}$$

where $s = 0,\ 1/2,\ 1,\ 3/2,\ 2, \ldots$ and $\sigma = -s, -s+1, \ldots, s$.

The total angular momentum is the sum of the orbital angular momentum and spin

$$\mathbf{j} = \mathbf{l} + \mathbf{s}. \tag{4.4}$$

As far as operators $\mathbf{l}$ and $\mathbf{s}$ are applied to the different arguments of wave function then they are commuting operators. The total angular momentum operator $\mathbf{j}$ obeys the same commutation relations as operators $\mathbf{l}$ and $\mathbf{s}$, because, as we have mentioned, the equations (4.1) are the general form of the commutation relations for arbitrary angular momentum operator.

4.1.2 Pauli matrix and spinors

The wave function of spin-1/2 particle is two-component, as well as $(2s+1) = 2$. It is convenient to take it in the form of the two-row column, called by spinor,

$$\psi(\mathbf{r}) = \begin{pmatrix} \psi_{1/2}(\mathbf{r}) \\ \psi_{-1/2}(\mathbf{r}) \end{pmatrix}. \tag{4.5}$$

Hence the spin projection operators are proportional to the Pauli matrices $\boldsymbol{\sigma} = (\sigma_x, \sigma_y, \sigma_z)$

$$\mathbf{s} = \frac{1}{2}\boldsymbol{\sigma}, \tag{4.6}$$

where

$$\sigma_x = \begin{pmatrix} 0 & 1 \\ 1 & 0 \end{pmatrix}, \quad \sigma_y = \begin{pmatrix} 0 & -i \\ i & 0 \end{pmatrix}, \quad \sigma_z = \begin{pmatrix} 1 & 0 \\ 0 & -1 \end{pmatrix}. \tag{4.7}$$

The Pauli matrices σ_α possess the following properties

$$\begin{gathered} \sigma_x^2 = \sigma_y^2 = \sigma_z^2 = I, \\ \sigma_x\sigma_y = i\sigma_z, \quad \sigma_y\sigma_z = i\sigma_x, \quad \sigma_z\sigma_x = i\sigma_y, \\ [\sigma_\alpha, \sigma_\beta] = 2i\sigma_\gamma, \end{gathered} \tag{4.8}$$

in the last equation the indexes (α, β, γ) are (x, y, z) or any sequence of them obtained by the even number of permutations of the indexes (x, y, z). The eigenfunctions of the operator σ_z corresponding to its eigenvalues $\sigma = \pm 1$ are

$$\sigma_z \begin{pmatrix} 1 \\ 0 \end{pmatrix} = (+1) \begin{pmatrix} 1 \\ 0 \end{pmatrix}, \quad \sigma_z \begin{pmatrix} 0 \\ 1 \end{pmatrix} = (-1) \begin{pmatrix} 0 \\ 1 \end{pmatrix}. \tag{4.9}$$

The equations (4.8) enable to derive the following identity

$$(\sigma\mathbf{A})(\sigma\mathbf{B}) = \mathbf{AB} + i\sigma[\mathbf{AB}]. \tag{4.10}$$

The components of the spinor wave function (4.5) are transformed at the reference frame rotations in the following way

$$\psi_1' = a\psi_1 + b\psi_2, \quad \psi_2' = c\psi_1 + d\psi_2,$$

or

$$\begin{pmatrix} \psi_1' \\ \psi_2' \end{pmatrix} = U \begin{pmatrix} \psi_1 \\ \psi_2 \end{pmatrix} = \begin{pmatrix} a & b \\ c & d \end{pmatrix} \begin{pmatrix} \psi_1 \\ \psi_2 \end{pmatrix}. \tag{4.11}$$

The elements of the rotation matrix U are in general case complex and depend on the angles of the reference frame rotation. As far as the matrix U defines the transformation rules for the particle wave functions, the elements of this matrix should satisfy the definite requirements.

Firstly, the matrix U should be the unitary matrix, because, in accordance with the probabilistic interpretation of the wave function, the bilinear combination

$$\psi^+(\mathbf{r})\,\psi(\mathbf{r}) = \begin{pmatrix} \psi_1^* & \psi_2^* \end{pmatrix} \begin{pmatrix} \psi_1 \\ \psi_2 \end{pmatrix} = \psi_1^*\psi_1 + \psi_2^*\psi_2$$

defines the probability for particle to be at specific spatial point. This probability should not depend on the reference frame

$$\psi'^+(\mathbf{r})\,\psi'(\mathbf{r}) = \psi^+(\mathbf{r})\,U^+U\psi(\mathbf{r}).$$

Hence

$$U^+U = I$$

or

$$U^+ = U^{-1}. \tag{4.12}$$

Secondly, the matrix U should be unimodal. This is a general condition for any matrix of rotations. It means

$$ad - bc = 1. \tag{4.13}$$

When the condition (4.13) holds the wave function normalization condition does not depend on the rotations of the reference frame

$$\int \psi'^+(\mathbf{r}')\,\psi'(\mathbf{r}')\,dV' = \int \psi^+(\mathbf{r})\,\psi(\mathbf{r})\,dV = 1.$$

Under the condition (4.13) the matrix U^{-1} is

$$U^{-1} = \begin{pmatrix} d & -b \\ -c & a \end{pmatrix}.$$

By equating it to the conjugated matrix U^+

$$U^+ = \begin{pmatrix} a^* & c^* \\ b^* & d^* \end{pmatrix},$$

we get

$$a = d^*, \quad b = -c^*. \tag{4.14}$$

It is seen that if the conditions (4.13) and (4.14) hold, then the four complex parameters a, b, c, d are the functions of the three real parameters. These three parameters are the three rotation angles that determine the reference frame rotation unambiguously.

Let us make an infinitesimally small rotation $\delta\varphi$ around the z-axis. The rotation matrix of this transformation is $1 + i\delta\varphi s_z$. The z projection of spin under this rotation remains invariable therefore the wave function $\psi(\sigma)$ takes the form $\psi(\sigma) + \delta\psi(\sigma)$, where

$$\delta\psi(\sigma) = i\delta\varphi s_z \psi(\sigma) = i\sigma\psi(\sigma)\,\delta\varphi.$$

Hence for the finite rotation angle φ we get

$$\psi'(\sigma) = \psi(\sigma)\exp(i\sigma\varphi). \tag{4.15}$$

If the rotation angle is equal to $\varphi = 2\pi$, then the wave function components are multiplied by factor of $\exp(i2\pi\sigma) = (-1)^{2\sigma}$ which is the same for each component σ at any spin s. For the spin-1/2 particle this factor is equal to -1. Thus the 2π-rotation brings the particle into the initial state at integer spin s and changes the sign of the particle wave function at half-integer spin.

It should be noted that, if the condition (4.13) holds, then the following bilinear combination is invariant

$$\psi_1'\varphi_2' - \psi_2'\varphi_1' = (ad - bc)(\psi_1\varphi_2 - \psi_2\varphi_1) = (\psi_1\varphi_2 - \psi_2\varphi_1),$$

this bilinear combination corresponds to the zero spin particle consisting of the two spin-1/2 particles. On the other hand, the condition (4.12) yields the following transformation

$$\psi_1'^*\psi_1' + \psi_2'^*\psi_2' = \psi_1^*\psi_1 + \psi_2^*\psi_2.$$

By comparing the last two transformations we can see that the components ψ_1^* and ψ_2^* are transformed as ψ_2 and $-\psi_1$ respectively. The same result follows from the transformation (4.11) directly. Indeed according to (4.11) we get

$$\psi_2'^* = a\psi_2^* + b(-\psi_1^*), \quad \psi_1'^* = c(-\psi_2^*) + d\psi_1^*.$$

This property of the spinor wave function is directly related to the symmetry of the wave equation with respect to the time reversal transformation.

4.1.3 Hamiltonian of Pauli equation

In classical electrodynamics both angular momentum and magnetic moment of the particle are described by the axial vectors. The angular momentum operator in quantum mechanics is also the axial vector, therefore the spin particle possessing the inner angular momentum will possesses the inner magnetic moment

$$\boldsymbol{\mu} = \frac{\mu}{s}\mathbf{s}. \tag{4.16}$$

Thus the Hamiltonian for the spin particle interacting with the electromagnetic field is

$$H = \frac{1}{2m_0}\left(\mathbf{p} - \frac{e}{c}\mathbf{A}\right)^2 + e\varphi - \boldsymbol{\mu}\mathbf{B}, \tag{4.17}$$

where

$$\mathbf{B} = \text{curl } \mathbf{A}.$$

The product of the two axial vectors $\boldsymbol{\mu}$ and $\mathbf{B}$ is invariant with respect to the space inversion transformation. Thus the Hamiltonian (4.17) for the particle moving in the spherically symmetrical potential, $\varphi(\mathbf{r}) = \varphi(r)$, is invariant with respect to the space inversion transformation. As we have shown in the previous chapters, it means that the parity operator, P, commutes with the Hamiltonian (4.17).

The total angular momentum of a spin particle is the sum of the orbital momentum and spin

$$\hbar\mathbf{j} = \hbar\mathbf{l} + \hbar\mathbf{s} = [\mathbf{rp}] + \frac{\hbar}{2}\boldsymbol{\sigma}. \tag{4.18}$$

The magnetic dipole interaction breaks the spherical symmetry of the Hamiltonian, but if the magnetic field is the axially symmetric then the total angular momentum projection onto the symmetry axis remains conservative. The vector potential of the uniform magnetic field is given by

$$\mathbf{A}_0 = \frac{1}{2}\left[\mathbf{B}_0 r\right].$$

By substituting this expression into the Hamiltonian (4.17), we get

$$H = \frac{\mathbf{p}^2}{2m_0} + U(r) + \frac{|e|}{2m_0c}\mathbf{B}_0[\mathbf{rp}] + \frac{e^2}{8m_0c^2}\left[\mathbf{B}_0\mathbf{r}\right]^2 + \frac{|e|\,\hbar}{m_0c}\mathbf{B}_0\mathbf{s}. \tag{4.19}$$

By taking into account the definition $[\mathbf{rp}] = \hbar\mathbf{l}$, the last equation can be rewritten in the following form

$$H = H_0 + \mu_B\left(\mathbf{l} + 2\mathbf{s}\right)\mathbf{B}_0 + \frac{e^2}{8m_0c^2}\left[\mathbf{B}_0\mathbf{r}\right]^2, \tag{4.20}$$

where H_0 is the Hamiltonian defined by the equation (2.2) and μ_B is the Bohr magneton:

$$\mu_B = \frac{|e|\,\hbar}{2m_0c}. \tag{4.21}$$

By assuming that the z-axis is directed along the magnetic field direction $\mathbf{B}_0 = \mathbf{e}_z B_0$ we can see that the z projection of the total angular momentum

$$j_z = l_z + s_z = -i\frac{\partial}{\partial\varphi} + \frac{1}{2}\sigma_z$$

commutes with the term of the Hamiltonian (4.20) proportional to the magnetic field

$$[j_z, l_z + 2s_z] = [j_z, j_z] + \frac{1}{4}[\sigma_z, \sigma_z] = 0.$$

The rest of the terms of Hamiltonian (4.20) do not depend on the spin operator and, hence, commute with the operator σ_z. The operator l_z commutes with H_0 and with the last term in the Hamiltonian (4.20), because this term does not depend on z coordinate. Thus, the Hamiltonian (4.20) commutes with the z projection of the total angular momentum operator.

We have seen that the time reversal invariance of the Schrödinger equation relates with the transformation $\psi^*(-t) \to \psi(t)$. But the magnetic moment projection, as well as the angular momentum projection, changes sign at the time reversal transformation, therefore the wave function of spin-1/2 particle should be transformed in accordance with the following rules: $\psi^*(1/2) \to \psi(-1/2)$ and $\psi^*(-1/2) \to -\psi(1/2)$. These rules are agree with the spinor wave function transformations considered in the previous subsection.

Thus, we have seen that the Hamiltonian (4.17) for the particle interacting with superposition of the Coulomb field (as a special case of the spherically symmetric potential $\varphi(\mathbf{r}) = \varphi(r)$) and axially symmetric magnetic field commutes with the parity operator and z projection of the total angular momentum operator. Therefore the energy, E, parity, p, and total angular momentum projection, $M = m + \sigma/2$, are the set of quantum numbers characterizing the eigenstates of this problem.

4.2 Geonium atom

The problem on the particle motion in uniform magnetic field provides the simplest model of the geonium atom. This model describes the cyclotron motion of a particle in the Penning trap and precession of its spin.

4.2.1 Electron motion in homogeneous magnetic field

Let us consider the problem on the electron motion in the homogeneous magnetic field. By using the Hamiltonian (4.19) we get the following eigenvalue problem

$$\left[-\frac{\hbar^2}{2m_0}\Delta - i\mu_{\mathrm{B}}B_0\frac{\partial}{\partial\varphi} + \frac{e^2B_0^2}{8m_0c^2}\rho^2 + \mu_{\mathrm{B}}B_0\sigma_z\right]\psi(\mathbf{r}) = E\psi(\mathbf{r}). \quad (4.22)$$

Recalling that the operators l_z and σ_z commute separately with the Hamiltonian (4.22) it is convenient to take the wave function in the following form

$$\psi(\mathbf{r}) = u_\sigma \exp(im\varphi + ik_z z) f_\sigma(\rho), \quad (4.23)$$

where the spinors u_σ are the eigenfunctions of operator σ_z:

$$\sigma_z u_\sigma = \sigma u_\sigma.$$

By substituting the wave function (4.23) into the equation (4.22) we get the following equation for the radial wave function $f(\rho)$

$$\frac{d^2 f_\sigma}{d\rho^2} + \frac{1}{\rho}\frac{df_\sigma}{d\rho} + \\ + \left(\frac{2m_0E}{\hbar^2} - k_z^2 - \frac{|e|B_0}{\hbar c}(m+\sigma) - \frac{m^2}{\rho^2} - \left(\frac{eB_0}{2\hbar c}\right)^2\rho^2\right) f_\sigma = 0. \quad (4.24)$$

By introducing the new variable

$$x = \frac{|e|B_0}{2\hbar c}\rho^2 = \kappa\rho^2,$$

and the new unknown function $R(x)$

$$f_\sigma(x) = x^{m/2}\exp\left(-\frac{x}{2}\right)R_\sigma(x)$$

we can transform the equation (4.24) to the equation for the confluent hypergeometric functions

$$xR'' + (m+1-x)R' - \left(\frac{m+1}{2} - \nu\right)R = 0, \quad (4.25)$$

where

$$\nu = \frac{\hbar c}{2|e|B_0}\left(\frac{2m_0E}{\hbar^2} - k_z^2 - \frac{|e|B_0}{\hbar c}(m+\sigma)\right).$$

By using the solutions of equation (4.25), the solution of the equation (4.24) can be chosen in the following form

$$f(\rho) = C\left(\kappa\rho^2\right)^{m/2}\exp\left(-\frac{\kappa\rho^2}{2}\right)F\left(\frac{m+1}{2} - \nu, m+1, \kappa\rho^2\right). \quad (4.26)$$

The solution (4.26) satisfies the boundary condition at $\rho \to \infty$ and it is not divergent at $\rho \to 0$ if the following condition holds

$$\frac{m+1}{2} - \nu_\sigma = -n, \tag{4.27}$$

where n is the non-negative integer, and m is arbitrary positive integer or negative integer obeying the condition

$$|m| \leq n. \tag{4.28}$$

The condition (4.27) results in the following equation for the energy spectrum

$$\begin{aligned} E_{n,M,k_z} &= 2\mu_{\text{B}} B_0 \left(n + m + \frac{1+\sigma}{2}\right) + \frac{\hbar^2 k_z^2}{2m_0} = \\ &= 2\mu_{\text{B}} B_0 \left(n + M + \frac{1}{2}\right) + \frac{\hbar^2 k_z^2}{2m_0}, \end{aligned} \tag{4.29}$$

where $M = m + \sigma/2$ is the z projection of the total angular momentum. It follows from the equation (4.29) that the states with the same magnitude of sum $n + M$ are degenerated.

The eigenfunction (4.23) is the product of the eigenfunctions of the angular momentum and spin. The operator of the total angular momentum projection, j_z is also integral of motion. The eigenfunctions of the this operator

$$j_z u_M = M u_M$$

are the following spinors

$$u_M = \begin{pmatrix} \exp(im\varphi) \\ \exp(i(m+1)\varphi) \end{pmatrix}. \tag{4.30}$$

Indeed,

$$j_z \begin{pmatrix} \exp(im\varphi) \\ \exp(i(m+1)\varphi) \end{pmatrix} = \left(m + \frac{1}{2}\right) \begin{pmatrix} \exp(im\varphi) \\ \exp(i(m+1)\varphi) \end{pmatrix}.$$

Hence, the superpositions of the eigenfunctions (4.23), corresponding the same energy eigenvalue, produce the eigenfunctions of the operator of total angular momentum projection.

As already mentioned, at the non-negative integer n, the confluent hypergeometric functions are coupled with the Laguerre polynomials, therefore the normalized wave functions can be written in the following form

$$\psi_{nm\sigma k_z}(\mathbf{r}) = \sqrt{\frac{\kappa}{\pi L}\frac{n!}{(n+m)!}}\,(-1)^n u_\sigma L_n^{(m)}\left(\kappa\rho^2\right)\left(\kappa\rho^2\right)^{m/2} \times \\ \times \exp\left(-\frac{\kappa\rho^2}{2}\right)\exp\left(im\varphi + ik_z z\right), \quad (4.31)$$

where L the spatial size of the region available for electron motion along the direction of the applied magnetic field.

The explicit form of the normalized radial wave function

$$f_{nm}(\rho) = \sqrt{\frac{2\kappa n!}{(n+m)!}}\,(-1)^n L_n^{(m)}\left(\kappa\rho^2\right)\left(\kappa\rho^2\right)^{m/2}\exp\left(-\frac{\kappa\rho^2}{2}\right)$$

for a number of the lower eigenstates is given below :

1 $n = 0,\ m = 0$

$$f_{00}(\rho) = \sqrt{2\kappa}\exp\left(-\frac{\kappa\rho^2}{2}\right);$$

2 $n = 1,\ m = 1$

$$f_{11}(\rho) = \kappa\rho\left(\kappa\rho^2 - 2\right)\exp\left(-\frac{\kappa\rho^2}{2}\right);$$

3 $n = 1,\ m = 0$

$$f_{10}(\rho) = \sqrt{2\kappa}\left(\kappa\rho^2 - 1\right)\exp\left(-\frac{\kappa\rho^2}{2}\right);$$

4 $n = 1,\ m = -1$

$$f_{1-1}(\rho) = \sqrt{2}\kappa\rho\exp\left(-\frac{\kappa\rho^2}{2}\right).$$

The spatial profiles of these functions are shown in the Fig. 4.1. The Fig. 4.2 shows the energy level diagram. As far as the magnitude of the possible negative projection of the orbital momentum is limited by the condition (4.28), it is natural to assume that at a given n the projection of the total orbital momentum lies into the interval $-n - 1/2 < M < < n + 1/2$.

The energy of the state, with the largest negative projection of the total angular momentum $M = -n - 1/2$, is equal to zero and the eigenfunctions of these states are

$$f_{n,-n} = \sqrt{\frac{2\kappa}{n!}}\left(\kappa\rho^2\right)^{n/2}\exp\left(-\frac{\kappa\rho^2}{2}\right). \quad (4.32)$$

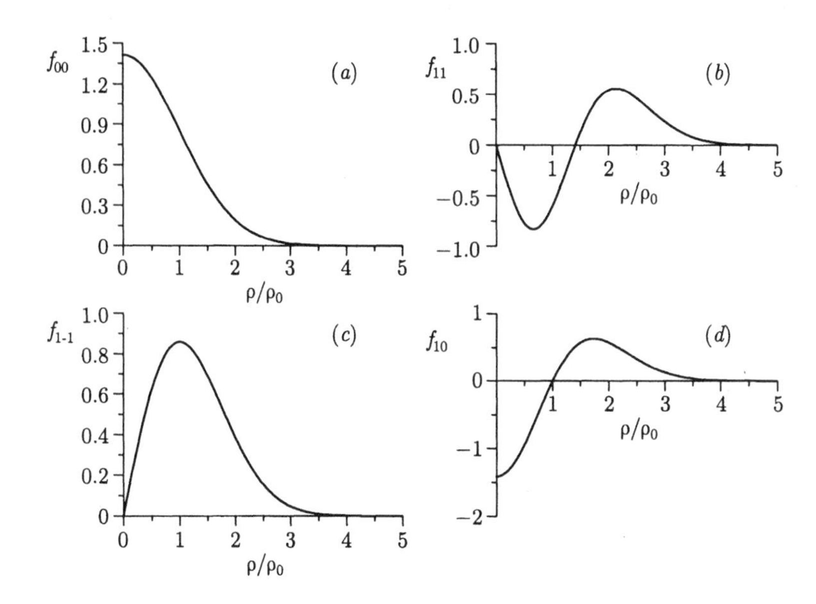

Figure 4.1. The radial wave functions of the geonium atom: (a) $n = 0$, $m = 0$; (b) $n = 1$, $m = 1$; (c) $n = 1$, $m = -1$; (d) $n = 1$, $m = 0$

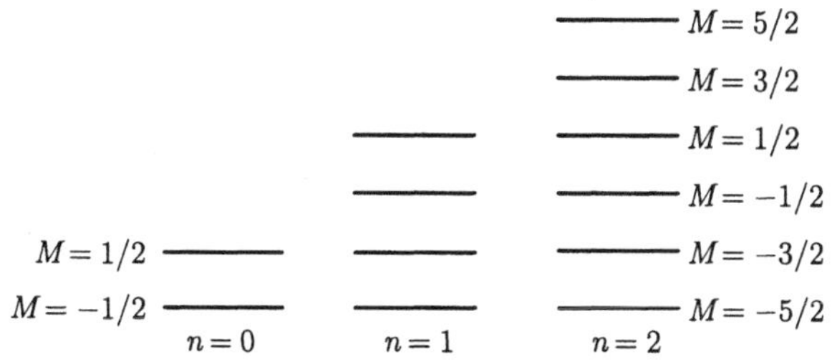

$M = 5/2$
$M = 3/2$
$M = 1/2$
$M = -1/2$
$M = 1/2$ $M = -3/2$
$M = -1/2$ $M = -5/2$
$n = 0$ $n = 1$ $n = 2$

Figure 4.2. The energy level diagram of the geonium atom

According to the equation (4.20) the total magnetic moment is

$$\mathbf{m} = -\mu_{\mathrm{B}} \left(\mathbf{l} + 2\mathbf{s} \right) . \tag{4.33}$$

In the state of the smallest energy the projection of the total magnetic moment is

$$m_z = \mu_{\mathrm{B}} \left(n + 1 \right) .$$

Thus in the state of the smallest energy the magnetic moment is directed along the applied magnetic field.

The state of the largest energy at a given n is the state with $M = n + 1/2$. The energy of this state is

$$E_{n,M=n+1/2} = 2\mu_{\rm B} B_0 (2n+1) = \hbar\omega_H (2n+1),$$

where $\omega_H = |e| B_0/(m_0 c)$ is the cyclotron resonance frequency. In this state the total magnetic moment is directed toward the applied magnetic field.

4.2.2 Strength of induced magnetic field

The orbital motion of electron results in the appearance of the induced magnetic field of response. It will be shown in chapter 6, that, for the case of the Pauli equation, the charge $\rho(\mathbf{r},t)$ and current $\mathbf{j}(\mathbf{r},t)$ density are defined by

$$\rho(\mathbf{r},t) = e\psi^{+}(\mathbf{r},t)\,\psi(\mathbf{r},t),$$

$$\mathbf{j}(\mathbf{r}) = \frac{ie\hbar}{2m_0}\left(\nabla\psi^{+}\cdot\psi - \psi^{+}\nabla\psi\right) - \frac{e^2}{m_0 c}\psi^{+}\mathbf{A}\psi + \frac{e\hbar}{2m_0}\mathrm{curl}\left(\psi^{+}\boldsymbol{\sigma}\psi\right).$$

By substituting the wave function (5.29), for the current density we get

$$\mathbf{j} = -\mathbf{e}_{\varphi}\mu_B c \frac{4\kappa^{3/2}}{\pi L}\frac{n!}{(n+M-1/2)!}\left(\kappa\rho^2\right)^M \exp\left(-\kappa\rho^2\right) L_n^{(M-1/2)}\left(\kappa\rho^2\right)\times \\ \times L_n^{(M+1/2)}\left(\kappa\rho^2\right),$$

where we have assumed that $k_z = 0$. It is seen from the last equation that the current density is equal to zero in the states of the smallest energy $M = -n - 1/2$.

In order to calculate the magnetic field of response we can use the Maxwell equation

$$\mathrm{curl}\,\mathbf{B} = \frac{4\pi}{c}\mathbf{j}.$$

By using the above equation for the current density we obtain

$$B_z = -\frac{4\kappa\mu_B}{L}\frac{n!}{(n+m)!}\int_{\kappa\rho^2}^{\infty}(m+x)\,x^{m-1}\exp(-x)\left(L_n^{(m)}(x)\right)^2 dx - \\ -\sigma\frac{4\kappa\mu_B}{L}\frac{n!}{(n+m)!}\left(\kappa\rho^2\right)^m \exp\left(-\kappa\rho^2\right)\left(L_n^{(m)}\left(\kappa\rho^2\right)\right)^2. \quad (4.34)$$

Particularly, for the states of $m = -n$ we have

$$\mathbf{B} = -\mathbf{e}_z B_0 (1+\sigma)\frac{1}{m_0 c^2}\frac{e^2}{L}\frac{1}{n!}\left(\kappa\rho^2\right)^n \exp\left(-\kappa\rho^2\right).$$

It is seen that the magnetic field of response is equal to zero in the states of the smallest energy $M = -n - 1/2$. This is quite natural, because, as we have mentioned above, the electron current density is zero in these states.

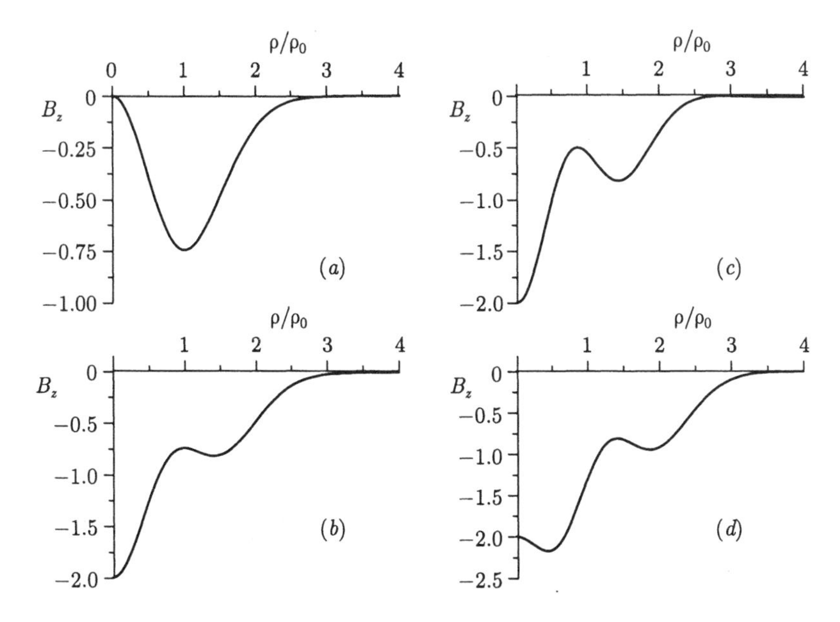

Figure 4.3. The spatial profile of the induced magnetic field produced by particle in the state: (*a*) $n = 0$, $m = 0$; (*b*) $n = 1$, $m = 1$; (*c*) $n = 1$, $m = -1$; (*d*) $n = 1$, $m = 0$

Fig. 4.3 shows the spatial profiles of the magnetic field for a number of eigenstates. It is seen that the induced magnetic field is directed opposite to the direction of the inducing magnetic field. It means that in the frames of theory based on the Pauli equation the electron response is diamagnetic. This is in complete agreement with the classical electromagnetic induction law, i.e. the induced currents tend to decrease the magnetic flux of the inducing magnetic field.

4.3 Hydrogen atom

4.3.1 Action for ensemble of non-relativistic spin-1/2 particles

There are the two moments only, that distinguish the Pauli equation from the Schrödinger equation. Firstly, the wave function of the Pauli equation is the spinor. Secondly, the Hamiltonian of the Pauli equation includes the interaction of the electron inner magnetic moment with the magnetic field. Therefore, the action for an ensemble of the spin-1/2 particles coupled by the electromagnetic field can be written in the form

$$S = \frac{1}{8\pi}\int\left[\left(\frac{1}{c}\frac{\partial \mathbf{A}}{\partial t}+\nabla\varphi\right)^2-(\text{curl }\mathbf{A})^2\right]dV\,dt+$$
$$+\sum_{\substack{i,j\\(i\neq j)}}\int\left[\psi_i^+\left(i\hbar\frac{\partial\psi_i}{\partial t}-q_i\varphi_j(\mathbf{r}_i)\psi_i\right)-\right.$$
$$-\frac{1}{2m_i}\left(i\hbar\nabla\psi_i^+-\frac{q_i}{c}\mathbf{A}_j(\mathbf{r}_i)\psi_i^+\right)\left(-i\hbar\nabla\psi_i-\frac{q_i}{c}\mathbf{A}_j(\mathbf{r}_i)\psi_i\right)+$$
$$\left.+\psi_i^+\boldsymbol{\mu}_i\mathbf{B}_j(\mathbf{r}_i)\psi_i\right]dV_i\,dt \quad (4.35)$$

The variation of this action with respect to ψ_i^+ and the field potentials results in the following equations

$$i\hbar\frac{\partial\psi_i}{\partial t}=\sum_{j(\neq i)}\left[\frac{1}{2m_i}\left(\mathbf{p}_i-\frac{q_i}{c}\mathbf{A}_j(\mathbf{r}_i,t)\right)^2+q_i\varphi_j(\mathbf{r}_i,t)-\boldsymbol{\mu}_i\mathbf{B}_j(\mathbf{r}_i,t)\right]\psi_i, \quad (4.36a)$$

$$\Delta\mathbf{A}-\frac{1}{c^2}\frac{\partial^2\mathbf{A}}{\partial t^2}=-\frac{4\pi}{c}\sum_i\mathbf{j}_i(\mathbf{r},t), \quad (4.36b)$$

$$\Delta\varphi-\frac{1}{c^2}\frac{\partial^2\varphi}{\partial t^2}=-4\pi\sum_i\rho_i(\mathbf{r},t), \quad (4.36c)$$

where

$$\mathbf{j}_i(\mathbf{r},t)=\frac{q_i}{2m_i}\left[\psi_i^+(\mathbf{r},t)\left(-i\hbar\nabla_i\psi_i(\mathbf{r},t)-\frac{q_i}{c}\sum_{j(\neq i)}\mathbf{A}_j(\mathbf{r},t)\psi_i(\mathbf{r},t)\right)+\right.$$
$$\left.+\left(i\hbar\nabla_i\psi_i^+(\mathbf{r},t)-\frac{q_i}{c}\sum_{j(\neq i)}\mathbf{A}_j(\mathbf{r},t)\psi_i^+(\mathbf{r},t)\right)\psi_i(\mathbf{r},t)\right]+$$
$$+c\,\text{curl}\left(\psi_i^+(\mathbf{r},t)\boldsymbol{\mu}_i\psi_i(\mathbf{r},t)\right), \quad (4.37a)$$

$$\rho_i(\mathbf{r},t)=q_i\psi_i^+(\mathbf{r},t)\psi_i(\mathbf{r},t). \quad (4.37b)$$

Since the field equations (4.36b), (4.36c) are linear equations, then the field potentials are the sums of potentials produced by the individual particles

$$\mathbf{A}(\mathbf{r},t)=\sum_i\mathbf{A}_i(\mathbf{r},t), \quad \varphi(\mathbf{r},t)=\sum_i\varphi_i(\mathbf{r},t).$$

Particularly if an ensemble of particles interacts with some external fields (usually produced by the external macroscopic bodies), then the integral field is the sum of the fields produced by the individual particles and external field

$$\mathbf{A}(\mathbf{r},t) = \mathbf{A}_{\text{ext}}(\mathbf{r},t) + \sum_i \mathbf{A}_i(\mathbf{r},t),$$
$$\varphi(\mathbf{r},t) = \varphi_{\text{ext}}(\mathbf{r},t) + \sum_i \varphi_i(\mathbf{r},t). \tag{4.38}$$

The generalized momenta canonically conjugate to the electromagnetic and matter fields are defined by

$$\mathbf{\Pi} = \frac{\partial L}{\partial \dot{\mathbf{A}}} = \frac{1}{4\pi c}\left(\frac{1}{c}\frac{\partial \mathbf{A}}{\partial t} + \nabla\varphi\right), \quad \pi_i = \frac{\partial L}{\partial \dot{\psi}_i} = i\hbar\psi_i^+.$$

The Hamiltonian function is

$$H = \mathbf{\Pi}\dot{\mathbf{A}} + \sum_i \pi_i\dot{\psi}_i - L.$$

The space-integral of the Hamiltonian function is the energy of a system

$$E = \frac{1}{8\pi}\int\left[\frac{1}{c^2}\left(\frac{\partial \mathbf{A}}{\partial t}\right)^2 + (\text{curl }\mathbf{A})^2 - (\nabla\varphi)^2\right]dV +$$
$$+ \sum_{\substack{i,j \\ (i\neq j)}} \int \psi_i^+\left[\frac{1}{2m_i}\left(\mathbf{p}_i - \frac{q_i}{c}\mathbf{A}_j(\mathbf{r}_i)\right)^2 + q_i\varphi_j(\mathbf{r}_i) - \boldsymbol{\mu}_i\mathbf{B}_j(\mathbf{r}_i)\right]\psi_i\, dV_i \tag{4.39}$$

where the electromagnetic field potentials obey the equations (4.36b), (4.36c).

The energy of electromagnetic field is defined as

$$E_f = \int H_f\, dV = \frac{1}{8\pi}\int\left(\mathbf{E}^2 + \mathbf{B}^2\right)dV,$$

where the electric and magnetic fields are

$$\mathbf{E} = -\frac{1}{c}\frac{\partial \mathbf{A}}{\partial t} - \nabla\varphi, \quad \mathbf{B} = \text{curl }\mathbf{A}.$$

By substituting the latter equalities into H_f, we can transform it to the following form

$$H_f = \frac{1}{8\pi}\left(\mathbf{E}^2 + \mathbf{B}^2\right) = \frac{1}{8\pi}\left[\frac{1}{c^2}\left(\frac{\partial \mathbf{A}}{\partial t}\right)^2 + (\text{curl }\mathbf{A})^2 - (\nabla\varphi)^2\right] +$$
$$+ \frac{1}{4\pi}\left(\frac{1}{c}\frac{\partial \mathbf{A}}{\partial t} + \nabla\varphi\right)\nabla\varphi.$$

In the previous chapter we have already used the following vectorial equalities

$$(\nabla\varphi)^2 = -\varphi\Delta\varphi + \operatorname{div}(\varphi\nabla\varphi),$$

$$\frac{1}{c}\operatorname{div}\left(\varphi\frac{\partial \mathbf{A}}{\partial t}\right) = \frac{1}{c}\frac{\partial \mathbf{A}}{\partial t}\nabla\varphi + \frac{1}{c}\varphi\frac{\partial}{\partial t}\operatorname{div}\mathbf{A} = \frac{1}{c}\frac{\partial \mathbf{A}}{\partial t}\nabla\varphi - \frac{1}{c^2}\varphi\frac{\partial^2\varphi}{\partial t^2}.$$

With the help of these equalities and equation (4.36c) the field energy becomes

$$E_f = \frac{1}{8\pi}\int\left(\mathbf{E}^2 + \mathbf{B}^2\right)dV - \sum_{\substack{i,j \\ (i\neq j)}} q_i \int \psi_i^+(\mathbf{r}_i)\,\varphi_j(\mathbf{r}_i)\,\psi_i(\mathbf{r}_i)\,dV_i.$$

By substituting this equation into the equation (4.39), we finally get

$$E = \frac{1}{8\pi}\int\left(\mathbf{E}^2 + \mathbf{B}^2\right)dV + \\ + \sum_{\substack{i,j \\ (i\neq j)}} \int \psi_i^+\left[\frac{1}{2m_i}\left(\mathbf{p}_i - \frac{q_i}{c}\mathbf{A}_j(\mathbf{r}_i)\right)^2 - \boldsymbol{\mu}_i\mathbf{B}_j(\mathbf{r}_i)\right]\psi_i\,dV_i. \quad (4.40)$$

Thus, the energy of an isolated system of particles is the sum of the energy of electromagnetic field, produced by these particles, the integral kinetic energy of their motion, and the integral energy of interaction of the particle magnetic moments with the magnetic field of a system.

It should be noted that equation (4.40) is gauge invariant. Indeed under the simultaneous transformation of the wave functions

$$\psi_i'(\mathbf{r},t) = \psi_i(\mathbf{r},t)\exp\left[\frac{i}{\hbar c}\sum_j q_j\chi_j(\mathbf{r},t)\right] \quad (4.41a)$$

and electromagnetic field potentials

$$\mathbf{A}'(\mathbf{r},t) = \mathbf{A}(\mathbf{r},t) + \sum_i \nabla\chi_i(\mathbf{r},t),$$
$$\varphi'(\mathbf{r},t) = \varphi(\mathbf{r},t) - \frac{1}{c}\sum_i \frac{\partial\chi_i(\mathbf{r},t)}{\partial t} \quad (4.41b)$$

the equation (4.40) remains invariable.

4.3.2 Orbital, spin-orbital, and spin-spin interactions

Let us consider the interaction of the two spin-1/2 particles. This problem enables us to model the hydrogen atom again, and, as a result, to clarify the new features of the hydrogenic spectra associated with the spin of electron and nucleus and originated from the interaction of spins with the intra-atomic magnetic field. The general formalism of the analysis of the two-particle problem was developed in the previous chapter. According to this formalism, the wave functions of the steady states of a system are determined by the variation of the energy functional (4.40). The global minimum gives us the wave function of the ground state, the subsequent steady states realize the local minima of the energy functional (4.40).

The energy given by (4.40) is functional of the particle wave functions and potentials of the electromagnetic field. On the other hand, the potentials of electromagnetic field are the functionals of the particle wave functions. Therefore we can vary the energy functional over the particle wave functions allowing for the constraint equations. Or, we can exclude the electromagnetic field potentials from the equation (4.40) (retaining only those part of the field energy that depends on the relative position of particles) and vary the energy functional with respect to the particle wave functions alone.

In steady state case the equations relating the electric and magnetic fields with the field potentials are

$$\mathbf{E} = -\nabla\varphi, \quad \mathbf{B} = \mathrm{curl}\,\mathbf{A}.$$

By using now the vectorial equalities

$$(\nabla\varphi)^2 = -\varphi\Delta\varphi + \mathrm{div}\,(\varphi\nabla\varphi),$$

$$(\mathrm{curl}\,\mathbf{A})^2 = \mathrm{div}\,[\mathbf{A}\,\mathrm{curl}\,\mathbf{A}] + \mathbf{A}\,\mathrm{curl}\,\,\mathrm{curl}\,\mathbf{A},$$

we get the following equation for the field energy depending on the relative position of particles

$$E_f = \frac{1}{8\pi}\int\left(\mathbf{E}^2 + \mathbf{B}^2\right)dV = -\frac{1}{8\pi}\int\left(\varphi\Delta\varphi + \mathbf{A}\Delta\mathbf{A}\right)dV = \\ = \frac{1}{2}\int\varphi\rho\,dV + \frac{1}{2c}\int\mathbf{A}\mathbf{j}\,dV. \quad (4.42)$$

Notice that in the latter equation we have used the Lorentz gauge condition. In steady-state case this condition is

$$\mathrm{div}\,\mathbf{A} = 0.$$

Taking into account that the particle could not interact with its own electromagnetic field for the hydrogen atom we should write

$$E_f = \frac{1}{2}\int \varphi_n(\mathbf{r}_e)\,\rho_e(\mathbf{r}_e)\,dV_e + \frac{1}{2}\int \varphi_e(\mathbf{r}_n)\,\rho_n(\mathbf{r}_n)\,dV_n + \\ + \frac{1}{2c}\int \mathbf{A}_n(\mathbf{r}_e)\,\mathbf{j}_e(\mathbf{r}_e)\,dV_e + \frac{1}{2}\int \mathbf{A}_e(\mathbf{r}_n)\,\mathbf{j}_n(\mathbf{r}_n)\,dV_n,$$

where ρ_a and $\mathbf{j}_a$ are defined by the equations (4.37a), (4.37b), $\varphi_b(\mathbf{r}_a)$ and $\mathbf{A}_b(\mathbf{r}_a)$ are the potentials produced by the particle $b = (n, e)$ at the position of the particle $a = (e, n)$.

By substituting the equation (4.42) into the equation (4.40) we get

$$E = \int \psi_e^+\left(\frac{\mathbf{p}_e^2}{2m_e} + \frac{1}{2}q_e\varphi_n(\mathbf{r}_e)\right)\psi_e\,dV_e + \int \psi_n^+\left(\frac{\mathbf{p}_n^2}{2m_n} + \frac{1}{2}q_n\varphi_e(\mathbf{r}_n)\right)\psi_n\,dV_n - \\ - \frac{q_e}{2m_e c}\int \psi_e^+\mathbf{A}_n(\mathbf{r}_e)\,\mathbf{p}_e\psi_e\,dV_e - \frac{q_n}{2m_n c}\int \psi_n^+\mathbf{A}_e(\mathbf{r}_n)\,\mathbf{p}_n\psi_n\,dV_n - \\ - \frac{1}{2}\int \psi_e^+\boldsymbol{\mu}_e\mathbf{B}_n(\mathbf{r}_e)\,\psi_e\,dV_e - \frac{1}{2}\int \psi_n^+\boldsymbol{\mu}_n\mathbf{B}_e(\mathbf{r}_n)\,\psi_n\,dV_n \quad (4.43)$$

where $\psi_e = \psi_e(\mathbf{r}_e)$ is electron wave function of the hydrogen atom, and $\psi_n = \psi_n(\mathbf{r}_n)$ is the nucleus wave function. Deriving of the equation (4.43) we have again used the Lorentz gauge condition

$$\mathrm{div}_a\,\mathbf{A}_b(\mathbf{r}_a) = 0. \quad (4.44)$$

In steady-state case the equations (4.36b), (4.36c) have the following integral solutions

$$\varphi_b(\mathbf{r}_a) = \int \frac{\rho_b(\mathbf{r}_b)}{|\mathbf{r}_a - \mathbf{r}_b|}\,dV_b,$$

$$\mathbf{A}_b(\mathbf{r}_a) = \frac{1}{c}\int \frac{\mathbf{j}_b(\mathbf{r}_b)}{|\mathbf{r}_a - \mathbf{r}_b|}\,dV_b.$$

By substituting here the equations (4.37a), (4.37b) we get

$$\varphi_b(\mathbf{r}_a) = \int \psi_b^+(\mathbf{r}_b)\,\frac{q_b}{|\mathbf{r}_a - \mathbf{r}_b|}\psi_b(\mathbf{r}_b)\;dV_b, \quad (4.45a)$$

$$\mathbf{A}_b(\mathbf{r}_a) = \frac{q_b}{c}\int \frac{\psi_b^+\mathbf{v}_b\psi_b}{|\mathbf{r}_a - \mathbf{r}_b|}\,dV_b + \int \frac{\mathrm{curl}\,(\psi_b^+\boldsymbol{\mu}_b\psi_b)}{|\mathbf{r}_a - \mathbf{r}_b|}\,dV_b, \quad (4.45b)$$

$$\mathbf{B}_b(\mathbf{r}_a) = -\frac{q_b}{c}\int \frac{\psi_b^+\,[\mathbf{r}_{ab}\mathbf{v}_b]\,\psi_b}{r_{ab}^3}\,dV_b + \int \psi_b^+\frac{3\mathbf{r}_{ab}\,(\boldsymbol{\mu}_b\mathbf{r}_{ab}) - \boldsymbol{\mu}_b r_{ab}^2}{r_{ab}^5}\psi_b\,dV_b, \quad (4.45c)$$

where

$$\mathbf{v}_b = \frac{1}{m_b}\left(\mathbf{p}_b - \frac{q_b}{c}\mathbf{A}_a(\mathbf{r}_b) - \frac{i\hbar\,|\mathbf{r}_a - \mathbf{r}_b|}{2}\nabla_b\frac{1}{|\mathbf{r}_a - \mathbf{r}_b|}\right). \tag{4.46}$$

Notice, that the last term in the equation (4.46) is appeared due to the use of the following transformation in the equations (4.45b), (4.45c): with the help of integration by parts the expression $(\mathbf{p}\psi)^+\psi - \psi^+\mathbf{p}\psi$ can be easily transformed to the following one, $\psi^+\mathbf{p}\psi$.

By substituting the equations (4.45a)–(4.45c) into the equation (4.43) we get

$$\begin{aligned} E = \int \psi_e^+\psi_n^+ \Big\{ &-\frac{\hbar^2}{2m_e}\Delta_e - \frac{\hbar^2}{2m_n}\Delta_n + \frac{q_e q_n}{r_{en}} - \\ &-\frac{q_e q_n}{2c^2 r_{en}}\left(\mathbf{v}_n\frac{\mathbf{p}_e}{m_e} + \mathbf{v}_n\frac{\mathbf{p}_n}{m_n}\right) - \frac{q_e}{m_e c}\frac{\boldsymbol{\mu}_n[\mathbf{r}_{en}\mathbf{p}_e]}{r_{en}^3} - \frac{q_n}{m_n c}\frac{\boldsymbol{\mu}_e[\mathbf{r}_{ne}\mathbf{p}_n]}{r_{en}^3} - \\ &-\frac{3(\boldsymbol{\mu}_e\mathbf{r}_{en})(\boldsymbol{\mu}_n\mathbf{r}_{en}) - \boldsymbol{\mu}_e\boldsymbol{\mu}_n r_{en}^2}{r_{en}^5}\Big\}\psi_e\psi_n\,dV_e\,dV_n \end{aligned} \tag{4.47}$$

where $\psi_e\psi_n$ is the direct product of the electron and nucleus spinor wave functions. The equation (4.47) is the desired form of the energy functional which is the functional of the particle wave functions only, and the potentials of the electromagnetic field were excluded with the help of the electromagnetic field equations.

Let us dwell on the physical meaning of the different terms in the equation (4.47).

(1) The sum of the first three terms coincides with equation for the Hamiltonian of the two spinless particles with the Coulomb interaction between them

$$H_0 = -\frac{\hbar^2}{2m_e}\Delta_e - \frac{\hbar^2}{2m_n}\Delta_n + \frac{q_e q_n}{r_{en}}. \tag{4.48}$$

The eigenvalues and eigenfunctions of the Hamiltonian (4.48) have been found in the Chapter 3.

(2) The forth term in (4.47) we have already met in the Chapter 3. With the help of the vectorial transformations

$$\begin{aligned} \mathbf{v}_n\mathbf{p}_e &= \frac{1}{r_{en}^2}\left\{[\mathbf{r}_{en}\mathbf{v}_n][\mathbf{r}_{en}\mathbf{p}_e] + (\mathbf{r}_{en}\mathbf{v}_n)(\mathbf{r}_{en}\mathbf{p}_e) + \frac{i\hbar}{m_n}(\mathbf{r}_{en}\mathbf{v}_e) + \frac{\hbar^2}{2m_n}\right\}, \\ \mathbf{v}_e\mathbf{p}_n &= \frac{1}{r_{en}^2}\left\{[\mathbf{r}_{en}\mathbf{v}_e][\mathbf{r}_{en}\mathbf{p}_n] + (\mathbf{r}_{en}\mathbf{v}_e)(\mathbf{r}_{en}\mathbf{p}_n) - \frac{i\hbar}{m_e}(\mathbf{r}_{en}\mathbf{v}_n) + \frac{\hbar^2}{2m_e}\right\}. \end{aligned} \tag{4.49}$$

and the use of the Lorentz gauge condition (4.44)

$$\operatorname{div}_a \mathbf{A}_b(\mathbf{r}_a) = -\frac{q_b}{c}\int \frac{\psi_b^+ \mathbf{r}_{ab}\mathbf{v}_b\psi_b}{r_{ab}^3}\, dV_b = 0. \tag{4.50}$$

this term can be transformed to the following form

$$H_{ll} = \frac{q_e q_n \hbar^2}{2m_e m_n c^2}\frac{\mathbf{l}_e\mathbf{l}_n + \mathbf{l}_n\mathbf{l}_e - 1}{r_{en}^3}, \tag{4.51}$$

where

$$\hbar\mathbf{l}_e = [\mathbf{r}_{en}\mathbf{p}_e], \quad \hbar\mathbf{l}_n = [\mathbf{r}_{ne}\mathbf{p}_n]. \tag{4.52}$$

The Hamiltonian (4.51) describes the interaction of the electron and nucleus orbital currents in the hydrogen atom. It should be noted again that the last term in the equation (4.51) is due to the fact that the operators $\mathbf{l}_e$ and $\mathbf{l}_n$ are the non-commuting operators.

(3) The next two terms in (4.47) are new ones and they describe the spin-orbital interaction: the nucleus magnetic moment interacts with the magnetic field resulted from the orbital motion of electron (i.e. electron orbital current), and vise versa

$$H_{ls} = -\frac{q_e\hbar}{m_e c}\frac{\boldsymbol{\mu}_n\mathbf{l}_e}{r_{en}^3} - \frac{q_n\hbar}{m_n c}\frac{\boldsymbol{\mu}_e\mathbf{l}_n}{r_{en}^3}. \tag{4.53}$$

(4) The last term in (4.47) is also new one and it describes the spin-spin interaction: the electron magnetic moment interacts with the magnetic field produced by the nucleus magnetic moment, and vise versa

$$H_{ss} = -\frac{3(\boldsymbol{\mu}_e\mathbf{r}_{en})(\boldsymbol{\mu}_n\mathbf{r}_{en}) - \boldsymbol{\mu}_e\boldsymbol{\mu}_n r_{en}^2}{r_{en}^5}. \tag{4.54}$$

Thus, one can see that the account for the electron spin results in appearance of the new mechanisms of particle interaction. Undoubtedly, these new interactions affect on the structure of the hydrogenic spectra.

4.3.3 Integrals of motion for hydrogen atom

The variation of the functional (4.47) with respect to the two-particle wave function $\Psi(\mathbf{r}_e, \mathbf{r}_n) = \psi_e(\mathbf{r}_e)\psi_n(\mathbf{r}_n)$ results in the following wave equation

$$(H_0 + H_{ll} + H_{ls} + H_{ss})\Psi = E\Psi. \tag{4.55}$$

As we have discussed earlier, to characterize completely the eiegenstates of the equation (4.55) we should find the operators commuting with the Hamiltonian of this equation. In the previous chapter it was shown, from the first principles, that the total angular momentum of an isolated ensemble of spinless particle is the integral of motion. Let us verify whether it is true for an ensemble of spin-1/2 particles. The total angular momentum of ensemble of particles possessing spin,

$$\mathbf{J} = \mathbf{L} + \mathbf{S}, \tag{4.56}$$

is the sum of total angular momentum of the orbital motion of particles

$$\hbar\mathbf{L} = [\mathbf{r}_e\mathbf{p}_e] + [\mathbf{r}_n\mathbf{p}_n]$$

and the total spin

$$\mathbf{S} = \mathbf{s}_e + \mathbf{s}_n. \tag{4.57}$$

In analysis of the hydrogen atom problem it is convenient to use the center-of-mass reference frame

$$\mathbf{r} = \mathbf{r}_e - \mathbf{r}_n, \quad \mathbf{R} = \frac{m_e\mathbf{r}_e + m_n\mathbf{r}_n}{M}$$

and

$$\mathbf{p} = -i\hbar\frac{\partial}{\partial\mathbf{r}}, \quad \mathbf{P} = -i\hbar\frac{\partial}{\partial\mathbf{R}}.$$

There are the following relationships between the momentum and angular momentum operators in the center-of-mass and laboratory reference frames

$$\begin{gathered}\mathbf{p} = \frac{m_n}{M}\mathbf{p}_e - \frac{m_e}{M}\mathbf{p}_n, \quad \mathbf{P} = \mathbf{p}_e + \mathbf{p}_n, \\ \hbar\mathbf{L} = [\mathbf{r}_e\mathbf{p}_e] + [\mathbf{r}_n\mathbf{p}_n] = [\mathbf{r}\mathbf{p}] + [\mathbf{R}\mathbf{P}].\end{gathered} \tag{4.58}$$

In the previous chapter we have shown that the commutation relations for the operator $\mathbf{L}$ and operators $\mathbf{l}_{e,n}$, defined by (4.52), are:

$$[L_\alpha, l_{e\beta}] = ie_{\alpha\beta\gamma}l_{e\gamma}, \quad [L_\alpha, l_{n\beta}] = ie_{\alpha\beta\gamma}l_{n\gamma}. \tag{4.59}$$

We have also shown that the operator of the total angular momentum of orbital motion $\mathbf{L}$ commutes with the Hamiltonian H_{ll}. Hence the total angular momentum operator $\mathbf{J}$ commutes with H_{ll} too.

It can be easily shown, with the help of the commutation relations (4.59) and commutation relations for spin operators, that the operator $\mathbf{J}$ commutes with the Hamiltonian of spin-orbital interaction, H_{ls}. Indeed

$$\begin{aligned}[J_\alpha, l_{e\beta}s_{n\beta}] &= [L_\alpha, l_{e\beta}]\, s_{n\beta} + l_{e\beta}\,[S_\alpha, s_{n\beta}] = \\ &= -ie_{\alpha\gamma\beta}l_{e\gamma}s_{n\beta} + ie_{\alpha\beta\gamma}l_{e\beta}s_{n\gamma} = 0.\end{aligned} \tag{4.60}$$

The spin operator $\mathbf{S}$ commutes with the last term in the Hamiltonian of spin-spin interaction, given by equation (4.54). One can see

$$[S_\alpha, s_{e\beta}s_{n\beta}] = [S_\alpha, s_{e\beta}]\, s_{n\beta} + s_{e\beta}\,[S_\alpha, s_{n\beta}] = \\ = -ie_{\alpha\gamma\beta}s_{e\gamma}s_{n\beta} + ie_{\alpha\beta\gamma}s_{e\beta}s_{n\gamma} = 0. \quad (4.61)$$

However, the operator $\mathbf{S}$ does not commute with the first term of the Hamiltonian (4.54)

$$[\mathbf{S}, (\mathbf{s}_e\mathbf{r})(\mathbf{s}_n\mathbf{r})] = i\,[\mathbf{r}\mathbf{s}_e]\,(\mathbf{r}\mathbf{s}_n) + i\,(\mathbf{r}\mathbf{s}_e)\,[\mathbf{r}\mathbf{s}_n]. \quad (4.62)$$

On the other hand, the commutator of the operator $\mathbf{L}$ with the same term is

$$[\mathbf{L}, (\mathbf{s}_e\mathbf{r})(\mathbf{s}_n\mathbf{r})] = i\,[\mathbf{s}_e\mathbf{r}]\,(\mathbf{s}_n\mathbf{r}) + i\,(\mathbf{s}_e\mathbf{r})\,[\mathbf{s}_n\mathbf{r}]. \quad (4.63)$$

Hence, the total angular momentum operator $\mathbf{J}$ commutes with the Hamiltonian of spin-spin interaction, H_{ss}.

Thus, the operator of the total angular momentum of the orbital motion $\mathbf{L}$ and operator of the total spin $\mathbf{S}$ do not separately commute with the Hamiltonian of the equation (4.55). The integral of motion of the equation (4.55) is the total angular momentum $\mathbf{J}$:

$$[\mathbf{J}, (H_0 + H_{ll} + H_{ls} + H_{ss})] = 0. \quad (4.64)$$

As far as the Hamiltonian of the equation (4.55) depends only on the radius vector $\mathbf{r} = \mathbf{r}_e - \mathbf{r}_n$, it is evidently, that the operator of the total momentum $\mathbf{P} = \mathbf{p}_e + \mathbf{p}_n$ is the integral of motion

$$[\mathbf{P}, (H_0 + H_{ll} + H_{ls} + H_{ss})] = 0. \quad (4.65)$$

Thus, the eigenfunctions of the equation (4.55) can be expressed in terms of the eigenfunctions of operators of the total momentum , total angular momentum, and projection of total angular momentum.

4.3.4 Angular dependency of hydrogenic wave functions

Let us consider the motionless hydrogen atom, $\mathbf{P} = 0$. In this case the Hamiltonian H of the equation (4.55) is slightly simplified

$$H = -\frac{\hbar^2}{2m_r}\Delta - \frac{Ze^2}{r} - \frac{4\mu_B\mu_N}{r^3}\mathbf{l}^2 + \frac{4\mu_B\mu_N}{r^3}\left(\gamma_e\mathbf{s}_e + \gamma_n\mathbf{s}_n\right)\mathbf{l} + \\ + \frac{4\mu_B\mu_N\gamma_e\gamma_n}{r^3}\left(3(\mathbf{s}_e\mathbf{e})\,(\mathbf{s}_n\mathbf{e}) - \mathbf{s}_e\mathbf{s}_n\right), \quad (4.66)$$

where

$$\boldsymbol{\mu}_e = \frac{\mu_e}{s}\mathbf{s}_e = -2\gamma_e\mu_B\mathbf{s}_e, \quad \boldsymbol{\mu}_n = \frac{\mu_n}{s}\mathbf{s}_n = 2\gamma_n\mu_N\mathbf{s}_n, \quad (4.67)$$

$\gamma_{e(n)}$ is the gyromagnetic ratio for the magnetic moment of electron (nucleus). The Bohr magneton μ_{B} and nuclear magneton μ_N are defined by the well known equations:

$$\mu_{\mathrm{B}} = -\frac{q_e \hbar}{2m_e c} = \frac{|e|\,\hbar}{2m_e c}, \quad \mu_N = \frac{q_n \hbar}{2m_n c} = \frac{Z\,|e|\,\hbar}{2m_n c}. \tag{4.68}$$

The wave function $\Psi(\mathbf{r}_e, \mathbf{r}_n)$ of the equation

$$H\Psi = E\Psi \tag{4.69}$$

is the second rank spinor

$$\Psi(\mathbf{r}_e, \mathbf{r}_n) = \psi_e(\mathbf{r}_e)\,\psi_n(\mathbf{r}_n) = \psi(\mathbf{r}, \mathbf{s}_e, \mathbf{s}_n), \tag{4.70}$$

i.e. the four row column. It is well known that the products of the two spinors are decompose into the two irreducible representations corresponding to the spin zero particle and spin one particle, respectively. If we use the products of eigenfunctions of operators $\sigma_{(e,n)z}$:

$$\sigma_z\,|\pm\rangle = \pm\,|\pm\rangle,$$

where $|\pm\rangle$ is the two row columns

$$|+\rangle = \begin{pmatrix} 1 \\ 0 \end{pmatrix}, \quad |-\rangle = \begin{pmatrix} 0 \\ 1 \end{pmatrix},$$

as the basis two-body wave functions, then the wave functions for these two representations can be written in the following form. The wave function of spin zero particle is

$$\chi^{(0)} = \frac{1}{\sqrt{2}}\left(|+\rangle_e\,|-\rangle_n - |-\rangle_e\,|+\rangle_n\right) = \frac{1}{\sqrt{2}}\begin{pmatrix} 0 \\ 1 \\ -1 \\ 0 \end{pmatrix}. \tag{4.71}$$

The rest three linear independent wave functions describe the spin one particle and correspond to three different projection of the total spin

$$\begin{aligned} \chi_1^{(1)} &= |+\rangle_e\,|+\rangle_n = \begin{pmatrix} 1 \\ 0 \\ 0 \\ 0 \end{pmatrix}, \\ \chi_0^{(1)} &= \frac{1}{\sqrt{2}}\left(|+\rangle_e\,|-\rangle_n + |-\rangle_e\,|+\rangle_n\right) = \frac{1}{\sqrt{2}}\begin{pmatrix} 0 \\ 1 \\ 1 \\ 0 \end{pmatrix}, \\ \chi_{-1}^{(1)} &= |-\rangle_e\,|-\rangle_n = \begin{pmatrix} 0 \\ 0 \\ 0 \\ 1 \end{pmatrix}. \end{aligned} \tag{4.72}$$

As it was shown in the previous chapters the eigenfunctions of the angular momentum operator, associated with the translational degrees of freedom, are the spherical harmonics. The angular dependency of the atomic wave functions, $\Psi(\mathbf{r}_e, \mathbf{r}_n)$, is determined by the rule of angular momenta coupling. For the spinless particle the angular wave function is the product of spherical harmonic and second rank spinor (4.71)

$$\Omega^{(s=0)}_{j,l,m} = Y_{lm}(\theta, \varphi)\,\chi^{(0)}. \tag{4.73}$$

For the spin one particle the angular wave functions, $\Omega^{(1)}_{j,l,m}$, are the series of products of the eigenfunctions of orbital momentum Y_{lm_1} and spin momentum $\chi^{(1)}_{m_2}$. The coefficients of series are the $3-j$ symbols

$$\Omega^{(s=1)}_{jlm} = (-1)^{l-1+m}\sqrt{2j+1}\sum_{m_1,m_2}\begin{pmatrix} l & 1 & j \\ m_1 & m_2 & -m \end{pmatrix} Y_{lm_1}(\theta,\varphi)\,\chi^{(1)}_{m_2}. \tag{4.74}$$

At a given value of l there are the three solutions differing in the value of the total angular momentum j

$$\Omega^{(s=1)}_{j=l+1,l,m} = \sqrt{\frac{(l+m)(l+m+1)}{2(l+1)(2l+1)}}Y_{l,m-1}\chi^{(1)}_1 + $$
$$+\sqrt{\frac{(l+m+1)(l-m+1)}{(l+1)(2l+1)}}Y_{l,m}\chi^{(1)}_0 + \sqrt{\frac{(l-m+1)(l-m)}{2(l+1)(2l+1)}}Y_{l,m+1}\chi^{(1)}_{-1},$$

$$\Omega^{(s=1)}_{j,l,m} = -\sqrt{\frac{(l+m)(l-m+1)}{2l(l+1)}}Y_{l,m-1}\chi^{(1)}_1 +$$
$$+\frac{m}{\sqrt{l(l+1)}}Y_{l,m}\chi^{(1)}_0 + \sqrt{\frac{(l+m+1)(l-m)}{2l(l+1)}}Y_{l,m+1}\chi^{(1)}_{-1}, \tag{4.75}$$

$$\Omega^{(s=1)}_{j=l-1,l,m} = \sqrt{\frac{(l-m+1)(l-m)}{2l(2l+1)}}Y_{l,m-1}\chi^{(1)}_1 -$$
$$-\sqrt{\frac{(l-m)(l+m)}{l(2l+1)}}Y_{l,m}\chi^{(1)}_0 + \sqrt{\frac{(l+m)(l+m+1)}{2l(2l+1)}}Y_{l,m+1}\chi^{(1)}_{-1}.$$

The eigenfunctions (4.73) and (4.75) are normalized by the condition

$$\int \Omega^{(s)+}_{jlm}\Omega^{(s')}_{j'l'm'}\sin\theta\, d\theta\, d\varphi = \delta_{jj'}\delta_{ll'}\delta_{mm'}\delta_{ss'}.$$

Thus, at given values of the total angular momentum j and its projection m the general solution of the equation (4.69) has the form

$$\Psi_{jm}(\mathbf{r}) = f(r)\,\Omega^{(0)}_{j,l=j,m}(\theta,\varphi) + \sum_{\sigma=-1}^{+1} g_{\sigma}(r)\,\Omega^{(1)}_{j,l=j-\sigma,m}(\theta,\varphi), \qquad (4.76)$$

where

$$\Omega^{(1)}_{j,l=j-1,m} = \sqrt{\frac{(j+m-1)(j+m)}{2j(2j-1)}}Y_{j-1,m-1}\chi^{(1)}_{1} + \\ + \sqrt{\frac{(j+m)(j-m)}{j(2j-1)}}Y_{j-1,m}\chi^{(1)}_{0} + \sqrt{\frac{(j-m)(j-m-1)}{2j(2j-1)}}Y_{j-1,m+1}\chi^{(1)}_{-1},$$

$$\Omega^{(1)}_{j,l=j,m} = -\sqrt{\frac{(j+m)(j-m+1)}{2j(j+1)}}Y_{j,m-1}\chi^{(1)}_{1} + \\ + \frac{m}{\sqrt{j(j+1)}}Y_{j,m}\chi^{(1)}_{0} + \sqrt{\frac{(j+m+1)(j-m)}{2j(j+1)}}Y_{j,m+1}\chi^{(1)}_{-1}, \qquad (4.77)$$

$$\Omega^{(1)}_{j,l=j+1,m} = \sqrt{\frac{(j-m+1)(j-m+2)}{2(j+1)(2j+3)}}Y_{j+1,m-1}\chi^{(1)}_{1} - \\ - \sqrt{\frac{(j-m+1)(j+m+1)}{(j+1)(2j+3)}}Y_{j+1,m}\chi^{(1)}_{0} + \\ + \sqrt{\frac{(j+m+1)(j+m+2)}{2(j+1)(2j+3)}}Y_{j+1,m+1}\chi^{(1)}_{-1}.$$

By substituting the expression (4.76) into the equation (4.69) and then integrating over the angular variables we can obtain the equations for the radial wave functions $f(r)$ and $g_{\sigma}(r)$. To do this we need to know the matrix elements of Hamiltonians of spin-orbital and spin-spin interactions.

Notice finally that the case of the zero value of the total angular momentum, $j=0$, is the special case, because, in this case, the two of the three linear independent solutions (4.77) are equal to zero. Indeed, at $s=1$ the state with $j=0$ can be obtained only if $l=1$. The non-zero solution is

$$\Omega^{(1)}_{010} = \frac{1}{\sqrt{3}}\left(Y_{1,-1}\chi^{(1)}_{1} - Y_{10}\chi^{(1)}_{0} + Y_{11}\chi^{(1)}_{-1}\right).$$

Hence the case of zero value of the total angular momentum should be considered separately.

4.3.5 Angular matrix elements of Hamiltonian of spin-orbital and spin-spin interactions

By applying operators $\mathbf{s}_{e,n}\mathbf{l}$ to the wave functions (4.73) and (4.77), we get

$$
\begin{gathered}
(\boldsymbol{\sigma}_e \mathbf{l})\, \Omega^{(0)}_{jlm} = \sqrt{j\,(j+1)}\,\Omega^{(1)}_{j,l=j,m}, \quad (\boldsymbol{\sigma}_n \mathbf{l})\, \Omega^{(0)}_{jlm} = -\sqrt{j\,(j+1)}\,\Omega^{(1)}_{j,l=j,m}, \\
(\boldsymbol{\sigma}_e \mathbf{l})\, \Omega^{(1)}_{j,l=j,m} = \sqrt{j\,(j+1)}\,\Omega^{(0)}_{jlm} - \Omega^{(1)}_{j,l=j,m}, \\
(\boldsymbol{\sigma}_n \mathbf{l})\, \Omega^{(1)}_{j,l=j,m} = -\sqrt{j\,(j+1)}\,\Omega^{(0)}_{jlm} - \Omega^{(1)}_{j,l=j,m}, \\
(\boldsymbol{\sigma}_e \mathbf{l})\, \Omega^{(1)}_{j,l=j-1,m} = (j-1)\,\Omega^{(1)}_{j,l=j-1,m}, \\
(\boldsymbol{\sigma}_n \mathbf{l})\, \Omega^{(1)}_{j,l=j-1,m} = (j-1)\,\Omega^{(1)}_{j,l=j-1,m}, \\
(\boldsymbol{\sigma}_e \mathbf{l})\, \Omega^{(1)}_{j,l=j+1,m} = -\,(j+2)\,\Omega^{(1)}_{j,l=j+1,m}, \\
(\boldsymbol{\sigma}_n \mathbf{l})\, \Omega^{(1)}_{j,l=j+1,m} = -\,(j+2)\,\Omega^{(1)}_{j,l=j+1,m}.
\end{gathered} \tag{4.78}
$$

These equations enable us to determine easily the angular matrix elements for the Hamiltonian of the spin-orbital interaction. It is seen that the spin-orbital interaction couples only the states with the zero projection of the total angular momentum, $m = 0$.

Particularly, for the diagonal elements of the spin-orbital interaction Hamiltonian we have

$$\langle s=0|\, H_{sp}\, |s=0\rangle = 0,$$

$$
\begin{aligned}
\langle j, l=j-1, m, s=1|\, H_{sp}\, |j, l=j-1, m, s=1\rangle = \\
= \frac{2\mu_{\mathrm{B}}\mu_N}{r^3}\,(\gamma_e + \gamma_n)\,(j-1)\,,
\end{aligned}
$$

$$\langle j, l=j, m, s=1|\, H_{sp}\, |j, l=j, m, s=1\rangle = -\frac{2\mu_{\mathrm{B}}\mu_N}{r^3}\,(\gamma_e + \gamma_n)\,,$$

$$
\begin{aligned}
\langle j, l=j+1, m, s=1|\, H_{sp}\, |j, l=j+1, m, s=1\rangle = \\
= -\frac{2\mu_{\mathrm{B}}\mu_N}{r^3}\,(\gamma_e + \gamma_n)\,(j+2)\,.
\end{aligned}
$$

To calculate the energy shifts due to the spin-orbital interaction exactly we should solve the equations for radial wave functions. However, in the frame of the perturbation theory methods, the first order corrections to the energy eigenvalues are determined by the quantum mechanical averages of the interaction Hamiltonian, i.e. we should average the function $1/r^3$ with the wave functions of Hamiltonian H_0. The radial wave functions of Hamiltonian H_0 depend solely on the quantum number l, therefore the obtained equations enable us to estimate the relative shifts of the different states with the same l, at least in the first order approximation. It is seen from the obtained equations that in the hydrogen atom, where the magnetic moments of electron and nucleus have the opposite signs, the spin-orbital interaction makes energetically more favorable the states with the smallest value of j at a given $l > 0$. The spin-orbital interaction tends to decrease the magnitude of the atomic magnetic moment and the value of the total angular momentum.

Before calculating the matrix elements of the Hamiltonian of spin-spin interaction, it is convenient to express it in terms of the total spin. It can be done in the following way. The total spin square is

$$\mathbf{S}^2 = \mathbf{s}_e^2 + \mathbf{s}_n^2 + 2(\mathbf{s}_e\mathbf{s}_n) = \frac{3}{2} + 2(\mathbf{s}_e\mathbf{s}_n)\,,$$

hence

$$(\mathbf{s}_e\mathbf{s}_n) = \frac{1}{2}\left(\mathbf{S}^2 - \frac{3}{2}\right).$$

By using the properties of the Pauli matrices we can write

$$(\mathbf{Se})^2 = (\mathbf{s}_e\mathbf{e})^2 + (\mathbf{s}_n\mathbf{e})^2 + 2(\mathbf{s}_e\mathbf{e})\,(\mathbf{s}_n\mathbf{e}) = \frac{1}{2} + 2(\mathbf{s}_e\mathbf{e})\,(\mathbf{s}_n\mathbf{e})\,,$$

hence

$$(\mathbf{s}_e\mathbf{e})\,(\mathbf{s}_n\mathbf{e}) = \frac{1}{2}\left[(\mathbf{Se})^2 - \frac{1}{2}\right].$$

Finally, for the Hamiltonian of spin-spin interaction we get

$$H_{ss} = \frac{2\mu_{\mathrm{B}}\mu_N\gamma_e\gamma_n}{r^3}\left(3(\mathbf{Se})^2 - \mathbf{S}^2\right). \tag{4.79}$$

It follows directly from the obtained equation that the energy of spin-spin interaction is equal to zero in spin zero state $\Omega^{(0)}$. For the spin one states we have

$$\mathbf{S}^2\Omega_{jlm}^{(1)} = s\,(s+1)\,\Omega_{jlm}^{(1)} = 2\Omega_{jlm}^{(1)}. \tag{4.80}$$

By applying the operator $(\mathbf{Se})$ to the wave functions (4.72) we get

$$(\mathbf{Se})\,\chi_1^{(1)} = -i\sqrt{\frac{4\pi}{3}}\left(Y_{10}\chi_1^{(1)} - Y_{1,+1}\chi_0^{(1)}\right),$$

$$(\mathbf{Se})\,\chi_0^{(1)} = -i\sqrt{\frac{4\pi}{3}}\left(Y_{1,-1}\chi_1^{(1)} - Y_{1,+1}\chi_{-1}^{(1)}\right),$$

$$(\mathbf{Se})\,\chi_{-1}^{(1)} = -i\sqrt{\frac{4\pi}{3}}\left(Y_{1,-1}\chi_0^{(1)} - Y_{10}\chi_{-1}^{(1)}\right),$$

where $Y_{lm}(\theta,\varphi)$ is the spherical harmonics. By using the obtained equations and the matrix elements for spherical harmonics

$$\langle l_1 m_1 | Y_{lm} | l_2 m_2 \rangle = (-1)^{m_1} i^{-l_1+l_2+l} \begin{pmatrix} l_1 & l & l_2 \\ -m_1 & m & m_2 \end{pmatrix} \begin{pmatrix} l_1 & l & l_2 \\ 0 & 0 & 0 \end{pmatrix} \times \\ \times \sqrt{\frac{(2l+1)(2l_1+1)(2l_2+1)}{4\pi}},$$

we get the following equations for the angular matrix elements of the first term in Hamiltonian (4.79)

$$\begin{gathered} \int \Omega^{(1)+}_{j,l=j-1,m} (\mathbf{Se})^2 \, \Omega^{(1)}_{j,l=j-1,m} \sin\theta \, d\theta \, d\varphi = \frac{j+1}{2j+1}, \\ \int \Omega^{(1)+}_{j,l=j,m} (\mathbf{Se})^2 \, \Omega^{(1)}_{j,l=j,m} \sin\theta \, d\theta \, d\varphi = 1, \\ \int \Omega^{(1)+}_{j,l=j+1,m} (\mathbf{Se})^2 \, \Omega^{(1)}_{j,l=j+1,m} \sin\theta \, d\theta \, d\varphi = \frac{j}{2j+1}, \\ \int \Omega^{(1)+}_{j,l=j+1,m} (\mathbf{Se})^2 \, \Omega^{(1)}_{j,l=j-1,m} \sin\theta \, d\theta \, d\varphi = -\frac{\sqrt{j(j+1)}}{2j+1}, \\ \int \Omega^{(1)+}_{j,l=j,m} (\mathbf{Se})^2 \, \Omega^{(1)}_{j,l=j-1,m} \sin\theta \, d\theta \, d\varphi = 0. \end{gathered} \tag{4.81}$$

It is seen that the spin-spin interaction couples only the states with the non-zero value of the total spin projection. The sum of the diagonal elements of (4.81) is equal to 2. Hence, accounting the equation (4.80), we can see that the sum of the diagonal elements of the Hamiltonian of spin-spin interaction $3(\mathbf{Se})^2 - \mathbf{S}^2$ is equal to zero. Besides, it follows directly from the definition of this operator.

With the help of the equations (4.81) we obtain the following form of the diagonal matrix elements of Hamiltonian of spin-spin interaction

$$\begin{gathered} \langle j, l=j-1, m | H_{ss} | j, l=j-1, m \rangle = -\frac{2\mu_B \mu_N \gamma_e \gamma_n}{r^3} \frac{j-1}{2j+1}, \\ \langle j, l=j, m | H_{ss} | j, l=j, m \rangle = \frac{2\mu_B \mu_N \gamma_e \gamma_n}{r^3}, \\ \langle j, l=j+1, m | H_{ss} | j, l=j+1, m \rangle = -\frac{2\mu_B \mu_N \gamma_e \gamma_n}{r^3} \frac{j+2}{2j+1}. \end{gathered}$$

Reminding the remarks, made with respect to the spin-orbital interaction, we can make some preliminary conclusions on the influence of the spin-spin interaction onto the hydrogen energy spectrum. It is seen from the above equations that the spin-spin interaction makes energetically more favorable the states with the smallest j at a given $l > 0$. The energy of spin-spin interaction is equal to zero in the antisymmetric state with the oppositely directed spins, and it is positive in the symmetric state with the oppositely directed spins. The spin-spin interaction tends to decrease the atomic magnetic moment and the total angular momentum of atom.

As we have mentioned above, the case of $j = 0$ requires the special consideration. In this special case for spin-orbital interaction we have

$$(\boldsymbol{\sigma}_e \mathbf{l})\, \Omega_0^{(0)} = 0, \quad (\boldsymbol{\sigma}_n \mathbf{l})\, \Omega_0^{(0)} = 0,$$
$$(\boldsymbol{\sigma}_e \mathbf{l})\, \Omega_{010}^{(1)} = -2\Omega_{010}^{(1)}, \quad (\boldsymbol{\sigma}_n \mathbf{l})\, \Omega_{010}^{(1)} = -2\Omega_{010}^{(1)}.$$

One can see that these equations follow directly from the appropriate equations (4.78). In this special case, the non-zero matrix element of spin-spin interaction Hamiltonian is

$$\langle 010| \left(3(\mathbf{Se})^2 - \mathbf{S}^2\right) |010\rangle = -1. \tag{4.82}$$

It is seen that this matrix element does not follow from the equations (4.81), because the equations (4.81) were obtained under assumption $j > 0$.

4.3.6 Equations for radial wave functions

Let us introduce the dimensionless energy E' and coordinate x:

$$E = \frac{\hbar^2 E'}{2m_r a_{\mathrm{B}}^2}, \quad x = \frac{r}{a_{\mathrm{B}}},$$

where the Bohr radius is

$$a_{\mathrm{B}} = \frac{\hbar^2}{m_r e^2}.$$

In dimensionless units the equation (4.69) becomes

$$\frac{d^2\Psi}{dx^2} + \frac{2}{x}\frac{d\Psi}{dx} + \left(E' + \frac{2Z}{x} - \frac{\mathbf{l}^2}{x^2}\right)\Psi +$$
$$+ \frac{2Z\alpha^2}{x^3}\frac{m_r}{M}\left(\mathbf{l}^2 - \gamma_e \mathbf{s}_e \mathbf{l} - \gamma_n \mathbf{s}_n \mathbf{l} - \frac{\gamma_e \gamma_n}{2}\left(3(\mathbf{Se})^2 - \mathbf{S}^2\right)\right)\Psi = 0, \tag{4.83}$$

where α is the fine structure constant.

One can see from the equation (4.83), that the energy of orbital, spin-orbital, and spin-spin interactions is smaller than the energy of the Coulomb interaction in the ratio

$$\alpha^2 \frac{m_e}{m_n} \approx 3 \cdot 10^{-8},$$

when the distance between the electron and nucleus is about the Bohr radius, $x \approx 1$. However, the energy of these hyperfine interactions increases at $x \to 0$ faster then the potential energy of Coulomb interaction or centrifugal energy. As a result the hyperfine interactions can, in principle, change the electron-nucleus interaction at small distances.

Now, the equations (4.78) and (4.81) enable us to get the equations for radial wave functions. By substituting the wave function (4.76) into the equation (4.83) and then averaging over the angular variables we obtain the equations for the functions $f(x)$ and $g_\sigma(x)$. We have already mentioned that the spin-orbital interaction couples the states with zero projection of the total spin, and spin-spin interaction couples the states with the non-zero projection of the total spin. Thus it is evident that the set of the four equations is decomposed into the two sets of the coupled equations:

for zero spin projection

$$\frac{d^2 f}{dx^2} + \frac{2}{x}\frac{df}{dx} + \left(E' + \frac{2Z}{x} - \frac{j(j+1)}{x^2}\right) f = \\ = -\frac{\beta}{x^3}\left[2j(j+1) f - \sqrt{j(j+1)}(\gamma_e - \gamma_n) g_0\right],$$

$$\frac{d^2 g_0}{dx^2} + \frac{2}{x}\frac{dg_0}{dx} + \left(E' + \frac{2Z}{x} - \frac{j(j+1)}{x^2}\right) g_0 = \\ = -\frac{\beta}{x^3}\left[(2j(j+1) + \gamma_e + \gamma_n - \gamma_e\gamma_n) g_0 - \sqrt{j(j+1)}(\gamma_e - \gamma_n) f\right] \tag{4.84}$$

and non-zero spin projection

$$\frac{d^2 g_1}{dx^2} + \frac{2}{x}\frac{dg_1}{dx} + \left(E' + \frac{2Z}{x} - \frac{j(j-1)}{x^2}\right) g_1 = \\ = -\frac{\beta}{x^3}\left[\left(2j(j-1) - (\gamma_e + \gamma_n)(j-1) + \gamma_e\gamma_n\frac{j-1}{2j+1}\right) g_1 - \\ - 3\gamma_e\gamma_n\frac{\sqrt{j(j+1)}}{2j+1} g_{-1}\right],$$

$$\frac{d^2 g_{-1}}{dx^2} + \frac{2}{x}\frac{dg_{-1}}{dx} + \left(E' + \frac{2Z}{x} - \frac{(j+1)(j+2)}{x^2}\right) g_{-1} = \\ = -\frac{\beta}{x^3}\left[\left(2(j+1)(j+2) + (\gamma_e + \gamma_n)(j+2) + \gamma_e\gamma_n\frac{j+2}{2j+1}\right) g_{-1} - \\ - 3\gamma_e\gamma_n\frac{\sqrt{j(j+1)}}{2j+1} g_1\right], \tag{4.85}$$

where

$$\beta = Z\alpha^2 \frac{m_r}{M}. \tag{4.86}$$

Notice, that the equations (4.84) and (4.85) have been used to derive the last equations.

4.3.7 Influence of orbital, spin-orbital, and spin-spin interactions on the energy spectrum of hydrogen atom

The obtained equations for radial wave functions are identical each other and they differ only in the magnitude of the incoming coefficients. However, as well as the analytical solutions of equations of this type are not known we shall use the perturbation theory to analyze the structure of the energy spectrum. The basis for applicability of the perturbation theory methods is in the smallness of the parameter β defined by the equation (4.86). The equations of the zero-order approximation are given by the left-hand-sides of the equations (4.84), (4.85). These equations depend solely on the quantum number l, therefore to systematize the levels of the corrected spectrum we can use the notations similar to the spectroscopic notations. The level with the principal quantum number n, total angular momentum j, orbital angular momentum $L = l$, and spin s is designated as

$$n^{(2s+1)}L_j. \tag{4.87}$$

The capital L shows the orbital angular momentum of atom, i.e. the sum of the electron and nucleus angular momenta, but not the orbital angular momentum of electron alone. It should be also noted that the total angular momentum j in (4.87) is referred to $\mathbf{J} = \mathbf{j}_e + \mathbf{j}_n$ while it is often referred to the total angular momentum of electron, which is $\mathbf{j}_e = \mathbf{l}_e + \mathbf{s}_e$.

As we have shown above the only operator $\mathbf{J}$ is integral of motion in general case, while the total orbital angular momentum $\mathbf{L}$ is not. But in zero order approximation, when the hyperfine interactions are neglected, the orbital angular momentum is integral of motion too. It gives a basis to introduce the notations (4.87) and explains the reason why these notations are widely used in spectroscopy.

To facilitate reading, the Table 4.1 gives the correspondence between the states (4.87) and states of different j and s.

One can see from the table that the S-state splits into the two sublevels, ${}^1S_0$ and ${}^3S_1$, the all other states split into the four sublevels, ${}^1L_j$, ${}^3L_{j-1}$, ${}^3L_j$, ${}^3L_{j+1}$.

The equations (4.84), (4.85) show that, in terms of Table 4.1, the spin-orbital interaction results in the coupling of n^1L_j and n'^3L_j states,

Table 4.1. Energy states of hydrogen atom

j	s	
	$s=0$	$s=1$
$j=0$	$^1S_0$	$^3P_0$
$j=1$	$^1P_1$	$^3S_1,\ ^3P_1,\ ^3D_1$
$j=2$	$^1D_2$	$^3P_2,\ ^3D_2,\ ^3F_2$
$j=3$	$^1F_3$	$^3D_3,\ ^3F_3,\ ^3G_3$

and spin-spin interaction couples the states $n^3(L-1)_j$ and $n'^3(L+1)_j$, where n and n' are the principal quantum numbers. It is also seen from the equations (4.84), (4.85) that, if the principle quantum numbers n and n' do not coincide, then the energy shift, due to the coupling of the unperturbed states, is the correction of the second order on the smallness parameter β. Indeed, in the frame of the perturbation theory methods the energies of the coupled states are given by

$$E^{(1,2)} = \frac{E_n + E_{n'} + \Delta E_n + \Delta E_{n'}}{2} \pm \pm \sqrt{\left(\frac{E_n - E_{n'} + \Delta E_n - \Delta E_{n'}}{2}\right)^2 + (\Delta E_{nn'})^2}, \quad (4.88)$$

where ΔE_n and $\Delta E_{nn'}$ are the partial shifts and coupling coefficients (i.e. the diagonal and non-diagonal elements of the Hamiltonian of hyperfine interactions), respectively. At $|E_n - E_{n'}| \gg \Delta E_{nn'}$, the contribution of the non-diagonal element is about $(\Delta E_{nn'})^2 / (E_n - E_{n'})$ and we can neglect them. At $n = n'$, the contribution of the non-diagonal elements becomes significant.

The radial wave functions of the equations (4.84), (4.85) in zero order approximation were calculated in Chapter 3 and Chapter 4. With the help of these wave functions, at $n = n'$, for matrix elements of $1/x^3$ we get the following formulas

$$\int_0^\infty \frac{1}{x^3} R_{nl}^2(x) x^2\, dx = \left(\frac{2Z}{n}\right)^3 \frac{1}{4l(l+1)(2l+1)},$$

$$\int_0^\infty \frac{1}{x^3} R_{nl}^2(x) R_{n,l+1}(x) x^2\, dx = \left(\frac{2Z}{n}\right)^3 \frac{\sqrt{(n-l-1)(n+l+1)}}{2n\,(2l+1)\,(2l+2)\,(2l+3)}, \quad (4.89)$$

$$\int_0^\infty \frac{1}{x^3} R_{nl}^2(x) R_{n,l+2}(x) x^2\, dx = 0.$$

Thus, the spin-orbital coupling of singlet and triplet states with the same l results in the corrections of the first order. The spin-spin coupling of the states $l = j - 1$ and $l' = j + 1$ contributes only into the second order corrections.

Consider firstly the equations (4.84) corresponding to the states of $m = 0$. At $j = 0$, the coupling coefficient is equal to zero, therefore the energy shifts are

$$\begin{aligned} \Delta E\left(n^1S_0\right) &= -\frac{\beta}{2}\left(\frac{2Z}{n}\right)^3 \mathrm{Ry}, \\ \Delta E\left(n^3P_0\right) &= -\beta\left(\frac{2Z}{n}\right)^3 \frac{4 + 2(\gamma_e + \gamma_n) + \gamma_e\gamma_n}{24}\mathrm{Ry}. \end{aligned} \tag{4.90}$$

At $j > 0$ the levels become coupled. The partial shifts and coupling coefficient in this case are

$$\begin{aligned} \Delta E\left(n^1J_j\right) &= -\beta\left(\frac{2Z}{n}\right)^3 \frac{1}{2(2j+1)}\mathrm{Ry}, \\ \Delta E\left(n^3J_j\right) &= -\beta\left(\frac{2Z}{n}\right)^3 \frac{2j\,(j+1) + \gamma_e + \gamma_n - \gamma_e\gamma_n}{4j\,(j+1)\,(2j+1)}\mathrm{Ry}, \\ \Delta E\left(n^1J_j \leftrightarrow n^3J_j\right) &= \beta\left(\frac{2Z}{n}\right)^3 \frac{\gamma_e - \gamma_n}{4(2j+1)\sqrt{j\,(j+1)}}\mathrm{Ry}, \end{aligned} \tag{4.91}$$

where the symbol L has been substituted by symbol J because at $m = 0$ we have $l = j$. By substituting these equations into the equation (4.88) we can easily get the corrected energy for the corresponding states.

For the states with $m \neq 0$, hence $j > 0$, in accordance with the equations (4.85) and (4.89), we get for partial shifts

$$\begin{aligned} \Delta E\left(n^3(J-1)_j\right) &= -\beta\left(\frac{2Z}{n}\right)^3 \frac{2j\,(2j+1) - (2j+1)\,(\gamma_e + \gamma_n) + \gamma_e\gamma_n}{4j\,(2j-1)\,(2j+1)}\mathrm{Ry}, \\ \Delta E\left(n^3(J+1)_j\right) &= \\ &= -\beta\left(\frac{2Z}{n}\right)^3 \frac{2(j+1)\,(2j+1) + (2j+1)\,(\gamma_e + \gamma_n) + \gamma_e\gamma_n}{4(j+1)\,(2j+1)\,(2j+3)}\mathrm{Ry}. \end{aligned} \tag{4.92}$$

The coupling coefficient for this case is exactly equal to zero due to the last integral in (4.89).

In accordance with the discussion given above, the equations (4.90) and (4.92) are directly defined the first order corrections to the shifts of the corresponding energy states. The energies of the coupled levels $n^1J_j \leftrightarrow n^3J_j$ are determined by the solutions of the equation (4.88), where the diagonal and non-diagonal elements of the Hamiltonian of hyperfine interactions are given by equation (4.91). It should be reminded that the Rydberg constant Ry has been defined above as

$$\mathrm{Ry} = \frac{m_r e^4}{2\hbar^2}, \tag{4.93}$$

therefore in final calculations it should be replaced by $\mathrm{Ry}\cdot(m_e+m_n)/m_n$.

The energy level diagram including $1S$, $2S$, and $2P$ states of hydrogen atom is shown in Fig. 4.4. By comparing the Fig. 4.4 and Fig. 1.1, one can see that the equation (4.83) yields the results for the singlet and triplet sublevels of S state that are qualitatively agree with the

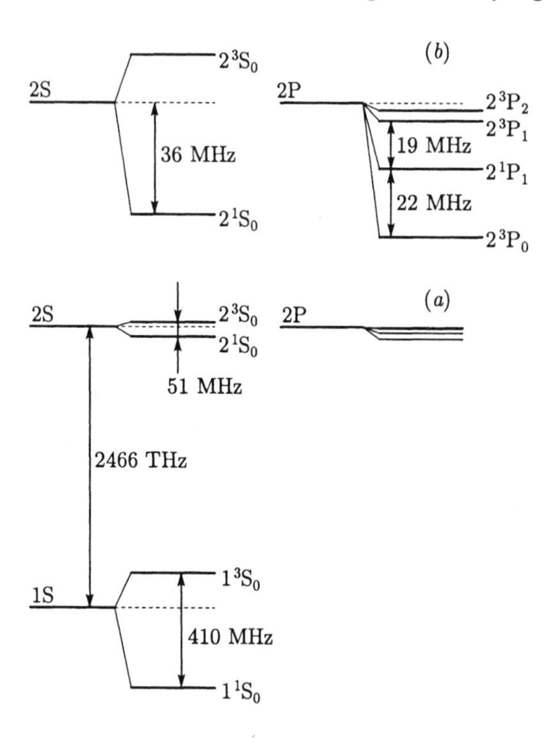

Figure 4.4. The Pauli equation corrections to the hydrogen atom spectrum: (*a*) $1S$, $2S$, $2P$ states; (*b*) $2S$, $2P$ (magnified)

experimental data. However, the numerical values of shifts differ from experimentally measured ones. The relative shifts of the states 2^3P_0 and 2^1P_1 with respect to state $2P$ of electron in Coulomb field are qualitatively agree with the experimental data, but their numerical values are again differ from the experimentally measured frequencies.

The shifts of states 2^3P_2 and 2^3P_1 differ from the experimental data not only in the numerical values but even in the direction of shift.

Thus, we can see that the incorporation of the idea on spin of electron has given significant improvements into the theory of hydrogenic spectra. The results of calculations are qualitatively agree with the results of the experimental measurements, but the numerical values of the transition frequencies remain still different from the experimental data. The further improvements in the theory were achieved with the help of relativistic equations of quantum mechanics.

Chapter 5

RELATIVISTIC EQUATION FOR SPIN ZERO PARTICLE

The first quantum relativistic equation was proposed by Klein, Fock, and Gordon [50–52] in 1926. It is well known now that this equation corresponds to the spin zero particle. This equation enables us to calculate the energy spectrum of the hydrogenic system consisting of the two spin zero particles. There are a lot of the nuclei of zero spin. Hence, if we substitute the electron in the hydrogen-like ion by the spin zero particle, we get a system describing by the Klein–Gordon–Fock equation. The most famous system of this type is the mesoatom, when the electron is substituted by the μ^- meson.

5.1 Klein–Gordon–Fock equation

The Klein–Gordon–Fock equation, or relativistic Schrödinger equation, for the case of a free particle is

$$\frac{\hbar^2}{c^2}\frac{\partial^2 \psi}{\partial t^2} = \left(\hbar^2 \Delta - M^2 c^2\right) \psi, \tag{5.1}$$

where M is the mass of a particle. It is seen that this equation has the relativistic invariant form, but it becomes more evident if we introduce the four dimensional radius vector

$$x_\mu = (\mathbf{r}, ict) \tag{5.2}$$

and four-dimensional momentum operator

$$p_\mu = \left(\mathbf{p}, -\frac{\hbar}{c}\frac{\partial}{\partial t}\right) = \left(-i\hbar\nabla, -\frac{\hbar}{c}\frac{\partial}{\partial t}\right) = -i\hbar\frac{\partial}{\partial x_\mu}. \tag{5.3}$$

If we apply the transformations (5.2) and (5.3) to the equation (5.1), it becomes

$$\left(p_\mu p_\mu + M^2 c^2\right) \psi = 0. \tag{5.4}$$

We use here and shall use further, the generally accepted convention that the double-recurring index means summation over this index.

It can be easily shown that the equation (5.1) results in the following continuity equation

$$\frac{\partial \rho}{\partial t} + \operatorname{div} \mathbf{j} = 0, \tag{5.5}$$

where

$$\rho(\mathbf{r}, t) = \frac{i\hbar}{2Mc^2}\left(\psi^* \frac{\partial \psi}{\partial t} - \frac{\partial \psi^*}{\partial t}\psi\right), \tag{5.6}$$

$$\mathbf{j}(\mathbf{r}, t) = -\frac{i\hbar}{2M}\left[\psi^* \nabla \psi - (\nabla \psi^*)\psi\right]. \tag{5.7}$$

The equation (5.5) can be also written in the relativistic invariant form

$$\frac{\partial j_\mu}{\partial x_\mu} = 0, \tag{5.8}$$

where $j_\mu = (\mathbf{j}, ic\rho)$.

One can see from the equation (5.6) that the time-component of the current density four-vector is not the certainly positively defined variable. Indeed the equation (5.1) is the second order differential equation with respect to the time derivative, therefore to determine unambiguously the initial state of the particle we should assign the initial values both the wave function $\psi(\mathbf{r}, 0)$ and its first time-derivative $\partial\psi(\mathbf{r}, 0)/\partial t$. Hence, if there are no any restrictions for these initial values then the initial value of $\rho(\mathbf{r}, 0)$ can be arbitrary, i.e. positive, negative, or zero. On the other hand, in accordance with the equation (5.6) the time-component of the current density plays the role of the density of probability for particle to be at a specific spatial point at a given moment of time. Therefore, in the frames of the probabilistic interpretation of the wave function, it must be positively defined. We have seen that this requirement holds for the probability density of Schrödinger equation. It is the uncertainty of the sign of $\rho(\mathbf{r}, t)$ that stimulated Paul Dirac to look for another quantum relativistic equation. It was assumed for a long time that the equation (5.1) was not applicable to describe any elementary particles, until it was not recognized the correctness of this equation for the case of the spin zero particles.

It becomes much easier to interpret the equation (5.1), if we introduce the electric charge and current density

$$\rho_e(\mathbf{r}, t) = \frac{iq\hbar}{2Mc^2}\left(\psi^* \frac{\partial \psi}{\partial t} - \frac{\partial \psi^*}{\partial t}\psi\right), \tag{5.9a}$$

$$\mathbf{j}_e(\mathbf{r}, t) = -\frac{iq\hbar}{2M}\left[\psi^* \nabla \psi - (\nabla \psi^*)\psi\right], \tag{5.9b}$$

where q is the elementary charge. In this case, by adjusting the sign of the time derivative $\partial\psi/\partial t$ and sign of charge, we can always satisfy the condition that the probability density, $\rho_e^{(\pm)}/(\pm|q|)$, is the positive defined value.

By concluding the introductory remarks, we can notice the following. The second order (with respect to the time derivative) differential equation has the two linear independent solutions. For example, the general solution of equation (5.1) for the case of free particle includes the two time-dependent functions

$$\psi(\mathbf{r},t) = \sum_{\mathbf{p}} \{C_1 \exp[-i(Et - \mathbf{pr})/\hbar] + C_2 \exp[i(Et - \mathbf{pr})/\hbar]\}, \tag{5.10}$$

where

$$E = \sqrt{\mathbf{p}^2c^2 + M^2c^4}. \tag{5.11}$$

The substitution of the equation (5.10) into (5.6) yields

$$\rho(\mathbf{r},t) = \frac{E}{Mc^2}\left(|\psi_1(\mathbf{r},t)|^2 - |\psi_2(\mathbf{r},t)|^2\right) = \frac{E}{Mc^2}\left(|C_1|^2 - |C_2|^2\right).$$

Thus, the probability densities, $\rho_{1,2}(\mathbf{r},t)$, corresponding to the two linear independent solutions, have the opposite signs

$$\rho_{1,2}(\mathbf{r},t) = \pm\frac{E}{Mc^2}|\psi_{1,2}(\mathbf{r})|^2.$$

It is seen that, if the time-dependent solutions are the complex functions, then the sign of probability density does not vary in time, and only in the case, when the time-dependent solution is a real function of time, the probability density could change the sign in the process of system evolution.

5.2 Interaction of zero spin particle with electromagnetic field

Whether the particle possesses the non-zero charge or not, we can understand only by studying (or observing) the process of the particle interaction with electromagnetic field. As we have discussed above, the general algorithm of obtaining the equations for particle interacting with the electromagnetic field from the free-particle equations consists in the replacement of the four-dimensional momentum operator by the generalized four-dimensional momentum operator

$$p_\mu \to -i\hbar\frac{\partial}{\partial x_\mu} - \frac{e}{c}A_\mu. \tag{5.12}$$

By applying the replacement (5.12) to the equation (5.1), we get

$$\frac{1}{c^2}\left(i\hbar\frac{\partial}{\partial t}-e\varphi\right)^2\psi=\left[\left(\mathbf{p}-\frac{e}{c}\mathbf{A}\right)^2+M^2c^2\right]\psi. \tag{5.13}$$

The covariant form of the equation (5.13) is

$$\left[\left(-i\hbar\frac{\partial}{\partial x_\mu}-\frac{e}{c}A_\mu\right)\left(-i\hbar\frac{\partial}{\partial x_\mu}-\frac{e}{c}A_\mu\right)+M^2c^2\right]\psi=0. \tag{5.14}$$

The covariant form demonstrates evidently, that the equation (5.14) is invariant with respect to the gauge transformation of the electromagnetic field potentials

$$A'_\mu=A_\mu+\frac{\partial\chi}{\partial x_\mu}$$

and wave function

$$\psi'(\mathbf{r},t)=\psi(\mathbf{r},t)\exp\left[\frac{ie}{\hbar c}\chi(\mathbf{r},t)\right].$$

Notice that the gauge transformation coincides with the gauge transformation for Schrödinger equation (see (2.13)).

If we compare the equation (5.13) with the Schrödinger equation, then we can see that the symmetry properties of equation (5.13) with respect to the orthogonal transformations of reference frame will be different of those for Schrödinger equation only in the case when the transformation includes the time axis. The equation (5.13) can be written in the relativistic invariant form (5.14), therefore its invariance with respect of Lorentz transformation is evident. Hence we really need not in the analysis of the symmetry properties of Klein–Gordon–Fock equation. There is only one moment worthy of attention. The orthogonal transformations are given by

$$x'_\mu=a_{\mu\nu}x_\nu \tag{5.15}$$

where matrix $a_{\mu\nu}$ obeys the condition

$$a_{\mu\nu}a_{\mu\lambda}=\delta_{\nu\lambda}. \tag{5.16}$$

The transformations (5.15), (5.16) remain invariable the spacetime interval, and they describe the Lorentz transformation, three-dimensional rotations, and space inversion. By applying transformations (5.15), (5.16) to the equation (5.14) we can see that the wave function is transformed in the following way

$$\psi(x)\rightarrow\psi'(x')=\lambda\psi(x), \tag{5.17}$$

where

$$|\lambda| = 1.$$

Particularly, at the space inversion transformation

$$\mathbf{r}' = -\mathbf{r}, \quad t' = t$$

for the parameter λ we have

$$\lambda^2 = 1,$$

because the twice repeated space inversion transformation is the identical transformation.

Hence, the wave function is even (at $\lambda_1 = +1$)

$$\psi(-\mathbf{r}, t) = \psi(\mathbf{r}, t),$$

or odd (at $\lambda_2 = -1$)

$$\psi(-\mathbf{r}, t) = -\psi(\mathbf{r}, t).$$

Thus, the wave function of the Klein–Gordon–Fock equation is either scalar or pseudoscalar function. As it follows from the discussion given in the beginning of the previous chapter the scalar and pseudoscalar wave functions describe the spinless particles.

5.3 Mesoatom

Let us consider the problem on zero spin particle motion in Coulomb field. The problem on the pion motion in the field of the heavy nuclei is an practical example. The bound state of pion and nucleus is mesoatom.

In the attracting Coulomb field the potential energy is

$$U(r) = e\varphi(r) = -\frac{Ze^2}{r}. \tag{5.18}$$

The substitution of equation (5.18) into the equation (5.13) results

$$\left[\Delta - \frac{M^2c^4 - E^2}{\hbar^2c^2} + \frac{2EZe^2}{\hbar c}\frac{1}{r} + \frac{Z^2e^4}{\hbar^2c^2}\frac{1}{r^2}\right]\psi(\mathbf{r}) = 0. \tag{5.19}$$

By taking into account that the equation (5.19) is spherically symmetric we can express the wave function in terms of the spherical harmonics

$$\psi(\mathbf{r}) = R_l(r)\, Y_{lm}(\theta, \varphi).$$

Due to orthogonality of spherical harmonics we get for the radial wave functions the following equation

$$\left[\frac{d^2}{dr^2} + \frac{2}{r}\frac{d}{dr} - \frac{l(l+1) - Z^2\alpha^2}{r^2} + \frac{2EZ\alpha}{\hbar c}\frac{1}{r} - \kappa^2\right] R_l(r) = 0, \tag{5.20}$$

where $\alpha = e^2/(\hbar c)$ is the fine structure constant, and

$$\kappa = \frac{\sqrt{M^2c^4 - E^2}}{\hbar c}. \tag{5.21}$$

The equation (5.20) is quite similar to the equation for radial functions (see (2.29)), that was obtained from the Schrödinger equation for electron moving in the Coulomb field. Therefore the solutions of the equation (5.20) are again expressed in terms of the confluent hypergeometric functions

$$R(r) = C_1 r^s \exp(-\kappa r) F\left(s + 1 - \frac{EZ\alpha}{\kappa\hbar c}, 2(s+1), 2\kappa r\right) + \\ + C_2 r^{-(s+1)} \exp(-\kappa r) F\left(-s - \frac{EZ\alpha}{\kappa\hbar c}, -2s, 2\kappa r\right) \tag{5.22}$$

where

$$s = \sqrt{(l + 1/2)^2 - Z^2\alpha^2} - 1/2. \tag{5.23}$$

In the considered case, the boundary conditions for the eigenvalue problem are the same as for the problem of electron motion in Coulomb field. The wave function should be finite at $r = 0$ and it should tend to zero at $r \to \infty$. The second term in (5.22) is divergent at $r \to 0$, therefore $C_2 = 0$. Similar to solution (2.30) of equation (2.29) the first term in (5.22) tends to zero at infinity, $r \to \infty$, when the first argument of the confluent hypergeometric function obeys the condition

$$s + 1 - \frac{EZ\alpha}{\kappa\hbar c} = -n_r, \tag{5.24}$$

where n_r is the non-negative integer. This condition yields the equation for the energy spectrum of the bound states. Notice, that, in analogy with the solution of the Schrödinger equation, it is convenient to introduce the principle quantum number

$$n = n_r + l + 1,$$

then the energy spectrum takes the form

$$E_{nl} = \frac{Mc^2(n + \delta_l)}{\sqrt{(n + \delta_l)^2 + Z^2\alpha^2}}, \tag{5.25}$$

where

$$\delta_l = s - l = \sqrt{(l + 1/2)^2 - Z^2\alpha^2} - (l + 1/2).$$

It is seen that the energy spectrum of the particle obeying the Klein–Gordon–Fock equation, in contrast to the energy spectrum of particle obeying the Schrödinger equation, includes the fine structure. Indeed the spectrum (5.25) depends on the two quantum numbers: principle quantum number n and angular momentum l. The energy distance between the levels with the same n and different l can be easily estimated in the case when $Z\alpha \ll 1$. In this case we have

$$\delta_l = \sqrt{(l+1/2)^2 - Z^2\alpha^2} - (l+1/2) \approx -\frac{Z^2\alpha^2}{2(l+1/2)^2}.$$

By expanding (5.25) in the series on $Z\alpha$, we get

$$E_{nl} = Mc^2\left[1 - \frac{Z^2\alpha^2}{2n^2} - \frac{Z^4\alpha^4}{2n^4}\left(\frac{n}{l+1/2} - \frac{3}{4}\right) + \ldots\right]. \tag{5.26}$$

If mass M coincides with electron mass then the first term in the expansion (5.26) is equal to

$$\Delta E_n^{(0)} = m_0c^2 - E_{nl} \approx \frac{Z^2 m_0 e^4}{2n^2\hbar^2} = \frac{Z^2}{n^2}\mathrm{Ry}.$$

Thus the first term in the expansion (5.26) coincides with the Bohr formula (2.33). The second term of expansion depends on the principle quantum number n and angular momentum l. Let us take, for example, states with $l = 1$ and $l = 0$ and the same n. For energy distance between them we have

$$\Delta E_{n0} - \Delta E_{n1} = \frac{Z^4\alpha^4 m_0 c^2}{2n^3}\left(2 - \frac{2}{3}\right) = \frac{4}{3}\frac{Z^4\alpha^2}{n^3}\mathrm{Ry}.$$

It is natural to compare energy of this splitting with the energy shift due to the hyperfine interactions. For example the energy shift of n^1S_0 states due to the hyperfine interactions is (see equation (4.90))

$$\Delta E\left(n^1S_0\right) = -\frac{4Z^4\alpha^2}{n^3}\frac{m_r}{m_e+m_n}\mathrm{Ry}.$$

One can see that the energy distance between the states with different l exceeds the energy of hyperfine splitting approximately in the ratio

$$\frac{\Delta E_{n0} - \Delta E_{n1}}{\Delta E\left(n^1S_0\right)} \sim \frac{m_n}{m_e}.$$

For the hydrogen atom this ratio is $m_p/m_e \approx 1836$. Fig. 5.1 shows in comparison the normalized energy spectra of the hydrogen atom $(\Delta E_{\mathrm{Sch}}/(m_e c^2\alpha^2))$ and mesoatom $(\Delta E_{\mathrm{KGF}}/(Mc^2\alpha^2))$.

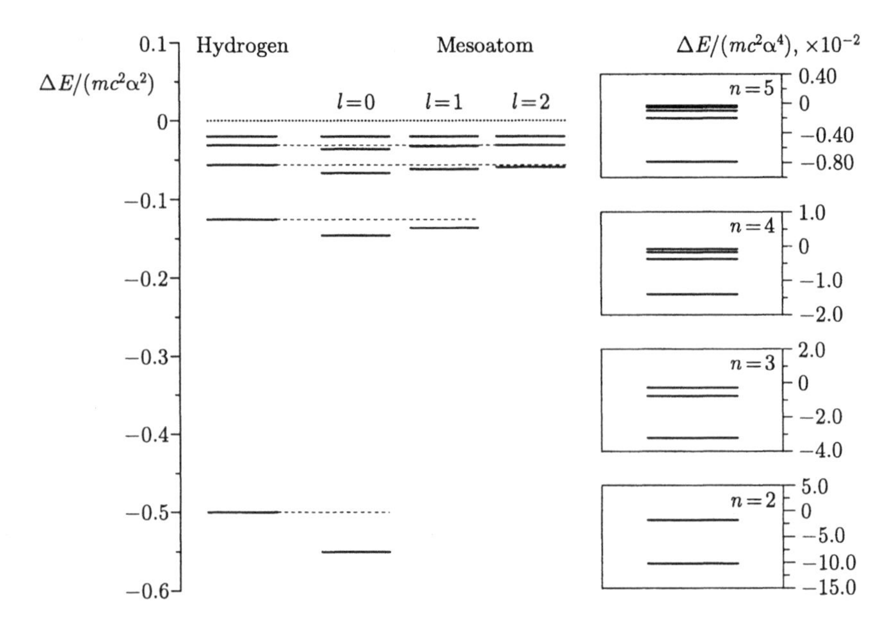

Figure 5.1. The normalized energy spectra of hydrogen and mesoatom. The sublevels of the mesoatom spectrum at the principle quantum number $n = 2, 3, 4, 5$ are shown in the magnified scale (right column)

The energy spectrum (5.26) does not coincide with the hydrogenic energy spectrum calculated by Sommerfeld [53] on the basis of Bohr quantization rules

$$\Delta E_{nl} = -\frac{m_0 c^2 Z^2 \alpha^2}{2(|k| + n_r)^2}\left[1 + \frac{Z^2\alpha^2}{(|k| + n_r)}\left(\frac{1}{|k|} - \frac{3}{4(|k| + n_r)}\right)\right], \qquad (5.27)$$

where $k = -(l+1), l$. The Sommerfeld spectrum coincides much better with the experimentally measured spectra than the spectrum (5.26). It was a serious basis to hesitate in the adequateness of the Klein–Gordon–Fock equation for description of electron. In the next chapter we shall see that the Dirac equation yielded the equation for hydrogenic spectrum, which is much closer to the experimentally measured spectra.

5.4 Wave functions

The normalized radial wave functions are

$$R_{nl}(r) = \sqrt{\frac{(2\kappa_{nl})^3 (n-l-1)!}{2(n+\delta_l)(n+l+2\delta_l)!}}\,(2\kappa_{nl} r)^{l+\delta} \exp(-\kappa_{nl} r) \times$$
$$\times L_{n-l-1}^{(2(l+\delta)+1)}(2\kappa_{nl} r), \qquad (5.28)$$

where

$$\kappa_{nl} = \frac{Z}{a_M} \frac{1}{\sqrt{(n+\delta_l)^2 + Z^2\alpha^2}}. \tag{5.29}$$

In analogy with the Bohr radius we have introduced in (5.29) the radius of orbit for π- meson atom:

$$a_M = \frac{\hbar^2}{Me^2}. \tag{5.30}$$

The radial wave functions given by equation (5.28) tends to zero at $r \to 0$, when $l > 0$. In the case of $l = 0$, the wave functions is singular at $r = 0$, because

$$\delta_0 = \sqrt{\frac{1}{4} - Z^2\alpha^2} - \frac{1}{2}$$

is negative. However at $Z\alpha \ll 1$ the singularity is weak and the difference between the wave functions (5.28) and non-relativistic wave

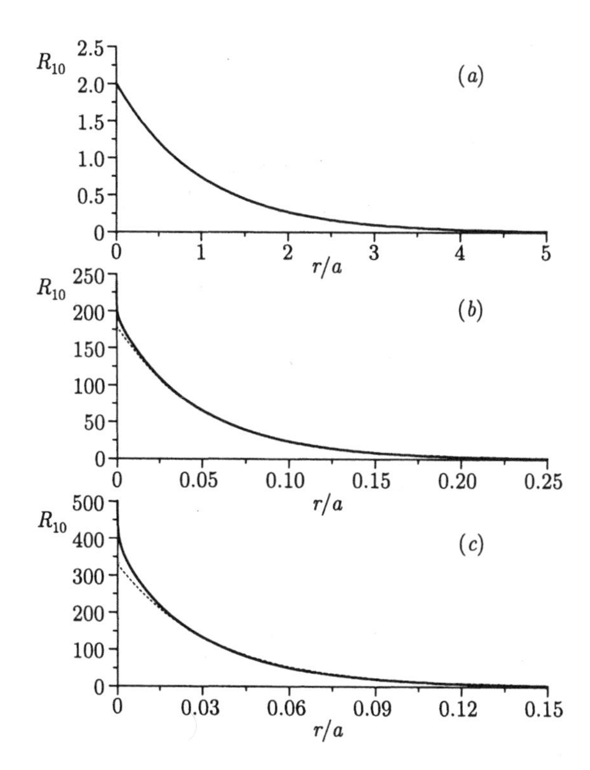

Figure 5.2. The radial wave functions of the KGF equation (solid lines) and Schrödinger equation (dashed lines) for the state of $n = 1$ and $l = 0$ at different nucleus charge $Z = 1$ (a), 20 (b), 30 (c)

functions (2.35) is small. For the nuclei with the high value of charge ($Z\alpha \to 1/2$) the singularity increases and the difference between the relativistic and non-relativistic wave functions becomes more significant. It follows from the equation (5.30) that the radius of orbit in π^--meson atom is smaller than the radius of orbit in hydrogen atom in the ratio of electron mass to meson mass. Hence, the π^- meson is much closer to the nucleus than electron in hydrogen atom. As a result the difference between the Coulomb potential and potential produced by the highly charged nucleus, having the finite size, becomes more significant. The substitution of Coulomb potential by some more realistic intra-atomic potential removes the problem of singularity of the wave function. It should be noted that the charge density, defined by $q(r) = R^2(r)\, r^2$, has no singularities at any l.

The Fig. 5.2 shows in comparison the wave functions (5.28) and non-relativistic wave functions (2.35) for the state of $n = 1$ and $l = 0$ at a different charge of nucleus $Z = 1$ (a), 20 (b), 30 (c). The dimensionless coordinate is normalized to a_M for π-meson atom and a_{B} for hydrogen atom. It is seen from the figure that the difference in the profiles of wave functions becomes visible only at $Z = 30$.

Table 5.1. The normalized radial matrix elements of transitions $1S \to nP$ for mesoatom

n	**2**	**3**	**4**	**5**	**6**	**7**	**8**
$\frac{Z}{a_M}\langle nP\vert\, r\,\vert 1S\rangle$	1.29015	0.51666	0.30457	0.20869	0.15513	0.12141	0.09849
n	**9**	**10**	**11**	**12**	**13**	**14**	**15**
$\frac{Z}{a_M}\langle nP\vert\, r\,\vert 1S\rangle$	0.08204	0.06974	0.06026	0.05276	0.0467	0.04172	0.03758
n	**16**	**17**	**18**	**19**	**20**	**21**	**22**
$\frac{Z}{a_M}\langle nP\vert\, r\,\vert 1S\rangle$	0.03408	0.03109	0.02851	0.02628	0.02432	0.02259	0.02106

The matrix elements of transitions $1S \to nP$ (i.e. transitions between the states $n_1 = 1, \quad l_1 = 0$ and $n_2 = n, \quad l_2 = 1$)

$$\langle nP|\, r|1S\rangle = \int_0^\infty R_{nP}(r) R_{1S}(r)\, r^3\, dr.$$

are given in Table 5.1. By comparing the data of Tabl. 3.1 and Tabl. 5.1 we can see that the difference in the magnitude of normalized matrix elements is about 10^{-4} for the lower states and tends to zero with the increase of the principle quantum number for the final state.

Chapter 6

DIRAC EQUATION

The intrinsic angular momentum or particle spin was introduced in the Chapter 4 on the basis of analogy between the spin and angular momentum associated with the translational degrees of freedom. As far as the wave function of the particle with the angular momentum l has the $(2l+1)$ components, corresponding to the different values of the angular momentum projections $-l \leq m \leq l$, then the wave function of the particle of spin s should have the $(2s+1)$ components. This number of components realizes the irreducible representation of the group of three-dimensional rotations. However, this analogy becomes incomplete when we turn to the equations of the relativistic theory. Indeed, the group of three-dimensional rotations is really a subgroup of the four-dimensional transformations. In classical physics, the rotations of the reference frame and the rotations of particle are equivalent transformations. In quantum mechanics, the elementary particle is a point object therefore the spin of the particle is completely associated with the rotations of the reference frame. Hence, the structure of the wave function should be adjusted with the group of the four-dimensional rotations. As a result, the wave function of the spin-1/2 particle became the four-component function or bispinor.

6.1 Dirac matrices

The Klein–Gordon–Fock equation has the relativistic invariant form. At the same time we have mentioned in Chapter 5 that the time component of the current density four-vector is not the certainly positively defined value. If we impose constraint on the structure of the particle many-component wave function, by demanding that the time component of the current density four-vector should be the bilinear combination of the wave function components of the following type

$$\sum_{\mu} \psi_{\mu}^{*}\psi_{\mu},$$

it results unambiguously that the equation for the wave function should be the differential equation of the first order with respect to the time derivative. To satisfy this condition we should factorize the differential operator of the Klein–Gordon–Fock equation. The required factorization can be made in the following way

$$\frac{1}{c^2}\frac{\partial^2}{\partial t^2} - \Delta + \frac{m_0^2 c^2}{\hbar^2} = \left(\frac{1}{c}\frac{\partial}{\partial t} + \boldsymbol{\alpha}\nabla + \beta\frac{i m_0 c}{\hbar}\right)\left(\frac{1}{c}\frac{\partial}{\partial t} - \boldsymbol{\alpha}\nabla - \beta\frac{i m_0 c}{\hbar}\right). \tag{6.1}$$

The self-conjugated matrices $\boldsymbol{\alpha}$ and β realizing the required factorization should obey the following conditions

$$\alpha_i\alpha_j + \alpha_j\alpha_i = 2\delta_{ij}, \quad \alpha_i\beta + \beta\alpha_i = 0, \quad \beta^2 = I. \tag{6.2}$$

It was shown by Paul Dirac that the matrices $\boldsymbol{\alpha}$ and β satisfying the all required conditions should be the four-dimensional matrices. One of the possible representations of the matrices α_i and β is

$$\boldsymbol{\alpha} = \begin{pmatrix} 0 & \boldsymbol{\sigma} \\ \boldsymbol{\sigma} & 0 \end{pmatrix}, \quad \beta = \begin{pmatrix} I & 0 \\ 0 & -I \end{pmatrix}, \tag{6.3}$$

where $\boldsymbol{\sigma}$ is two-dimensional Pauli matrices introduced and discussed in the Chapter 4.

Thus, the Hamiltonian of the Klein–Gordon–Fock equation which is the differential operator of the second order with respect to the time derivative is factorized in the product of the two differential operators of the first order with respect to the time derivative. Notice that if the wave function is a solution of the first order differential equation then it should be the solution of the second order differential equation too. By using the identical transformation

$$\frac{1}{c}\frac{\partial}{\partial t} + \boldsymbol{\alpha}\mathbf{p} + i\beta\frac{m_0 c}{\hbar} = \frac{i}{\hbar c}\left(-i\hbar\frac{\partial}{\partial t} + c\left(\boldsymbol{\alpha}\mathbf{p}\right) + \beta m_0 c^2\right),$$

the factorized equation can be written in the following form

$$i\hbar\frac{\partial\Psi}{\partial t} = \left(c\boldsymbol{\alpha}\mathbf{p} + \beta m_0 c^2\right)\Psi. \tag{6.4}$$

The Hamiltonian of the Dirac equation (6.4) is

$$H_D = c\left(\boldsymbol{\alpha}\mathbf{p}\right) + \beta m_0 c^2. \tag{6.5}$$

The Hamiltonian (6.5) includes the four-dimensional matrices, hence the wave function of the equation (6.4) should be the four row column.

It can be easily shown that the Dirac equation (6.4) results in the following continuity equation

$$\frac{\partial \rho}{\partial t} + div\mathbf{j} = 0, \tag{6.6}$$

where

$$\rho = e\Psi^{+}\Psi, \quad \mathbf{j} = ec\Psi^{+}\boldsymbol{\alpha}\Psi. \tag{6.7}$$

The second possible equation, resulted from the factorization (6.1), is

$$i\hbar\frac{\partial \Psi'}{\partial t} = -\left(c\left(\boldsymbol{\alpha}\mathbf{p}\right) + \beta m_0 c^2\right)\Psi'. \tag{6.8}$$

It is seen that this equation differs from the equation (6.4) only in the sign of the time derivative. As far as the charge density is the time component of the current density four-vector then we get the following equation for the current density four-vector of the particle obeying the equation (6.8)

$$\rho = -e\Psi'^{+}\Psi', \quad \mathbf{j} = ec\Psi'^{+}\boldsymbol{\alpha}\Psi'. \tag{6.9}$$

By comparing the equations (6.7) and (6.9) we can see that the equations (6.4) and (6.8) correspond to the oppositely charged particles.

Let us write the four-component wave function Ψ of the equation (6.4) in the form of the bispinor wave function

$$\Psi = \begin{pmatrix} \varphi \\ \chi \end{pmatrix}, \tag{6.10}$$

where $\varphi = \begin{pmatrix} \varphi_1 \\ \varphi_2 \end{pmatrix}$ and $\chi = \begin{pmatrix} \chi_1 \\ \chi_2 \end{pmatrix}$ are the three-dimensional spinors. By applying the transformation (6.10) to the equation (6.3) we get for the spinors φ and χ the following coupled set of equations

$$i\hbar\frac{\partial \varphi}{\partial t} = c\boldsymbol{\sigma}\mathbf{p}\chi + m_0c^2\varphi, \tag{6.11a}$$

$$i\hbar\frac{\partial \chi}{\partial t} = c\boldsymbol{\sigma}\mathbf{p}\varphi - m_0c^2\chi. \tag{6.11b}$$

By applying the Hermitian conjugation to the equation (6.11a) we get

$$i\hbar\frac{\partial \varphi^{+}}{\partial t} = c\mathbf{p}\chi^{+}\boldsymbol{\sigma} - m_0c^2\varphi^{+}.$$

It is seen that the spinor χ obeys the equation which is quite similar to the equation for the Hermitian conjugate spinor φ. Thus it should be anticipated that the transformation properties of the spinor χ with respect to the orthogonal transformations will coincide with the transformation properties of the spinor φ^+ and differ from the transformation properties of the spinor φ.

Let us turn again to the properties of the three-dimensional spinors. In the Chapter 4 we have shown that the matrix of the three-dimensional rotations $U = \begin{pmatrix} a & b \\ c & d \end{pmatrix}$ should be the unimodal and unitary matrix. The conditions of the unimodality and unitarity of the transformation matrix leave only the three real independent parameters from the eight possible ones. There are only three rotation angles that are required to specify unambiguously the three-dimensional rotations. But if we need in the invariance with respect to the four-dimensional rotations we should reject one of the two above mentioned conditions which are applied to the transformation matrix. Earlier we have already mentioned that the unimodality condition $ad - bc = 1$ is the common property of any arbitrary rotations. This condition holds for the three-dimensional rotations and for the Lorentz transformation as well. The unitarity condition applied to the transformation matrix U means that the bilinear combination $\psi'^+\psi' = \psi^+U^+U\psi = \psi^+\psi$ is a scalar. However, this bilinear combination is the time component of the current density four-vector. Hence it could not be the scalar in the extended group of four-dimensional rotations. The condition of the unimodality imposes only two constraints on the eight real parameters of the transformation matrix U. The left six parameters assign the six rotation angles in the four-dimensional reference frame. It is seen from the equations (4.11)–(4.14) that the transformation properties of the spinors and Hermitian conjugate spinors are not identical. It is this property that provides the linear independency of the spinors of the bispinor wave function (6.10).

6.2 Covariant form of the Dirac equation

Before turn to the study of the transformation properties of the Dirac equation (6.4) it is convenient to transform it to the symmetric form. Indeed the time and space derivatives in the equation (6.4) are not symmetric, the space derivative is multiplied by the spin operator and the time derivative is multiplied by constant. If we multiply the both sides of the equation (6.4) by the matrix β then the equation becomes symmetric, because the term, defining the rest energy of electron, takes the pure scalar form. However, it is more convenient to introduce the following matrices

$$\boldsymbol{\gamma} = -i\beta\boldsymbol{\alpha} = i\begin{pmatrix} 0 & -\boldsymbol{\sigma} \\ \boldsymbol{\sigma} & 0 \end{pmatrix}, \quad \gamma_4 = \beta. \tag{6.12}$$

It is seen from the equations (6.12) that the matrices γ_μ ($\mu = 1, 2, 3, 4$) are the self-conjugated matrices and they obey the following commutation relations

$$\gamma_\mu\gamma_\nu + \gamma_\nu\gamma_\mu = 2\delta_{\mu\nu}. \tag{6.13}$$

By multiplying both sides of the equation (6.4) by the factor $-i\beta/c$, we get

$$(\gamma_\mu p_\mu - im_0c)\,\Psi = 0. \tag{6.14}$$

As in Chapter 5 we have used here the following notations for the four-vector of coordinate $x_\mu = (\mathbf{r}, ict)$ and the four-momentum operator

$$p_\mu = -i\hbar\frac{\partial}{\partial x_\mu}.$$

The equation for the Dirac adjoint wave function

$$\bar{\Psi} = \Psi^+\gamma_4, \tag{6.15}$$

is

$$\bar{\Psi}\,(\gamma_\mu p_\mu + im_0c) = 0. \tag{6.16}$$

It is seen from the definition of Dirac adjoint function (6.15) that $\bar{\Psi}$ is a row vector, but not a column as Ψ. Hence the Dirac adjoint wave function $\bar{\Psi}$ is multiplied by the spin matrices γ_μ from the left, as a result the differential operators, acting on the Dirac adjoint function, are on the right side of it.

There is additional convenience to introduce the matrices γ_μ and the Dirac adjoint function $\bar{\Psi}$ because the continuity equation takes the clear relativistic invariant form

$$\frac{\partial j_\mu}{\partial x_\mu} = 0,$$

where the current density four-vector j_μ is

$$j_\mu = (\mathbf{j}, ic\rho) = \left(ec\Psi^+\boldsymbol{\alpha}\Psi, ice\Psi^+\Psi\right) = iec\bar{\Psi}\gamma_\mu\Psi. \tag{6.17}$$

It should be noted that the equations (6.12) give the standard representation of the matrices γ_μ. In principle, the matrices γ_μ are the arbitrary four by four matrices. Indeed with the help of the unitary transformation of the wave function, $\Psi_U = U\Psi$, the equation (6.12) can be transformed to the following form

$$\left(\gamma_\mu^{(U)} p_\mu - im_0c\right)\Psi_U = 0,$$

where the commutation relations for matrices $\gamma_\mu^{(U)} = U\gamma_\mu U^{-1}$ coincide with that given by equations (6.13)

$$\gamma_\mu^{(U)}\gamma_\nu^{(U)} + \gamma_\nu^{(U)}\gamma_\mu^{(U)} = U\left(\gamma_\mu\gamma_\nu + \gamma_\nu\gamma_\mu\right)U^{-1} = 2\delta_{\mu\nu}.$$

To obtain the equation for the particle interacting with the electromagnetic field we can use the standard replacement of the four-momentum p_μ by the generalized four-momentum

$$p_\mu \to -i\hbar\nabla_\mu - \frac{e}{c}A_\mu, \tag{6.18}$$

where $A_\mu = (\mathbf{A}, i\varphi)$ is the four vector of field, the spatial component of which is the vector potential $\mathbf{A}(\mathbf{r}, t)$ and the time component is the scalar potential $\varphi(\mathbf{r}, t)$ of the electromagnetic field.

6.3 Symmetry properties of the Dirac equation with respect to the orthogonal transformations

In the relativistic case, the orthogonal transformations are the transformations of the coordinate four-vector $x_\mu = (\mathbf{r}, ict)$:

$$x'_\mu = a_{\mu\nu}x_\nu + a_\mu, \tag{6.19}$$

where the transformation matrix $a_{\mu\nu}$ obeys the condition

$$a_{\mu\nu}a_{\mu\lambda} = \delta_{\nu\lambda}. \tag{6.20}$$

The transformations (6.19), (6.20) remain invariable the spacetime interval $\Delta x_\mu = x_{2\mu} - x_{1\mu}$, since

$$\Delta x'_\mu \Delta x'_\mu = a_{\mu\nu}a_{\mu\lambda}\Delta x_\nu \Delta x_\lambda = \Delta x_\nu \Delta x_\nu.$$

The transformations (6.19), (6.20) include the discrete (space inversion and time-reversal) and continuous (spacetime translations, three- and four-dimensional rotations) transformations.

Notice, that the components of the matrix γ_μ are the numbers, which remain invariable under the coordinate transformations (6.19), (6.20), and the four-momentum operator is transformed in the following way

$$p'_\mu = \frac{\partial x_\nu}{\partial x'_\mu}p_\nu = a_{\mu\nu}p_\nu.$$

It is seen that the free particle Hamiltonian (6.5) is invariant with respect to the infinitesimally small spacetime translation, $a_\mu = \delta x_\mu$.

Similar to the non-relativistic case the generator of this transformation is the four-momentum operator $p_\mu = -i\hbar(\partial/\partial x_\mu)$. Thus, we can assume now that a_μ in the equation (6.19) is equal to zero and consider further the transformations due to the matrix $a_{\mu\nu}$ only.

If we apply the transformations (6.19), (6.20) to the equation (6.14) it becomes

$$\left(\gamma_\mu p'_\mu - im_0c\right)\Psi'\left(x'\right) = 0. \tag{6.21}$$

The difference between the wave function in the transformed reference frame, $\Psi'(x')$, and the wave function in the initial reference frame, $\Psi(x)$, is due to both the transformation of its arguments and the transformation associated with the column vector manner of the wave function.

As we have mentioned in the previous chapters the equation is symmetric with respect to the transformations (6.19), (6.20) if there is such a matrix, S,

$$\Psi'\left(x'\right) = S\Psi\left(x\right), \tag{6.22}$$

that transforms the equation (6.21) to the initial unprimed form given by (6.14).

By applying the transformation (6.22) to the equation (6.21), we get

$$\left(\gamma_\mu a_{\mu\nu} p_\nu - im_0c\right) S\Psi\left(x\right) = 0.$$

Multiplying the last equation by the matrix S^{-1} from the left, we finally get

$$\left(S^{-1}\gamma_\mu S a_{\mu\nu} p_\nu - im_0c\right)\Psi = 0. \tag{6.23}$$

It should be noted the the matrix S is applied to the components of the bespinor wave function, while the matrix $a_{\mu\nu}$ is applied to the coordinate indexes, therefore these two matrices are commuting ones.

Thus we can see that the equations (6.23) and (6.14) coincide if the following condition holds

$$S^{-1}\gamma_\mu S a_{\mu\nu} = \gamma_\nu. \tag{6.24}$$

The obtained equation yields the relationships between the matrices γ_ν and $\gamma'_\mu = S^{-1}\gamma_\mu S$, hence, we can calculate the explicit form of the matrix S. If we shall use the orthogonality condition (6.20) then the set of equations (6.24) can be rewritten in the form

$$S^{-1}\gamma_\mu S = a_{\mu\nu}\gamma_\nu. \tag{6.25}$$

When we deal with the continuous transformations it is convenient to start with the infinitesimally small orthogonal transformation

$$a_{\mu\nu} = \delta_{\mu\nu} + \varepsilon_{\mu\nu}, \tag{6.26}$$

where $\varepsilon_{\mu\nu}$ is the infinitesimally small tensor of the second rank. Similar to the non-relativistic case, the tensor $\varepsilon_{\mu\nu}$ should be an antisymmetric tensor to satisfy the orthogonality condition (6.20) applied to the matrices $a_{\mu\nu}$. Indeed by substituting the equation (6.26) into the equation (6.20) we get

$$(\delta_{\mu\nu} + \varepsilon_{\mu\nu})(\delta_{\mu\lambda} + \varepsilon_{\mu\lambda}) = \delta_{\nu\lambda} + (\varepsilon_{\lambda\nu} + \varepsilon_{\nu\lambda}) + \ldots = \delta_{\nu\lambda}.$$

Hence, $\varepsilon_{\lambda\nu} = -\varepsilon_{\nu\lambda}$.

At the infinitesimally small transformation (6.26), the matrix S differs from the identity matrix by a small component proportional to the tensor $\varepsilon_{\mu\nu}$

$$S_{\alpha\beta} = \delta_{\alpha\beta} + \frac{1}{2} C^{\mu\nu}_{\alpha\beta} \varepsilon_{\mu\nu}. \tag{6.27}$$

If the transformation (6.27) is applied to the equation (6.25), we get the following equation for the generator of transformation $C^{\mu\nu}_{\alpha\beta}$:

$$\frac{1}{2}\left(\gamma_\mu C^{\alpha\beta} - C^{\alpha\beta}\gamma_\mu\right)\varepsilon_{\alpha\beta} = \varepsilon_{\mu\beta}\gamma_\beta$$

or

$$\left[\frac{1}{2}\left(\gamma_\mu C^{\alpha\beta} - C^{\alpha\beta}\gamma_\mu\right) - \delta_{\mu\alpha}\gamma_\beta\right]\varepsilon_{\alpha\beta} = 0.$$

The solution of the last equation is

$$C^{\alpha\beta} = \frac{1}{2}\gamma_\alpha\gamma_\beta. \tag{6.28}$$

Thus, for the infinitesimally small continuous transformations, the matrix S is defined by

$$S = I + \frac{1}{4}\varepsilon_{\mu\nu}\gamma_\mu\gamma_\nu. \tag{6.29}$$

6.3.1 Three-dimensional rotations

The matrix of reference frame rotation by the angle θ around the z axis is

$$a^{(z)}_{\mu\nu} = \begin{pmatrix} \cos\theta & \sin\theta & 0 & 0 \\ -\sin\theta & \cos\theta & 0 & 0 \\ 0 & 0 & 1 & 0 \\ 0 & 0 & 0 & 1 \end{pmatrix}. \tag{6.30}$$

Hence, the matrix of the infinitesimally small rotation by the angle $\delta\theta$ around the z axis is

$$\varepsilon_{\mu\nu} = \begin{pmatrix} 0 & \delta\theta & 0 & 0 \\ -\delta\theta & 0 & 0 & 0 \\ 0 & 0 & 0 & 0 \\ 0 & 0 & 0 & 0 \end{pmatrix}.$$

By substituting the last equation into the equation (6.29), we get

$$S_z(\delta\theta) = I + \frac{1}{4}\delta\theta\left(\gamma_1\gamma_2 - \gamma_2\gamma_1\right) = I + \frac{i}{2}\delta\theta\,\Sigma_3.$$

As we have mentioned above, the generator of the transformation of the infinitesimally small rotations is the intrinsic angular momentum operator. Therefore, the matrix Σ_3, having the form

$$\Sigma_3 = \begin{pmatrix} \sigma_3 & 0 \\ 0 & \sigma_3 \end{pmatrix}, \tag{6.31}$$

relates with the spin projection operator. By rotating the reference frame around other spatial axes we can easily get the general equation

$$S_R(\delta\boldsymbol{\theta}) = I + \frac{i}{2}\delta\boldsymbol{\theta}\,\boldsymbol{\Sigma}, \tag{6.32}$$

where the matrix $\boldsymbol{\Sigma}$ is

$$\boldsymbol{\Sigma} = \begin{pmatrix} \boldsymbol{\sigma} & 0 \\ 0 & \boldsymbol{\sigma} \end{pmatrix}. \tag{6.33}$$

To generalize the equation (6.32) for the case of any finite rotation angle we need in the operator of powers of $(\mathbf{n}\boldsymbol{\Sigma})$, where $\mathbf{n} = (n_1, n_2, n_3)$ is the arbitrary unit three-dimensional vector. The square of the operator $(\mathbf{n}\boldsymbol{\Sigma})$ is

$$(\mathbf{n}\boldsymbol{\Sigma})^2 = \begin{pmatrix} (\mathbf{n}\boldsymbol{\sigma})^2 & 0 \\ 0 & (\mathbf{n}\boldsymbol{\sigma})^2 \end{pmatrix} = \sum_{i,j} n_i n_j \begin{pmatrix} \sigma_i\sigma_j & 0 \\ 0 & \sigma_i\sigma_j \end{pmatrix} \tag{6.34}$$

By taking into account the commutation relations for the Pauli matrices, $\sigma_i\sigma_j + \sigma_j\sigma_i = 2\delta_{ij}$, we get

$$(\mathbf{n}\boldsymbol{\Sigma})^2 = I. \tag{6.35}$$

Thus, the all even powers of the operator $(\mathbf{n}\boldsymbol{\Sigma})$ are the identity operator, $(\mathbf{n}\boldsymbol{\Sigma})^{2n} = I$, and the all odd powers are $(\mathbf{n}\boldsymbol{\Sigma})^{2n+1} = (\mathbf{n}\boldsymbol{\Sigma})$. Finally, for the matrix of rotation by the finite angle θ around the axis of $\mathbf{n}$ we get

$$S_R = \exp\left(\frac{i}{2}\boldsymbol{\theta}\boldsymbol{\Sigma}\right) = \cos\frac{\theta}{2} + i\,(\mathbf{n}\boldsymbol{\Sigma})\sin\frac{\theta}{2}. \tag{6.36}$$

It follows from the equation (6.36) that under the rotation by the angle 2π the transformed wave function does not coincide with the initial wave function, it takes the opposite sign

$$\Psi'(x') = S_R(2\pi)\,\Psi(x) = -\Psi(x).$$

Thus, the transformation of the three-dimensional rotations, for the bispinor wave function of the Dirac equation, is realized by the matrix $\mathbf{\Sigma}$. Therefore in the Dirac theory the spin operator is defined by

$$\mathrm{s} = \frac{\hbar}{2}\mathbf{\Sigma} = \frac{\hbar}{2}\begin{pmatrix} \boldsymbol{\sigma} & 0 \\ 0 & \boldsymbol{\sigma} \end{pmatrix}. \tag{6.37}$$

6.3.2 Lorentz transformation

The matrix of the Lorentz transformation is

$$a_{\mu\nu}^{(L)} = \begin{pmatrix} \cos\varphi & 0 & 0 & \sin\varphi \\ 0 & 1 & 0 & 0 \\ 0 & 0 & 1 & 0 \\ -\sin\varphi & 0 & 0 & \cos\varphi \end{pmatrix}, \tag{6.38}$$

where $\tan\varphi = iv/c$. The matrix (6.38) describes the transformation to the reference frame moving along the x axis with the velocity v with respect to the initial reference frame.

The matrix of the infinitesimally small Lorentz transformation is

$$\varepsilon_{\mu\nu} = \begin{pmatrix} 0 & 0 & 0 & i\delta v/c \\ 0 & 0 & 0 & 0 \\ 0 & 0 & 0 & 0 \\ -i\delta v/c & 0 & 0 & 0 \end{pmatrix}$$

By substituting the last equation into the equation (6.29) and applying the equalities

$$\gamma_1\gamma_4 = -\gamma_4\gamma_1 = i\alpha_1,$$

we get

$$S_x(\delta v) = I - \frac{1}{2}\frac{\delta v}{c}\alpha_1. \tag{6.39}$$

By making the similar transformations with the remaining spatial axes we can easily get for the arbitrary vector $\delta\mathbf{v} = \mathbf{n}\delta v$ the following matrix

$$S_L(\delta\mathbf{v}) = I - \frac{1}{2}\frac{\delta\mathbf{v}}{c}\boldsymbol{\alpha}. \tag{6.40}$$

Notice that

$$(\mathbf{n}\boldsymbol{\alpha})^2 = \begin{pmatrix} (\mathbf{n}\boldsymbol{\sigma})^2 & 0 \\ 0 & (\mathbf{n}\boldsymbol{\sigma})^2 \end{pmatrix}.$$

Hence, similar to (6.34), we get $(\mathbf{n}\boldsymbol{\alpha})^2 = I$. Thus, the Lorentz transformation, at the arbitrary finite velocity $\mathbf{v} = \mathbf{n}v$, is realized by the matrix

$$S_L(\mathbf{v}) = \cosh\left(\frac{1}{2}\tanh^{-1}\frac{v}{c}\right) - (\mathbf{n}\boldsymbol{\alpha})\sinh\left(\frac{1}{2}\tanh^{-1}\frac{v}{c}\right). \tag{6.41}$$

6.3.3 Space inversion

The matrix of the space inversion transformation is

$$a_{\mu\nu}^{(P)} = \begin{pmatrix} -1 & 0 & 0 & 0 \\ 0 & -1 & 0 & 0 \\ 0 & 0 & -1 & 0 \\ 0 & 0 & 0 & 1 \end{pmatrix}, \tag{6.42}$$

The space inversion is the discrete transformation, therefore to determine the explicit form of the transformation matrix S we should directly solve the equations (6.24) or the equivalent equations (6.25). By substituting the matrix $a_{\mu\nu}$ given by (6.42) into the equations (6.25) we get

$$\gamma_i S = -S\gamma_i, \quad \gamma_4 S = S\gamma_4, \tag{6.43}$$

where $i = 1, 2, 3$. It can be easily seen that the solution of the equations (6.43) is

$$S_P = \lambda\gamma_4, \tag{6.44}$$

where λ is the arbitrary constant. The double space inversion transformation can be considered as an identical transformation. In this case, we get the following equation for the constant λ

$$\lambda^2 = 1$$

the solutions of which are

$$\lambda_1 = 1, \quad \lambda_2 = -1. \tag{6.45}$$

However, we have seen above that, for the bispinor wave function, the rotation by the angle 2π is not the identical transformation. Hence if we assume that the double space inversion is equivalent to the rotation by the angle 2π we get the following equation for the constant λ

$$\lambda^2 = -1.$$

Hence

$$\lambda_3 = i, \quad \lambda_4 = -i. \tag{6.46}$$

The choice of the value of constant λ, among its four possible values, depends on the internal parity of a particle.

6.3.4 Time reversal

When we discussed the symmetry properties of the Schrödinger equation we have pointed out that, as the Schrödinger equation is the first order differential equation with respect to time derivative, so the time reversal transformation should inevitable include the complex conjugated wave function. The Dirac equation is also the differential equation of the first order with respect to time derivative. Hence, the time reversal transformation will also include the Hermitian or Dirac adjoint.

Let us write the equation (6.14) in the form

$$\left[\gamma_4\left(-\frac{\hbar}{c}\frac{\partial}{\partial t}\right)+\boldsymbol{\gamma}\mathbf{p}-im_0c\right]\Psi(t)=0.$$

By applying the time reversal transformation and complex conjugation to this equation, we get

$$\left[\gamma_4^*\left(\frac{\hbar}{c}\frac{\partial}{\partial t}\right)-\boldsymbol{\gamma}^*\mathbf{p}+im_0c\right]\Psi^*(-t)=0. \tag{6.47}$$

We look for the transformation

$$\Psi^*(-t)=S_T\Psi(t),$$

which converts the equation (6.47) to (6.14). Multiplying the equation (6.47) by matrix $-S_T^{-1}$ from the left we get

$$\left[S_T^{-1}\gamma_4^*S_T\left(-\frac{\hbar}{c}\frac{\partial}{\partial t}\right)+S_T^{-1}\boldsymbol{\gamma}^*S_T\mathbf{p}-im_0c\right]\Psi(t)=0.$$

Thus the transformation matrix S_T should satisfy the equations

$$S_T\gamma_4=\gamma_4^*S_T,\quad S_T\boldsymbol{\gamma}=\boldsymbol{\gamma}^*S_T. \tag{6.48}$$

In the standard representation of the matrices γ_μ (see eq. (6.12)) the solution of the equations (6.48) is

$$S_T=\lambda_T\gamma_3\gamma_1, \tag{6.49}$$

where λ_T is the constant of the unit modulus, $|\lambda_T|=1$.

6.3.5 Charge conjugation

As we have discussed above, the standard replacement (6.18) transforms the equation for the free particle into the equation for the particle interacting with the electromagnetic field

$$\left[\gamma_4\left(-\frac{\hbar}{c}\frac{\partial}{\partial t}-iq\varphi\right)+\boldsymbol{\gamma}\left(\mathbf{p}-\frac{q}{c}\mathbf{A}\right)-im_0c\right]\Psi=0. \tag{6.50}$$

The charge conjugation transformation defines the symmetry properties of equation with respect to the replacement $q\to -q$. Making this replacement in the equation (6.50), we get

$$\left[\gamma_4\left(-\frac{\hbar}{c}\frac{\partial}{\partial t}+iq\varphi\right)+\boldsymbol{\gamma}\left(\mathbf{p}+\frac{q}{c}\mathbf{A}\right)-im_0c\right]\Psi_C=0. \tag{6.51}$$

The complex conjugation of the equation (6.50) yields the following equation

$$\left[\gamma_4^*\left(-\frac{\hbar}{c}\frac{\partial}{\partial t}+iq\varphi\right)-\boldsymbol{\gamma}^*\left(\mathbf{p}+\frac{q}{c}\mathbf{A}\right)+im_0c\right]\Psi^*=0. \tag{6.52}$$

We should find such a matrix, S_C,

$$\Psi^* = S_C\Psi_C,$$

that transforms the equation (6.52) to the equation (6.51). Multiplying the equation (6.52) by the matrix $-S_C^{-1}$ from the left we get the following equations for the matrix S_C

$$S_C\gamma_4 = -\gamma_4^* S_C, \quad S_C\boldsymbol{\gamma} = \boldsymbol{\gamma}^* S_C. \tag{6.53}$$

In the standard representation of the matrices γ_μ the solution of the equations (6.53) is

$$S_C = \lambda_C\gamma_2. \tag{6.54}$$

where λ_C is the constant of the unit modulus, $|\lambda_C| = 1$.

6.3.6 *CPT* invariance

By summarizing the results of the last three subsections we write the space inversion, time reversal, and charge conjugation transformations all together:

a) space inversion

$$\Psi(-\mathbf{r},t) = \lambda_P\gamma_4\Psi(\mathbf{r},t) \quad \text{or} \quad \Psi(\mathbf{r},t) = \lambda_P^*\gamma_4\Psi(-\mathbf{r},t)\,; \tag{6.55}$$

b) time reversal

$$\Psi^*(\mathbf{r},-t) = \lambda_T\gamma_3\gamma_1\Psi(\mathbf{r},t) \quad \text{or} \quad \Psi(\mathbf{r},t) = \lambda_T^*\gamma_1\gamma_3\Psi^*(\mathbf{r},-t)\,; \tag{6.56}$$

c) charge conjugation

$$\Psi^*(\mathbf{r},t) = \lambda_C\gamma_2\Psi_C(\mathbf{r},t) \quad \text{or} \quad \Psi_C(\mathbf{r},t) = \lambda_C^*\gamma_2\Psi^*(\mathbf{r},t)\,. \tag{6.57}$$

The combined transformation can be written as follows

$$\begin{aligned} T\,\Psi(\mathbf{r},t) &= \lambda_T^*\gamma_1\gamma_3\Psi^*(\mathbf{r},-t)\,, \\ PT\,\Psi(\mathbf{r},t) &= \lambda_P^*\lambda_T^*\gamma_4\gamma_1\gamma_3\Psi^*(-\mathbf{r},-t)\,, \\ CPT\,\Psi(\mathbf{r},t) &= \lambda_C^*\lambda_P^*\lambda_T^*\gamma_2\gamma_4\gamma_1\gamma_3\Psi(-\mathbf{r},-t) = -\lambda_C^*\lambda_P^*\lambda_T^*\gamma_5\Psi(-\mathbf{r},-t) \end{aligned} \tag{6.58}$$

where

$$\gamma_5 = \gamma_1\gamma_2\gamma_3\gamma_4 = -\begin{pmatrix} 0 & I \\ I & 0 \end{pmatrix}. \tag{6.59}$$

In the frames of the Dirac theory there is some freedom in the choice of the coefficient of the combined CPT transformation, $\lambda_{CPT} = -\lambda_C^* \lambda_P^* \lambda_T^*$. Its value is determined by the internal symmetry of the particle. It is usually assumed that $\lambda_{CPT} = i$.

Thus, the CPT theorem can be formulated in the following way: any solution describing the particle motion in the external electromagnetic field has the counterpartner solution describing the space-inverted and time-reversed motion of the antiparticle.

6.4 Free particle

So we have seen in section 6.1 that the second order in space and time differential operator of the Klein–Gordon–Fock equation for the free particle is factorized into the product of the two operators, which are the first order in space and time differential operators. It is evident that any solution of the first order differential equation is at the same time the solution of the second order differential equation. The two first order differential equations are

$$i\hbar\frac{\partial \Psi_1}{\partial t} = \left(c\left(\boldsymbol{\alpha}\mathbf{p}\right) + \beta m_0 c^2\right)\Psi_1, \tag{6.60a}$$

$$i\hbar\frac{\partial \Psi_2}{\partial t} = -\left(c\left(\boldsymbol{\alpha}\mathbf{p}\right) + \beta m_0 c^2\right)\Psi_2. \tag{6.60b}$$

We have seen that the equation for the wave function Ψ_1 results in the following equation for the current density four-vector

$$j_\mu^{(1)} = (\mathbf{j}, ic\rho) = \left(ec\Psi_1^+\boldsymbol{\alpha}\Psi_1, ice\Psi_1^+\Psi_1\right), \tag{6.61}$$

while the equation for the wave function Ψ_2 yields the current density four-vector for the particle of the opposite charge

$$j_\mu^{(2)} = \left(ec\Psi_2^+\boldsymbol{\alpha}\Psi_2, -ice\Psi_2^+\Psi_2\right). \tag{6.62}$$

In the steady-state case, the wave functions are $\Psi_{1,2}(\mathbf{r}, t) = \Psi_{1,2}(\mathbf{r}) \times \times \exp(-iE_{1,2}t/\hbar)$. It is seen that the steady-state equations (6.60a) and (6.60b) coincide when $E_2 = -E_1$. Taking into account the equations (6.61) and (6.62) we can assume that the equation (6.60a) describes the particle and the equation (6.60b) describes the antiparticle. Hence, it is seen that the solutions corresponding to the particle and antiparticle have the opposite sign of energy. This will help us in systematization of the solutions of the equations (6.60a), (6.60b).

6.4.1 Plane waves

Let us consider the equation for particle (6.60a). The equations for the spinors φ and χ of the bispinor wave function Ψ (see (6.10)) are

$$i\hbar\frac{\partial\varphi}{\partial t} = c\boldsymbol{\sigma}\mathbf{p}\chi + m_0c^2\varphi, \quad i\hbar\frac{\partial\chi}{\partial t} = c\boldsymbol{\sigma}\mathbf{p}\varphi - m_0c^2\chi. \tag{6.63}$$

We can exclude one of the spinors from the coupled set of equations (6.63). For example, by excluding spinor χ, for the spinor φ we get

$$\left(\Delta - \frac{1}{c^2}\frac{\partial^2}{\partial t^2} - \frac{m_0^2c^2}{\hbar^2}\right)\varphi = 0. \tag{6.64}$$

The general solutions of the equation (6.64) for spinor φ, and similar equation for spinor χ, are

$$\begin{aligned}
\varphi(\mathbf{r},t) = &\left[A_{\mathbf{p}}\exp\left(i\frac{\mathbf{pr}}{\hbar}\right) + A_{-\mathbf{p}}\exp\left(-i\frac{\mathbf{pr}}{\hbar}\right)\right]\exp\left(-i\frac{E_pt}{\hbar}\right) + \\
&+\left[B_{\mathbf{p}}\exp\left(-i\frac{\mathbf{pr}}{\hbar}\right) + B_{-\mathbf{p}}\exp\left(i\frac{\mathbf{pr}}{\hbar}\right)\right]\exp\left(i\frac{E_pt}{\hbar}\right), \\
\chi(\mathbf{r},t) = &\left[C_{\mathbf{p}}\exp\left(i\frac{\mathbf{pr}}{\hbar}\right) + C_{-\mathbf{p}}\exp\left(-i\frac{\mathbf{pr}}{\hbar}\right)\right]\exp\left(-i\frac{E_pt}{\hbar}\right) + \\
&+\left[D_{\mathbf{p}}\exp\left(-i\frac{\mathbf{pr}}{\hbar}\right) + D_{-\mathbf{p}}\exp\left(i\frac{\mathbf{pr}}{\hbar}\right)\right]\exp\left(i\frac{E_pt}{\hbar}\right),
\end{aligned} \tag{6.65}$$

where

$$E_p = \sqrt{m_0^2c^4 + p^2c^2}. \tag{6.66}$$

and $A_{\pm\mathbf{p}}, B_{\pm\mathbf{p}}, C_{\pm\mathbf{p}}, D_{\pm\mathbf{p}}$ are the constants. These constants are really coupled, because not all of the solutions of the equation (6.64) are the solutions of the coupled equations (6.63). By substituting the solutions (6.65) into the equations (6.63) we get

$$C_{\pm\mathbf{p}} = \pm\frac{c(\boldsymbol{\sigma}\mathbf{p})}{E_p + m_0c^2}A_{\pm\mathbf{p}}, \quad B_{\pm\mathbf{p}} = \mp\frac{c(\boldsymbol{\sigma}\mathbf{p})}{E_p + m_0c^2}D_{\pm\mathbf{p}}.$$

The solutions (6.65) correspond to the free particle of a given energy E_p. We assume that particle makes a one-dimensional motion in finite volume V with the ideal boundaries. To account for the three-dimensional motion we shall further make a summation over the all directions $\mathbf{p}$.

It is evident from the equation (6.65) that both equations (6.60a) and (6.60b) have the solutions of positive and negative energy. Thus in the case of free particle there is no necessity to solve the equation for antiparticle (6.60b), because the general solution of the equation (6.60a) includes the antiparticle solutions as well. The solutions of the equation (6.60a) of the positive and negative energy are respectively

$$\Psi^{(+)}(\mathbf{r},t) = \left[A_{\mathbf{p}} u_{\mathbf{p}}^{(+)} \exp\left(i\frac{\mathbf{pr}}{\hbar}\right) + A_{-\mathbf{p}} u_{-\mathbf{p}}^{(+)} \exp\left(-i\frac{\mathbf{pr}}{\hbar}\right)\right] \exp\left(-i\frac{E_p t}{\hbar}\right),$$

$$\Psi^{(-)}(\mathbf{r},t) = \left[D_{\mathbf{p}} u_{\mathbf{p}}^{(-)} \exp\left(-i\frac{\mathbf{pr}}{\hbar}\right) + D_{-\mathbf{p}} u_{-\mathbf{p}}^{(-)} \exp\left(i\frac{\mathbf{pr}}{\hbar}\right)\right] \exp\left(i\frac{E_p t}{\hbar}\right), \tag{6.67}$$

where

$$u_{\pm\mathbf{p}}^{(+)} = \begin{pmatrix} w_{\pm\mathbf{p}} \\ \pm\dfrac{c(\boldsymbol{\sigma}\mathbf{p})}{E_p + m_0 c^2} w_{\pm\mathbf{p}} \end{pmatrix}, \quad u_{\pm\mathbf{p}}^{(-)} = \begin{pmatrix} \mp\dfrac{c(\boldsymbol{\sigma}\mathbf{p})}{E_p + m_0 c^2} w'_{\pm\mathbf{p}} \\ w'_{\pm\mathbf{p}} \end{pmatrix}, \tag{6.68}$$

here $w_{\pm\mathbf{p}}$ and $w'_{\pm\mathbf{p}}$ are the arbitrary two-dimensional spinors which obey the normalization conditions $w_{\pm\mathbf{p}}^{+} w_{\pm\mathbf{p}} = 1$. It is seen that the spinors, at positive and negative energy solutions, are orthogonal

$$\left(u_{\pm\mathbf{p}}^{(+)}\right)^{+} u_{\pm\mathbf{p}}^{(-)} = \mp w_{\pm\mathbf{p}}^{+} \frac{c(\boldsymbol{\sigma}\mathbf{p})}{E_p + m_0 c^2} w'_{\pm\mathbf{p}} \pm w_{\pm\mathbf{p}}^{+} \frac{c(\boldsymbol{\sigma}\mathbf{p})}{E_p + m_0 c^2} w'_{\pm\mathbf{p}} = 0.$$

In the particle rest frame, $\mathbf{p} = 0$, the bispinors $u^{(\pm)}$ take the form

$$u^{(+)} = \begin{pmatrix} w \\ 0 \end{pmatrix}, \quad u^{(-)} = \begin{pmatrix} 0 \\ w' \end{pmatrix}. \tag{6.69}$$

So, in the particle rest frame one of spinors in the bispinor wave function became zero. The solution $u^{(+)}$, having the non-zero upper spinor and zero lower spinor, corresponds to particle, the solution $u^{(-)}$ corresponds to antiparticle.

6.4.2 Helicity

It is seen from the equations of the previous subsection that the general solution of the positive energy E_p depends on the two spinors $w_{\pm\mathbf{p}}$. It is quite natural, because the energy eigenvalue E_p is degenerated with respect to the two directions of momentum $\pm\mathbf{p}$. Let us discuss how we can choose the explicit form of the spinors $w_{\pm\mathbf{p}}$.

The momentum operator commutes with the Hamiltonian of the equation (6.60), hence the momentum is the conservative value. The orbital momentum operator $\hbar\mathbf{l} = [\mathbf{r}\mathbf{p}]$ and the spin operator $\mathbf{s} = \frac{\hbar}{2}\boldsymbol{\Sigma}$ do not commute with the Hamiltonian H_D of the equation (6.60)

$$[\mathbf{l}, H_D] = ic\,[\boldsymbol{\alpha}\mathbf{p}], \quad [\boldsymbol{\Sigma}, H_D] = -2ic\,[\boldsymbol{\alpha}\mathbf{p}].$$

The commuting operator is the operator of the total angular momentum

$$\hbar\mathbf{j} = \hbar\mathbf{l} + \frac{\hbar}{2}\mathbf{\Sigma}. \tag{6.70}$$

As far as $(\mathbf{jp}) = (\mathbf{\Sigma p})/2$, then the conservation of the total angular momentum results in the conservation of the helicity. The helicity is the projection of the spin on the direction of the momentum $(\mathbf{\Sigma n})$, where $\mathbf{n} = \mathbf{p}/p$. We can check it directly

$$[(\mathbf{\Sigma n}), (\boldsymbol{\alpha}\mathbf{n})] = 2i\alpha_k e_{kij} n_i n_j = 0, \quad [(\mathbf{\Sigma n}), \beta] = 0.$$

So,

$$[(\mathbf{\Sigma p}), H_D] = 0. \tag{6.71}$$

If in the initial state the spin and momentum are parallel each other, then it is convenient as the linear independent spinors $w_{\pm\mathbf{p}}$ to take the eigenfunctions of the equation

$$\sigma_z w^{(\sigma)} = \sigma w^{(\sigma)},$$

where $\mathbf{e}_z = \mathbf{p}/p$. These two spinors are

$$w^{(\sigma=+1)} = \begin{pmatrix} 1 \\ 0 \end{pmatrix}, \quad w^{(\sigma=-1)} = \begin{pmatrix} 0 \\ 1 \end{pmatrix}. \tag{6.72}$$

In general case, when a particle moves in the direction defined by the angles θ and φ in the spin state reference frame, the linear independent spinors are the eigenfunctions of the eigenvalue problem

$$(\boldsymbol{\sigma}\mathbf{n})\, w^{(\sigma)} = \sigma w^{(\sigma)},$$

which are

$$w_{\mathbf{p}}^{(\sigma=+1)} = \begin{pmatrix} \exp\left(-i\frac{\varphi}{2}\right)\cos\frac{\theta}{2} \\ \exp\left(i\frac{\varphi}{2}\right)\sin\frac{\theta}{2} \end{pmatrix}, \quad w_{\mathbf{p}}^{(\sigma=-1)} = \begin{pmatrix} -\exp\left(-i\frac{\varphi}{2}\right)\sin\frac{\theta}{2} \\ \exp\left(i\frac{\varphi}{2}\right)\cos\frac{\theta}{2} \end{pmatrix}. \tag{6.73}$$

There are the following relationships between the spinors $w_{-\mathbf{p}}^{(\sigma)}$ and $w_{\mathbf{p}}^{(\sigma)}$

$$w_{-\mathbf{p}}^{(\sigma=+1)} = i w_{\mathbf{p}}^{(\sigma=-1)}, \quad w_{-\mathbf{p}}^{(\sigma=-1)} = i w_{\mathbf{p}}^{(\sigma=+1)}. \tag{6.74}$$

Hence, the general solution of the equation (6.60a) can be written in the following form

$$\Psi(\mathbf{r}, t) = \\ = \sum_{\mathbf{p}} \sum_{\sigma=\pm 1} \left[A_{\mathbf{p},\sigma} u_{\mathbf{p},\sigma}^{(+)} \exp\left(-i\frac{E_p t - \mathbf{pr}}{\hbar}\right) + B_{\mathbf{p},\sigma} u_{\mathbf{p},\sigma}^{(-)} \exp\left(i\frac{E_p t - \mathbf{pr}}{\hbar}\right)\right], \tag{6.75}$$

$$\bar{\Psi}(\mathbf{r},t) = = \sum_{\mathbf{p}} \sum_{\sigma=\pm 1} \left[A^*_{\mathbf{p},\sigma} \bar{u}^{(+)}_{\mathbf{p},\sigma} \exp\left(i\frac{E_p t - \mathbf{pr}}{\hbar}\right) + B^*_{\mathbf{p},\sigma} \bar{u}^{(-)}_{\mathbf{p},\sigma} \exp\left(-i\frac{E_p t - \mathbf{pr}}{\hbar}\right)\right], \tag{6.76}$$

where

$$u^{(+)}_{\mathbf{p},\sigma} = \sqrt{\frac{E_p + m_0 c^2}{2E_p}} \begin{pmatrix} w^{(\sigma)}_{\mathbf{p}} \\ \frac{c(\boldsymbol{\sigma}\mathbf{p})}{E_p + m_0 c^2} w^{(\sigma)}_{\mathbf{p}} \end{pmatrix}, \quad u^{(-)}_{\mathbf{p},\sigma} = \sqrt{\frac{E_p + m_0 c^2}{2E_p}} \begin{pmatrix} -\frac{c(\boldsymbol{\sigma}\mathbf{p})}{E_p + m_0 c^2} w^{(\sigma)}_{\mathbf{p}} \\ w^{(\sigma)}_{\mathbf{p}} \end{pmatrix}. \tag{6.77}$$

The spinors $u^{(\lambda=\pm 1)}_{\mathbf{p},\sigma}$ are normalized by the following condition

$$u^{(\lambda)+}_{\mathbf{p},\sigma} u^{(\lambda')}_{\mathbf{p},\sigma'} = \delta_{\lambda\lambda'}\delta_{\sigma\sigma'}. \tag{6.78}$$

Thus, at a given magnitude of the energy, we have the three binary quantum numbers to classify the particle states. They are the energy $E = \pm E_p$, momentum $\pm\mathbf{p}$, and helicity $\sigma = \pm 1$. Therefore the linear independent solutions can be chosen in the two equivalent forms: (1) the four combinations of the different energy $E = \pm E_p$ and helicity $\sigma = = \pm 1$ solutions, at a given momentum $\mathbf{p}$; (2) the four combinations of the different energy $E = \pm E_p$ and momentum $\pm\mathbf{p}$ solutions, at a given helicity σ.

6.4.3 Particle and antiparticle

In the frames of the quantum field theory formalism the linear independent solutions can be chosen in the following way: the solution at coefficient $A_{\mathbf{p},\sigma}$ in equation (6.75) is associated with the electron, and the solution at coefficient $B^*_{\mathbf{p},\sigma}$ in equation (6.76) is associated with the positron. However, it is more convenient to choose both electron and positron solutions having the positive energy. In this case, the electron solution is given by the positive energy part of the wave function (6.75). The positron solution is given by the positive energy part of the charge conjugated wave function $\Psi_C = S_C^{-1}\Psi^* = \lambda_C^*\gamma_2\Psi^*$:

$$\Psi_C = i\lambda_C^* \sum_{\mathbf{p}} \sum_{\sigma=\pm 1} \left[A^*_{\mathbf{p},\sigma} \sigma u^{(-)}_{-\mathbf{p},\sigma} \exp\left(i\frac{E_p t - \mathbf{pr}}{\hbar}\right) - B^*_{\mathbf{p},\sigma} \sigma u^{(+)}_{-\mathbf{p},\sigma} \exp\left(-i\frac{E_p t - \mathbf{pr}}{\hbar}\right)\right]. \tag{6.79}$$

The wave function normalization condition follows, as usual, from the continuity equation

$$\frac{\partial \rho}{\partial t} + \operatorname{div} \mathbf{j} = 0,$$

where the current density $\mathbf{j}$ and charge density ρ are defined by the equation (6.7). By integrating the continuity equation over the volume V we get

$$\frac{d}{dt} \int_V \rho(\mathbf{r}, t) dV = - \oint_S \mathbf{j}(\mathbf{r}, t) d\mathbf{S},$$

where S is the boundary surface of the volume V.

If the wave function obeys the boundary condition

$$\Psi(\mathbf{r})|_{r \to \infty} = 0, \tag{6.80}$$

then the continuity equation generates the charge conservation law

$$\int_V \rho(\mathbf{r}, t)\, dV = \text{const},$$

where the charged density is integrated over the infinite volume. The charge conservation law determines the normalization condition for the wave functions of the bound states

$$\int \Psi^+(\mathbf{r})\, \Psi(\mathbf{r})\, dV = 1, \tag{6.81}$$

The wave functions of the continuous spectrum are normalized by the condition (see Chapter 2)

$$\int \Psi_{\mathbf{p}}^+(\mathbf{r}, t)\, \Psi_{\mathbf{p}'}(\mathbf{r}, t)\, dV = (2\pi\hbar)^3\, \delta(\mathbf{p} - \mathbf{p}'). \tag{6.82}$$

By applying the normalization condition to the wave function (6.75) we get

$$\sum_{\mathbf{p}} \sum_{\sigma = \pm 1} \left(|A_{\mathbf{p},\sigma}|^2 + |B_{\mathbf{p},\sigma}|^2 \right) = 1. \tag{6.83}$$

Thus the normalization conditions (6.81) or (6.82) specify both the charge and the integral number of the particles.

6.4.4 Spherical waves

The free-particle Dirac equation (6.60) is the rotationally invariant. Hence, the solutions of this equation may be classified according to their total angular momentum j and parity. For a given j, the wave function may be expanded in the spherical harmonics $Y_{l,m}$ with the orbital angular momentum $l = j + 1/2$ and $l = j - 1/2$, but the definite parity $(-1)^{j \mp 1/2}$. As already shown, the upper and lower spinors of the bispinor wave function (6.10) have the opposite parity, hence, we can only have $l = j \mp 1/2$ in φ and $l = j \pm 1/2$ in χ. The usual rules of the angular-momentum addition then yield the following two linear independent spherical spinors

$$\Omega_{j=l+1/2,l,m} = \begin{pmatrix} \sqrt{\dfrac{j+m}{2j}} Y_{l,m-1/2} \\ \sqrt{\dfrac{j-m}{2j}} Y_{l,m+1/2} \end{pmatrix}, \qquad \Omega_{j=l-1/2,l,m} = \begin{pmatrix} -\sqrt{\dfrac{j-m+1}{2j+2}} Y_{l,m-1/2} \\ \sqrt{\dfrac{j+m+1}{2j+2}} Y_{l,m+1/2} \end{pmatrix}. \tag{6.84}$$

The spherical spinors (6.84) are orthonormalized

$$\int \Omega^{+}_{jlm} \Omega_{j'l'm'} \sin\theta \, d\theta \, d\varphi = \delta_{jj'} \delta_{ll'} \delta_{mm'}. \tag{6.85}$$

The relativistic total angular momentum is

$$\hbar \mathbf{j} = \hbar \mathbf{l} + \frac{\hbar}{2} \mathbf{\Sigma},$$

hence the bispinor wave function of the state with the total angular momentum j and its projection m is

$$\Psi_{jm}(\mathbf{r}) = \begin{pmatrix} f_1(r)\, \Omega_{j,l=j-1/2,m}(\theta,\varphi) \\ g_1(r)\, \Omega_{j,l=j+1/2,m}(\theta,\varphi) \end{pmatrix} + \begin{pmatrix} f_2(r)\, \Omega_{j,l=j+1/2,m}(\theta,\varphi) \\ g_2(r)\, \Omega_{j,l=j-1/2,m}(\theta,\varphi) \end{pmatrix}. \tag{6.86}$$

We have shown above that, in the case of the free particle, the coupled set of equations (6.63) can be transformed to the second order differential equations for spinors φ and χ in separate

$$\left(\Delta - \frac{1}{c^2} \frac{\partial^2}{\partial t^2} - \frac{m_0^2 c^2}{\hbar^2} \right) \varphi = 0, \quad \left(\Delta - \frac{1}{c^2} \frac{\partial^2}{\partial t^2} - \frac{m_0^2 c^2}{\hbar^2} \right) \chi = 0.$$

By substituting here the wave functions in the form

$$\Psi(\mathbf{r},t) = \begin{pmatrix} f_l(r)\, \Omega_{jlm} \\ g_{l+1}(r)\, \Omega_{j,l+1,m} \end{pmatrix} \exp\left(-i \frac{Et}{\hbar}\right),$$

we get for the radial wave functions the following equations

$$\frac{d^2 f_l}{dr^2} + \frac{2}{r}\frac{df_l}{dr} + \left(\frac{E^2 - m_0^2 c^4}{\hbar^2 c^2} - \frac{l(l+1)}{r^2}\right) f_l = 0,$$
$$\frac{d^2 g_{l+1}}{dr^2} + \frac{2}{r}\frac{dg_{l+1}}{dr} + \left(\frac{E^2 - m_0^2 c^4}{\hbar^2 c^2} - \frac{(l+1)(l+2)}{r^2}\right) g_{l+1} = 0. \tag{6.87}$$

The solutions of the equations (6.87), satisfying the boundary conditions at $r \to \infty$, are

$$f_l(r) = A j_l(\kappa r), \quad g_{l+1}(r) = B j_{l+1}(\kappa r), \tag{6.88}$$

where A and B are the constants, $\kappa = \sqrt{E^2 - m_0^2 c^4}/(\hbar c)$, and $j_l(x)$ is the spherical Bessel function. Thus, the radial wave functions are the spherical Bessel functions.

To find the relationships between the coefficients A and B in (6.88), we should substitute the equations (6.88) into the equations (6.63). It is convenient to use the following transformation

$$(\boldsymbol{\sigma}\mathbf{r})(\boldsymbol{\sigma}\mathbf{p}) = \mathbf{r}\mathbf{p} + i\boldsymbol{\sigma}[\mathbf{r}\mathbf{p}] = -i\hbar r\frac{\partial}{\partial r} + i\hbar\left(\mathbf{j}^2 - \mathbf{l}^2 - \frac{3}{4}\right).$$

Hence,

$$(\boldsymbol{\sigma}\mathbf{r})(\boldsymbol{\sigma}\mathbf{p}) j_l(r)\Omega_{jlm} = i\hbar\kappa r j_{l+1}(r)\Omega_{jlm},$$
$$(\boldsymbol{\sigma}\mathbf{r})(\boldsymbol{\sigma}\mathbf{p}) j_{l+1}(r)\Omega_{j,l+1,m} = -i\hbar\kappa r j_l(r)\Omega_{j,l+1,m}.$$

If the last transformations are applied to the equation (6.63), we get

$$\left(E - m_0 c^2\right) A\sigma_r\Omega_{jlm} = -i\hbar\kappa c B\Omega_{j,l+1,m},$$
$$\left(E + m_0 c^2\right) B\Omega_{j,l+1,m} = i\hbar\kappa c A\sigma_r\Omega_{jlm}.$$

Notice, that there is the following relationship between the spinors Ω_{jlm} and $\Omega_{jl'm}$:

$$(\mathbf{n}_r\boldsymbol{\sigma})\Omega_{jlm} = i^{l'-l}\Omega_{jl'm}. \tag{6.89}$$

This equation can be easily derived with the help of the explicit form of the matrix $\sigma_r = (\mathbf{n}_r\boldsymbol{\sigma})$. So, we get finally the following equation for the wave function

$$\Psi_{l=j-1/2}(\mathbf{r}, t) = C\begin{pmatrix} \sqrt{E + m_0 c^2}\, j_l(\kappa r)\Omega_{jlm} \\ -\sqrt{E - m_0 c^2}\, j_{l+1}(\kappa r)\Omega_{j,l+1,m} \end{pmatrix}\exp\left(-i\frac{Et}{\hbar}\right), \tag{6.90}$$

where C is the normalization constant.

The second linear independent solution (see equation (6.86)) can be derived in the similar way

$$\Psi_{l=j+1/2}(\mathbf{r},t) = C\begin{pmatrix} \sqrt{E+m_0c^2}\,j_{l+1}(\kappa r)\,\Omega_{j,l+1,m} \\ -\sqrt{E-m_0c^2}\,j_l(\kappa r)\,\Omega_{jlm} \end{pmatrix}\exp\left(-i\frac{Et}{\hbar}\right). \tag{6.91}$$

According to the definition (6.84) the spinors Ω_{jlm} are transformed under the space inversion in the following way

$$\Omega_{jlm}(-\mathbf{n}_r) = (-1)^l\,\Omega_{jlm}(\mathbf{n}_r).$$

Hence,

$$\Psi_{l=j-1/2}(-\mathbf{r}) = (-1)^l\begin{pmatrix} \sqrt{E+m_0c^2}\,j_l(\kappa r)\,\Omega_{jlm}(\mathbf{n}_r) \\ \sqrt{E-m_0c^2}\,j_{l+1}(\kappa r)\,\Omega_{j,l+1,m}(\mathbf{n}_r) \end{pmatrix} \tag{6.92}$$

and

$$\Psi_{l=j+1/2}(-\mathbf{r}) = (-1)^{l+1}\begin{pmatrix} \sqrt{E+m_0c^2}\,j_{l+1}(\kappa r)\,\Omega_{j,l+1,m}(\mathbf{n}_r) \\ \sqrt{E-m_0c^2}\,j_l(\kappa r)\,\Omega_{jlm}(\mathbf{n}_r) \end{pmatrix}. \tag{6.93}$$

Thus, as we have already mentioned, the the upper and lower spinors of the bispinor wave functions (6.90) and (6.91) have the opposite parity.

6.5 Particle interaction with electromagnetic field

The Dirac equation for a particle interacting with the electromagnetic field is

$$\left(\gamma_\mu\left(p_\mu - \frac{e}{c}A_\mu\right) - im_0c\right)\Psi(\mathbf{r},t) = 0. \tag{6.94}$$

Before start with the analysis of this equation it is helpful to make the following comments. We have already mentioned that there is a close connection between the Dirac equation for the free particle and the Klein–Gordon–Fock equation. Indeed,

$$(\gamma_\mu p_\mu + im_0c)(\gamma_\nu p_\nu - im_0c) = \frac{1}{2}(\gamma_\mu\gamma_\nu p_\mu p_\nu + \gamma_\nu\gamma_\mu p_\nu p_\mu) + m_0^2c^2.$$

As long as the components of the four-momentum operator $p_\mu = -i\hbar\partial/\partial x_\mu$ commute with each other, then, with the help of the commutation relations (6.13), we finally get

$$(\gamma_\mu p_\mu + im_0c)(\gamma_\nu p_\nu - im_0c) = p_\mu p_\mu + m_0^2c^2 = \frac{\hbar^2}{c^2}\frac{\partial^2}{\partial t^2} - \hbar^2\Delta + m_0^2c^2.$$

It is seen that the right-hand-side of the last equation is the operator of the Klein–Gordon–Fock equation for the case of free particle.

However, the components of the generalized four-momentum operator

$$P_\mu = p_\mu - \frac{e}{c} A_\mu$$

do not commute with each other. The commutation rules for them are

$$\left[\left(- i\hbar \frac{\partial}{\partial x_\mu} - \frac{e}{c} A_\mu\right), \left(- i\hbar \frac{\partial}{\partial x_\nu} - \frac{e}{c} A_\nu\right)\right] = \frac{i\hbar e}{c}\left(\frac{\partial A_\nu}{\partial x_\mu} - \frac{\partial A_\mu}{\partial x_\nu}\right) = \frac{ie\hbar}{c} F_{\mu\nu},$$

where $F_{\mu\nu}$ is the electromagnetic field tensor. As a result, the product of the two first order differential operators does not coincide with the operator of the Klein–Gordon–Fock equation for a particle interacting with the electromagnetic field. Instead of that, we have

$$(\gamma_\mu P_\mu + i m_0 c)(\gamma_\nu P_\nu - i m_0 c) =$$
$$= -\frac{1}{c^2}\left(i\hbar \frac{\partial}{\partial t} - e\varphi\right)^2 + \left(\mathbf{p} - \frac{e}{c}\mathbf{A}\right)^2 + m_0^2 c^2 - \frac{e\hbar}{c} \boldsymbol{\Sigma}\mathbf{B} + i\frac{e\hbar}{c} \boldsymbol{\alpha}\mathbf{E}. \quad (6.95)$$

Thus, there are the two additional terms in the right-hand-side of the equation (6.95) with respect to the operator of the Klein–Gordon–Fock equation for particle interacting with the electromagnetic field.

In analysis of the solutions of the Dirac equation for the case of free particle we have already used the following technique. The coupled set of equations for the upper and lower spinors of the bispinor wave function is transformed into the second order differential equation for each of the spinors in separate. Of course, not all of the solutions of these two second order differential equations are the solutions of the Dirac equation, because the upper and lower spinors, of the desired bispinor wave function, are not independent. However, the substitution of the obtained solutions into the Dirac equation enable us to relate the upper and lower spinor and find the solution for the bispinor wave function of the Dirac equation.

In the analysis of the Dirac equation for the particle interacting with the electromagnetic field it is also helpful often to find initially the solutions of the second order differential equation

$$\left(\gamma_\mu \left(p_\mu - \frac{e}{c} A_\mu\right) + i m_0 c\right)\left(\gamma_\nu \left(p_\nu - \frac{e}{c} A_\nu\right) - i m_0 c\right)\Phi = 0. \quad (6.96)$$

To exclude the unnecessary solutions of the second order differential equation (6.96) we can use the following technique. If the function $\Phi(\mathbf{r}, t)$ is a solution of the equation (6.96), then the solution of the Dirac equation (6.94) is defined by

$$\Psi(\mathbf{r}, t) = \left(\gamma_\mu \left(p_\mu - \frac{e}{c} A_\mu\right) + i m_0 c\right)\Phi(\mathbf{r}, t). \quad (6.97)$$

Indeed, it is seen that the function $\Psi(\mathbf{r}, t)$ satisfies the Dirac equation (6.94).

6.5.1 Pauli equation

With the application of the definitions (6.12) the equations (6.94) reads

$$i\hbar\frac{\partial\Psi}{\partial t}=\left[c\boldsymbol{\alpha}\left(\mathbf{p}-\frac{e}{c}\mathbf{A}\right)+\beta m_0c^2+U\right]\Psi, \tag{6.98}$$

where we have introduced the potential energy $U(\mathbf{r},t)=e\varphi(\mathbf{r},t)$ to avoid the confusion of the electromagnetic field scalar potential $\varphi(\mathbf{r},t)$ and the upper spinor of the bispinor wave function (6.10).

Let us start with the case when the particle kinetic energy is much smaller than its rest energy. In this case it is convenient to use the following transformation of the wave function

$$\Psi(\mathbf{r},t)=\Psi'(\mathbf{r},t)\exp\left(-i\frac{m_0c^2t}{\hbar}\right). \tag{6.99}$$

The substitution of the equation (6.99) into the equation (6.98) results in the following coupled set of equations for the spinors φ' and χ'

$$\left(i\hbar\frac{\partial}{\partial t}-U\right)\varphi'=c\boldsymbol{\sigma}\mathbf{P}\chi', \tag{6.100}$$

$$\left(i\hbar\frac{\partial}{\partial t}+2m_0c^2-U\right)\chi'=c\boldsymbol{\sigma}\mathbf{P}\varphi', \tag{6.101}$$

where $\mathbf{P}=\mathbf{p}-\frac{e}{c}\mathbf{A}=-i\hbar\nabla-\frac{e}{c}\mathbf{A}$. By neglecting the time derivative in the equation (6.101), we get

$$\chi'=\frac{c(\boldsymbol{\sigma}\mathbf{P})}{2m_0c^2-U}\varphi'.$$

The substitution of the last equation into the equation (6.100) results in the closed equation for the spinor φ'

$$i\hbar\frac{\partial\varphi'}{\partial t}=\left[\frac{c^2(\boldsymbol{\sigma}\mathbf{P})^2}{2m_0c^2-U}+U\right]\varphi'. \tag{6.102}$$

With the help of the identity (4.10) we get

$$(\boldsymbol{\sigma}\mathbf{P})(\boldsymbol{\sigma}\mathbf{P})=\mathbf{P}^2+i\boldsymbol{\sigma}\left[\left(\mathbf{p}-\frac{e}{c}\mathbf{A}\right)\left(\mathbf{p}-\frac{e}{c}\mathbf{A}\right)\right]=\left(\mathbf{p}-\frac{e}{c}\mathbf{A}\right)^2-\frac{e\hbar}{c}\boldsymbol{\sigma}\,\mathrm{curl}\,\mathbf{A}.$$

Hence in the case $U\ll m_0c^2$ the equation (6.102) becomes

$$i\hbar\frac{\partial\varphi'}{\partial t}=\left[\frac{1}{2m_0}\left(\mathbf{p}-\frac{e}{c}\mathbf{A}\right)^2+U-\frac{e\hbar}{2m_0c}\boldsymbol{\sigma}\mathbf{B}\right]\varphi'. \tag{6.103}$$

It is seen that the obtained equation coincides with the Pauli equation (see section 4.1.3), where the electron magnetic moment is

$$\boldsymbol{\mu} = \frac{e\hbar}{2m_0 c}\boldsymbol{\sigma} = \frac{e\hbar}{m_0 c}\mathbf{s}. \tag{6.104}$$

In the non-relativistic approximation the current density four vector is

$$\rho = e\left(\varphi^+\varphi + \chi^+\chi\right) \approx e\varphi'^+\varphi', \tag{6.105}$$

$$\mathbf{j} = ec\left(\varphi^+\boldsymbol{\sigma}\chi + \chi^+\boldsymbol{\sigma}\varphi\right) \approx \frac{e}{2m_0}\left(\varphi'^+\boldsymbol{\sigma}\left(\boldsymbol{\sigma}\mathbf{P}\right)\varphi' + \left(\mathbf{P}\varphi'^+\boldsymbol{\sigma}\right)\boldsymbol{\sigma}\varphi'\right) =$$
$$= \frac{ie\hbar}{2m_0}\left(\nabla\varphi'^+ \cdot \varphi' - \varphi'^+\nabla\varphi'\right) - \frac{e^2}{m_0 c}\varphi'^+\mathbf{A}\varphi' + \frac{e\hbar}{2m_0}\operatorname{curl}\left(\varphi'^+\boldsymbol{\sigma}\varphi'\right). \tag{6.106}$$

6.5.2 Non-relativistic approximation

To derive the equation accounting for the highest order of the non-relativistic approximation ($v \ll c$) we can use the following technique proposed in [54]. Let us write the equation (6.101) in the form

$$\chi' = \left(2m_0c^2 - U + i\hbar\frac{\partial}{\partial t}\right)^{-1} c\left(\boldsymbol{\sigma}\mathbf{P}\right)\varphi',$$

where the inverse operator is defined by the following power series

$$\chi' = \frac{c}{2m_0c^2 - U}\sum_{n=0}^{\infty}\left(\frac{-i\hbar}{2m_0c^2 - U}\frac{\partial}{\partial t}\right)^n\left(\boldsymbol{\sigma}\mathbf{P}\right)\varphi'. \tag{6.107}$$

The equivalence of the inverse operator to the series (6.107) can be shown in the following way. Let us consider the equation

$$\frac{dx}{dt} + ax = f\left(t\right).$$

The solution of this equation can be written in the following form

$$x\left(t\right) = \exp\left(-at\right)\int_0^t f\left(z\right)\exp(az)\,dz =$$
$$= \frac{1}{a}f\left(t\right) - \frac{1}{a}\exp\left(-at\right)\int_0^t \frac{df}{dz}\exp(az)\,dz - \frac{\exp\left(-at\right)}{a}f\left(0\right) =$$
$$= \frac{1}{a}f\left(t\right) - \frac{1}{a^2}\frac{df}{dt} + \frac{1}{a^2}\exp\left(-at\right)\int_0^t \frac{d^2f}{dz^2}\exp(az)\,dz -$$
$$- \exp\left(-at\right)\left(\frac{1}{a}f\left(0\right) - \frac{1}{a^2}\frac{df}{dt}\left(0\right)\right) = \ldots \tag{6.108}$$

By continuing further the obtained chain, we can see that the series (6.108) coincides with (6.107). In the equation (6.108) we can remove the terms depending on the initial conditions, because in our case $a = -i2m_0c^2/\hbar$ and we must omit these terms in non-relativistic approximation, even if the initial conditions are non-zero.

The substitution of series (6.107) into the equation (6.100) produces the closed equation for spinor φ'. If we take into account the only first two terms of series, we get

$$i\hbar\frac{\partial\psi}{\partial t} = \left\{\frac{1}{2m_1}\left(\mathbf{p} - \frac{e}{c}\mathbf{A}\right)^2 + U - \frac{e\hbar}{2m_1 c}\boldsymbol{\sigma}\mathbf{B} - \frac{e\hbar c^2}{\left(2m_0c^2 - U\right)^2}\left[\mathbf{E}\left(\mathbf{p} - \frac{e}{c}\mathbf{A}\right)\right] - \right.$$
$$- \frac{e\hbar^2c^2}{2}\,\mathrm{div}\left(\frac{\mathbf{E}}{(2m_0c^2 - U)^2}\right) + i\frac{e\hbar^2c^2}{2}\boldsymbol{\sigma}\,\mathrm{curl}\left(\frac{1}{(2m_0c^2 - U)^2}\frac{1}{c}\frac{\partial\mathbf{A}}{\partial t}\right) +$$
$$\left. + \hbar\boldsymbol{\sigma}\left[\nabla\frac{1}{2m_1}\left(\mathbf{p} - \frac{e}{c}\mathbf{A}\right)\right]\right\}\psi, \quad (6.109)$$

where

$$\psi(\mathbf{r},t) = \left(1 + \boldsymbol{\sigma}\mathbf{P}\frac{c^2}{\left(2m_0c^2 - U\right)^2}\boldsymbol{\sigma}\mathbf{P}\right)\varphi'(\mathbf{r},t),$$
$$\frac{1}{m_1} = \frac{1}{m_0}\left(1 - \left(\frac{U}{2m_0c^2 - U}\right)^2\right). \quad (6.110)$$

The equation (6.109) generates the following equations for components of the current density four vector

$$\rho = e\psi^+\psi,$$

$$\mathbf{j} = \frac{ie\hbar}{2m_1}\left(\nabla\psi^+ \cdot \psi - \psi^+\nabla\psi\right) - \frac{e^2}{m_1c}\psi^+\mathbf{A}\psi - \frac{e^2c^2\hbar}{\left(2m_0c^2 - U\right)^2}\psi^+\left[\mathbf{E}\boldsymbol{\sigma}\right]\psi. \quad (6.111)$$

In the case $U \ll m_0c^2$, the Hamiltonian of the equation (6.109) is simplified and takes the form

$$H = \frac{1}{2m_0}\left(\mathbf{p} - \frac{e}{c}\mathbf{A}\right)^2 + U - \frac{e\hbar}{2m_0c}\boldsymbol{\sigma}\mathbf{H} - \frac{e\hbar}{4m_0^2c^2}\boldsymbol{\sigma}\left[\mathbf{E}\left(\mathbf{p} - \frac{e}{c}\mathbf{A}\right)\right] -$$
$$- \frac{e\hbar^2}{8m_0^2c^2}\,\mathrm{div}\,\mathbf{E} - i\frac{e\hbar^2}{8m_0^2c^2}\boldsymbol{\sigma}\,\mathrm{curl}\,\mathbf{E}. \quad (6.112)$$

Thus we can see that the Hamiltonian of the equation (6.109) includes the extra terms in comparison with the Hamiltonian of the Pauli equation. Hence the structure of the hydrogenic spectra will differ from that calculated in the Chapter 4 on the basis of the Pauli equation. However we do not present here the analysis of the equation (6.109), because the Dirac equation for the problem of the electron motion in the Coulomb field has the analytically tractable solution. The Hamiltonian (6.112) is useful in the interpretation of the results of the exact analytical calculations.

6.5.3 Motion in Coulomb field

Energy spectrum

Let us consider the problem on the electron motion in the Coulomb field

$$U(r) = -\frac{Ze^2}{r},$$

where Z is the charge of nucleus the hydrogenlike atom.

As we have mentioned in the beginning of this section, we can find initially the solutions of the second order differential equation (6.96), then the solutions of the Dirac equation are defined by the equation (6.97). The second order differential equation for the considered problem is

$$\left[\Delta - \kappa^2 + \frac{2EZ\alpha}{\hbar c}\frac{1}{r} + \frac{Z^2\alpha^2}{r^2} + i\frac{Z\alpha}{r^2}\alpha_r\right]\Phi = 0, \tag{6.113}$$

where

$$\kappa = \frac{\sqrt{m_0^2c^4 - E^2}}{\hbar c}, \qquad \alpha_r = (\mathbf{n}_r\boldsymbol{\alpha}), \tag{6.114}$$

and we have introduced the fine structure constant $\alpha = e^2/(\hbar c)$. Let us write the wave function of the second order differential equation in the form

$$\Phi = \begin{pmatrix} \xi \\ \eta \end{pmatrix}, \tag{6.115}$$

then for spinors ξ and η we get the following coupled equations

$$\begin{aligned} \left[\Delta - \kappa^2 + \frac{2EZ\alpha}{\hbar c}\frac{1}{r} + \frac{Z^2\alpha^2}{r^2}\right]\xi &= -i\frac{Z\alpha}{r^2}\sigma_r\eta, \\ \left[\Delta - \kappa^2 + \frac{2EZ\alpha}{\hbar c}\frac{1}{r} + \frac{Z^2\alpha^2}{r^2}\right]\eta &= -i\frac{Z\alpha}{r^2}\sigma_r\xi. \end{aligned} \tag{6.116}$$

The Coulomb potential is the spherically symmetric, hence in accordance with the analysis given in subsection 6.4.4 the spinors ξ and η are:

a) for $j = l + 1/2$:

$$\xi_1(\mathbf{r}) = p_1(r)\,\Omega_{j,l=j-1/2,m}(\theta,\varphi), \quad \eta_1(\mathbf{r}) = q_1(r)\,\Omega_{j,l=j+1/2}(\theta,\varphi); \tag{6.117}$$

b) for $j = l - 1/2$:

$$\xi_2(\mathbf{r}) = p_2(r)\,\Omega_{j,l=j+1/2}(\theta,\varphi)\,, \quad \eta_2(\mathbf{r}) = q_2(r)\,\Omega_{j,l=j-1/2,m}(\theta,\varphi)\,. \tag{6.118}$$

Notice, that the equations (6.116) are symmetric with respect to spinors ξ and η, therefore we need not in the analysis of the equations (6.116) for the cases (6.117) and (6.118) separately. If we have got the solutions for the case $j = l + 1/2$, then the transposition of the upper and lower spinors in the wave function (6.115) generates the solutions for the case $j = l - 1/2$. By substituting the equations (6.117) into the equation (6.116), we get the following equations for the radial wave functions $p\,(r)$ and $q\,(r)$

$$\begin{aligned} \left[\frac{d^2}{dr^2} + \frac{2}{r}\frac{d}{dr} - \kappa^2 + \frac{2EZ\alpha}{\hbar c}\frac{1}{r} + \frac{Z^2\alpha^2 - l\,(l+1)}{r^2}\right] p &= -\frac{Z\alpha}{r^2} q, \\ \left[\frac{d^2}{dr^2} + \frac{2}{r}\frac{d}{dr} - \kappa^2 + \frac{2EZ\alpha}{\hbar c}\frac{1}{r} + \frac{Z^2\alpha^2 - (l+1)\,(l+2)}{r^2}\right] q &= \frac{Z\alpha}{r^2} p. \end{aligned} \tag{6.119}$$

The solutions of the coupled set of equations (6.119) can be easily found with the help of the solutions of the following equation

$$\frac{d^2 f}{dx^2} + \frac{2}{x}\frac{df}{dx} - \left(a - \frac{b}{x} - \frac{c}{x^2}\right) f = 0.$$

The last equation coincides with the equation for the radial wave functions of the Schrödinger equation for particle moving in the Coulomb field, and the solution of this equation, which is finite at $x \to 0$, is

$$\begin{aligned} f\,(x) = \exp\left(-\sqrt{a}x + \frac{\sqrt{1-4c}-1}{2}\ln x\right) \times \\ \times F\left(\frac{1+\sqrt{1-4c}}{2} - \frac{b}{2\sqrt{a}}, 1+\sqrt{1-4c}, 2\sqrt{a}x\right), \end{aligned} \tag{6.120}$$

where $F\,(p, q, z)$ is the confluent hypergeometric function. It is seen that we can assume that the radial wave functions $p\,(r)$ and $q\,(r)$ are

$$\begin{aligned} p\,(r) &= Ar^{\nu-1}\exp\left(-\kappa r\right) F\left(\nu - \frac{EZ\alpha}{\hbar c\kappa}, 2\nu, 2\kappa r\right), \\ q\,(r) &= Br^{\nu-1}\exp\left(-\kappa r\right) F\left(\nu - \frac{EZ\alpha}{\hbar c\kappa}, 2\nu, 2\kappa r\right), \end{aligned} \tag{6.121}$$

where

$$\nu = \frac{1}{2}\left(1 + \sqrt{1-4c}\right), \tag{6.122}$$

and A and B are the arbitrary constants. By substituting the equations (6.121) into the equations (6.119) we get the coupled set of the algebraic equations for coefficients A and B

$$\begin{aligned}\left(Z^2\alpha^2 - l\,(l+1) - c\right)A + Z\alpha B = 0,\\ Z\alpha A - \left(Z^2\alpha^2 - (l+1)\,(l+2) - c\right)B = 0.\end{aligned} \tag{6.123}$$

The condition of existence of the non-trivial solutions of the algebraic equations (6.123) yields the following two values for the coefficient c:

$$c_{1,2} = -\,(l+1)^2 + Z^2\alpha^2 \pm \sqrt{(l+1)^2 - Z^2\alpha^2}. \tag{6.124}$$

Hence,

$$\nu_1 = \sqrt{\left(j+\frac{1}{2}\right)^2 - Z^2\alpha^2}, \quad B_1 = -\varsigma A_1 \tag{6.125}$$

and

$$\nu_2 = 1 + \sqrt{\left(j+\frac{1}{2}\right)^2 - Z^2\alpha^2}, \quad A_2 = -\varsigma B_2, \tag{6.126}$$

where

$$\varsigma = \frac{Z\alpha}{\sqrt{(j+1/2)^2 - Z^2\alpha^2} + j + 1/2}. \tag{6.127}$$

Thus, in the case when $j = l + 1/2$, the two linear independent solutions of the second order differential equation (6.114) are

$$\begin{aligned}\Phi_1(\mathbf{r}) = \begin{pmatrix}\Omega_{j,l=j-1/2,m}\\ -\varsigma\,\Omega_{j,l=j+1/2,m}\end{pmatrix} R_j^{(1)}(r),\\ \Phi_2(\mathbf{r}) = \begin{pmatrix}-\varsigma\,\Omega_{j,l=j-1/2,m}\\ \Omega_{j,l=j+1/2,m}\end{pmatrix} R_j^{(2)}(r),\end{aligned} \tag{6.128}$$

where

$$R_j^{(1,2)}(r) = r^{\nu_{1,2}-1}\exp(-\kappa r)\,F\left(\nu_{1,2} - \frac{EZ\alpha}{\hbar c\kappa}, 2\nu_{1,2}, 2\kappa r\right). \tag{6.129}$$

The solutions (6.128) satisfy the boundary condition at $r \to 0$. The radial wave function (6.129) satisfies the boundary condition at $r \to \infty$ when the following condition holds

$$\nu_{1,2} - \frac{EZ\alpha}{\hbar c\kappa} = -n_r^{(1,2)}, \tag{6.130}$$

where $n_r^{(1,2)}$ is the non-negative integers. The equation (6.130) yields the following equation for the energy spectrum

$$E_{1,2} = \frac{m_0c^2}{\sqrt{1 + \left(\dfrac{Z\alpha}{n_r^{(1,2)} + \nu_{1,2}}\right)^2}}. \tag{6.131}$$

It is seen from the equations (6.125) and (6.126), that the obtained spectrum is degenerated. Indeed, by introducing the following notations

$$\nu_1 = \nu, \quad \nu_2 = \nu + 1, \quad n_r^{(1)} = n_r, \quad n_r^{(2)} = n_r - 1,$$

we get the general equation for the energy spectrum

$$E_{n_r j} = \frac{m_0 c^2 (n_r + \nu)}{\sqrt{(n_r + \nu)^2 + (Z\alpha)^2}}, \tag{6.132}$$

and the following equations for the wave functions

$$\begin{aligned} \Phi_1(\mathbf{r}) &= \begin{pmatrix} \Omega_{j,l=j-1/2,m} \\ -\varsigma\Omega_{j,l=j+1/2,m} \end{pmatrix} r^{\nu-1} \exp(-\kappa r)\, F(-n_r, 2\nu, 2\kappa r), \\ \Phi_2(\mathbf{r}) &= \begin{pmatrix} -\varsigma\Omega_{j,l=j-1/2,m} \\ \Omega_{j,l=j+1/2,m} \end{pmatrix} r^{\nu} \exp(-\kappa r)\, F(1 - n_r, 2\nu + 2, 2\kappa r). \end{aligned} \tag{6.133}$$

As we have mentioned above we need not in separate analysis of the case of $j = l - 1/2$. The energy spectrum is again defined by the equation (6.132), and the wave functions follow from the equations (6.133) with the help of transposition of the upper and lower spinors in these equations

$$\begin{aligned} \Phi_3(\mathbf{r}) &= \begin{pmatrix} \Omega_{j,l=j+1/2,m} \\ -\varsigma\Omega_{j,l=j-1/2,m} \end{pmatrix} r^{\nu} \exp(-\kappa r)\, F(1 - n_r, 2\nu + 2, 2\kappa r), \\ \Phi_4(\mathbf{r}) &= \begin{pmatrix} -\varsigma\Omega_{j,l=j+1/2,m} \\ \Omega_{j,l=j-1/2,m} \end{pmatrix} r^{\nu-1} \exp(-\kappa r)\, F(-n_r, 2\nu, 2\kappa r). \end{aligned} \tag{6.134}$$

It should be reminded here that there are the superfluous or unnecessary solutions among the solutions of the second order differential equation (6.113). These solutions are the solutions of the equation (6.113), but they are not the solutions of the Dirac equation

$$(\gamma_\mu p_\mu - i m_0 c)\, \Psi = 0. \tag{6.135}$$

We have mentioned above, that the general technique, to obtain the Dirac equation solutions from the solutions of the second order differential equation, is in the following. If Φ is a solution of the second order differential equation, then the solution of the Dirac equation is defined by

$$\Psi = (\gamma_\mu p_\mu + i m_0 c)\, \Phi = \left[\frac{i}{c}\gamma_4 (E - U) + \boldsymbol{\gamma}\mathbf{p} + i m_0 c\right] \Phi.$$

In the case when $j = l + 1/2$, the general solution of the second order differential equation is

$$\Phi(\mathbf{r}) = A\Phi_1(\mathbf{r}) + B\Phi_2(\mathbf{r}). \tag{6.136}$$

The equations for $\Phi_{1,2}$ can be symmetrized with the help of the recurrence relations for the confluent hypergeometric functions:

$$\begin{aligned}
&2\nu F(-n_r, 2\nu, 2\kappa r) = \\
&\quad = -n_r F(1 - n_r, 2\nu + 1{,}2\kappa r) - (n_r + 2\nu) F(-n_r, 2\nu + 1{,}2\kappa r), \\
&2\kappa r F(1 - n_r, 2\nu + 2{,}2\kappa r) = \\
&= (2\nu + 1) F(1 - n_r, 2\nu + 1{,}2\kappa r) - (2\nu + 1) F(-n_r, 2\nu + 1{,}2\kappa r).
\end{aligned} \tag{6.137}$$

If we apply the equations (6.137) to the equation (6.136) we can easily find the coefficients A and B that realize the equality (6.135). The resultant wave function is

$$\Psi_{njm}(\mathbf{r}) = \begin{pmatrix} f_{nj}(r)\,\Omega_{j,l=j-1/2,m} \\ g_{nj}(r)\,\Omega_{j,l=j+1/2,m} \end{pmatrix}, \tag{6.138}$$

where

$$\begin{aligned}
f_{nj}(r) = &\frac{(2\kappa)^{3/2}}{\Gamma(2\nu + 1)} \frac{E}{2m_0c^2} (2\kappa r)^{\nu - 1} \exp(-\kappa r) \times \\
&\times \sqrt{\frac{\Gamma(2\nu + n_r + 1)}{\Gamma(n_r + 1)} \frac{(m_0c^2 + E)}{(n_r + \nu)(m_0c^2(n_r + \nu) + E(j + 1/2))}} \times \\
&\times \left[\left(\frac{Z\alpha m_0 c}{\hbar\kappa} + j + \frac{1}{2}\right) F(-n_r, 2\nu + 1{,}2\kappa r) - n_r F(1 - n_r, 2\nu + 1{,}2\kappa r)\right],
\end{aligned} \tag{6.139}$$

$$\begin{aligned}
g_{nj}(r) = -&\frac{(2\kappa)^{3/2}}{\Gamma(2\nu + 1)} \frac{E}{2m_0c^2} (2\kappa r)^{\nu - 1} \exp(-\kappa r) \times \\
&\times \sqrt{\frac{\Gamma(2\nu + n_r + 1)}{\Gamma(n_r + 1)} \frac{(m_0c^2 - E)}{(n_r + \nu)(m_0c^2(n_r + \nu) + E(j + 1/2))}} \times \\
&\times \left[\left(\frac{Z\alpha m_0 c}{\hbar\kappa} + j + \frac{1}{2}\right) F(-n_r, 2\nu + 1{,}2\kappa r) + n_r F(1 - n_r, 2\nu + 1{,}2\kappa r)\right].
\end{aligned} \tag{6.140}$$

The wave function (6.138)–(6.140) is normalized by the condition

$$\int \Psi^+_{njm}(\mathbf{r})\,\Psi_{njm}(\mathbf{r})\,dV = 1.$$

The energy spectrum (6.132) depends on the two quantum numbers n_r and j. Hence, in contrast to the Bohr formula the spectrum (6.132) describes the fine structure, i.e. the splitting of the $nP_{1/2}$ and $nP_{3/2}$ levels. The first terms of expansion of the spectrum (6.132) in powers $(Z\alpha)^n$ are

$$\Delta E_{nj} = m_0c^2 - E_{nj} = \\ = m_0c^2 \left[\frac{Z^2\alpha^2}{2n^2} + \frac{Z^4\alpha^4}{2n^4\,(j+1/2)} \left(n - \frac{3}{4}\left(j + \frac{1}{2} \right) \right) \right] + \ldots, \quad (6.141)$$

where n is the principle quantum number, which is defined by

$$n = n_r + j + \frac{1}{2}.$$

It is seen that the equation (6.141) coincides with the Sommerfeld formula (5.27).

The spectrum (6.125) differs from the Bohr formula in the dependency on the charge Z of the ion. It is seen from the equation (6.125), that there is the critical charge Z_0, defined by

$$Z_0\alpha = j + \frac{1}{2}.$$

It is impossible in the frames of the Dirac theory to consider the Coulomb field with $Z > Z_0$. At $Z < Z_0$, the dependency of the parameter ν on Z

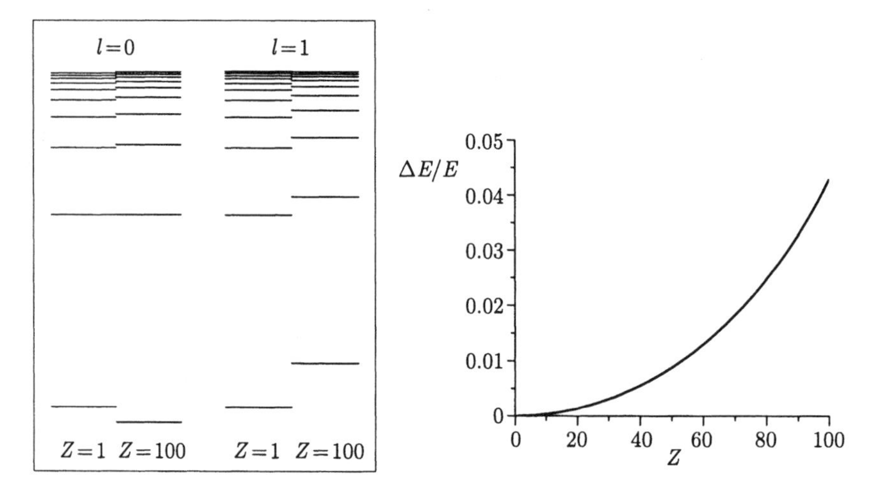

Figure 6.1. The relative energy shift of the $2P_{1/2}$ and $2P_{3/2}$ levels as a function of the ion charge Z. The insert shows the normalized spectra of hydrogen and ion of charge $Z = 100$ for $l = 0$ and $l = 1$

changes the magnitude of the relative shifts of the different states. The Fig. 6.1 shows in comparison the spectra given by equation (6.132) for hydrogen and for ion of the charge Z=100. The energy of states are normalized by the energy of $1s$ state E_{nl}/E_{1s}. The states of $2 \leq n \leq 10$ are only depicted. It is seen from the figure that the $2s$ state of ion lies below the $2s$ state of hydrogen (i.e. the relative energy of this state of ion is smaller than the relative energy of the same state of hydrogen). At $n > 3$, the relative energy of the ns states of ion is higher than the relative energy of the appropriate states of hydrogen. At $l > 0$, the relative energy of the all states of ion is higher than the relative energy of hydrogen, and the energy shift exceeds significantly the energy shift at $l = 0$. The graph in Fig. 6.1 shows the relative energy shift,

$$\frac{E\left(2p_{3/2}\right) - E\left(2p_{1/2}\right)}{E\left(1s_{1/2}\right)},$$

as a function of the ion charge. It is seen that at $Z > 50$ the energy shift of $2p_{3/2}$ and $2p_{1/2}$ states is about a few percents of the ground state energy.

Wave functions

To compare the wave functions (6.139), (6.140) with the wave functions of the non-relativistic Schrödinger equation, given by the equation (2.35), it is convenient to introduce the dimensionless coordinate

$$x = \frac{r}{a_{\mathrm{B}}}, \tag{6.142}$$

where a_{B} is the Bohr radius. Let us introduce additionally the effective principle quantum number

$$N = \sqrt{(n_r + \nu)^2 + (Z\alpha)^2}. \tag{6.143}$$

Let us consider the case of $j = l + 1/2$. In this case, at $Z\alpha \ll 1$, the parameter ν is approximately equal to

$$\nu = \sqrt{(l+1)^2 - (Z\alpha)^2} \approx l + 1 - \frac{(Z\alpha)^2}{2(l+1)}, \tag{6.144}$$

then for the effective principle quantum number N we get

$$N \approx n_r + l + 1 - \frac{(Z\alpha)^2 n_r}{2(n_r + l + 1)(l + 1)}.$$

It is seen that in the limiting case of $Z\alpha \to 0$ the quantum number N approaches to the principle quantum number, $n = n_r + l + 1$, of the Schrödinger theory.

By substituting the equations (6.142), (6.143) into the equations (6.139), (6.140), we get

$$f_{n_r l}(x) = \frac{1}{2N\Gamma(2\nu+1)} \times$$
$$\times \sqrt{\frac{\Gamma(n_r+2\nu+1)}{\Gamma(n_r+1)} \frac{N+n_r+\nu}{N+l+1}} \left(\frac{2Z}{N}\right)^{\nu+1/2} x^{\nu-1} \exp\left(-\frac{Zx}{N}\right) \times$$
$$\times \left((N+l+1) F\left(-n_r, 2\nu+1, \frac{2Zx}{N}\right) - n_r F\left(1-n_r, 2\nu+1, \frac{2Zx}{N}\right)\right), \tag{6.145}$$

$$g_{n_r l}(x) = -\frac{1}{2N\Gamma(2\nu+1)} \times$$
$$\times \sqrt{\frac{\Gamma(n_r+2\nu+1)}{\Gamma(n_r+1)} \frac{N-n_r-\nu}{N+l+1}} \left(\frac{2Z}{N}\right)^{\nu+1/2} x^{\nu-1} \exp\left(-\frac{Zx}{N}\right) \times$$
$$\times \left((N+l+1) F\left(-n_r, 2\nu+1, \frac{2Zx}{N}\right) + n_r F\left(1-n_r, 2\nu+1, \frac{2Zx}{N}\right)\right). \tag{6.146}$$

At $l > 0$, there is almost complete coincidence between the wave function (6.145) and the radial wave function $R_{nl}(x)$ of the Schrödinger

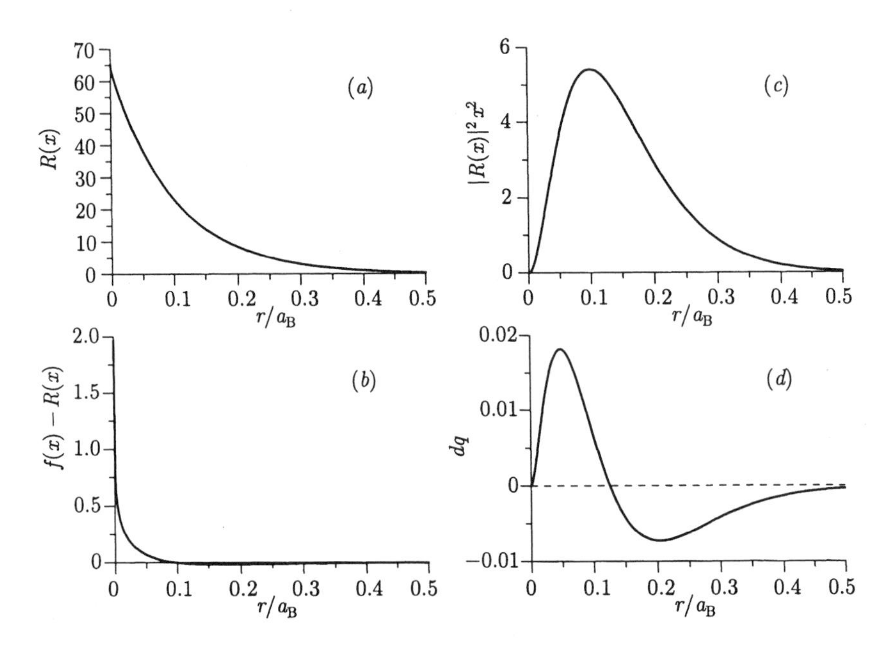

Figure 6.2. The comparison of the Dirac and Schrödinger wave functions for the $1S$ state of the ion of charge $Z = 10$

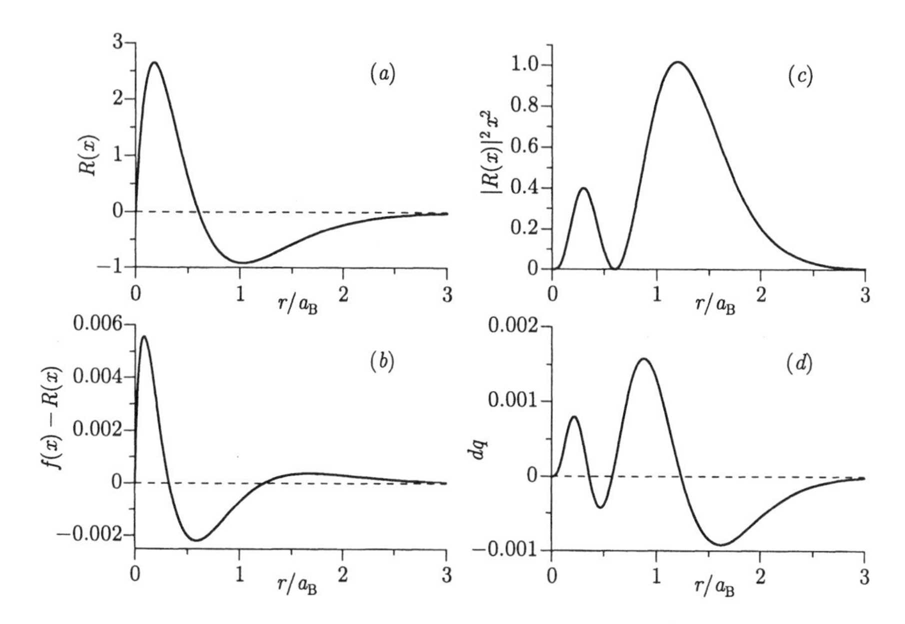

Figure 6.3. The comparison of the Dirac and Schrödinger wave functions for the $1P$ state of the ion of charge $Z = 10$

equation. At the same time, there is some difference between the radial wave functions of the s states. Indeed, the confluent hypergeometric function has the following asymptotical form at $x \to 0$: $F(a, b, x)|_{x \to 0} = 1$. Hence, at $x \to 0$ we get the following asymptotical expression for the wave function (6.145)

$$f(x) \sim g(x) \sim x^{-(Z\alpha)^2/2}. \tag{6.147}$$

We can see that the wave function $\Psi(x)$ is divergent at $x \to 0$. The divergency is weak, when $Z\alpha \ll 1$. The Fig. 6.2 shows in comparison the wave function $R(x)$ (curve (a)) and difference $f(x) - R(x)$ (curve (b)) for the $1s$ state of the hydrogenlike ion of charge $Z = 10$. The curves (c) and (d) illustrate the charge density distribution $q(x) = R^2(x)\, x^2$ and the difference $dq = \left(\Psi^+(x)\, \Psi(x) - R^2(x)\right) x^2$. The charge density distribution is not divergent, and the ratio of dq to q is $dq/q \leq 3 \cdot 10^{-3}$. The Fig. 6.3 shows the wave function $R(x)$ (curve (a)), the difference $f(x) - R(x)$ (curve (b)), charge density distribution $q(x) = R^2(x)\, x^2$ (curve (c)), and the difference $dq = \left(\Psi^+(x)\, \Psi(x) - R^2(x)\right) x^2$ (curve (d)) for the $2p$ state of the hydrogenlike ion of charge $Z = 10$. The wave functions are not divergent at $l > 0$, and the ratio of dq/q is about of the same order of magnitude as for the s states.

Continuous spectrum

The states of the energy $E > m_0c^2$ correspond to the continuous spectrum. In this case the parameter κ becomes pure imagine, and it is convenient to introduce the new parameter

$$k = \frac{\sqrt{E^2 - m_0^2c^4}}{\hbar c} = i\kappa. \tag{6.148}$$

The two linear independent solutions of the second order differential equation (6.114) are still given by the equation (6.128), where we should make the following replacement $\kappa \to -ik$, so

$$\begin{aligned} R^{(1)}(r) &= (2kr)^{\nu-1}\exp(ikr)F(\nu - i\eta, 2\nu, -i2kr), \\ R^{(2)}(r) &= (2kr)^{\nu}\exp(ikr)F(\nu + 1 - i\eta, 2\nu + 2, -i2kr), \end{aligned} \tag{6.149}$$

where

$$\eta = \frac{EZ\alpha}{c\hbar k}. \tag{6.150}$$

With the help of the recurrence relations (6.137) and the Kummer transformation $F(a, b, z) = \exp(z) F(b - a, b, -z)$, the wave functions (6.139) and (6.140) are transformed to the following form

$$\begin{aligned} f(r) &= C\sqrt{E + m_0c^2}(2kr)^{\nu-1}\times \\ &\quad \times \operatorname{Im}\{\exp[i(kr + \xi)]F(\nu - i\eta, 2\nu + 1, -i2kr)\}, \\ g(r) &= C\sqrt{E - m_0c^2}(2kr)^{\nu-1}\times \\ &\quad \times \operatorname{Re}\{\exp[i(kr + \xi)]F(\nu - i\eta, 2\nu + 1, -i2kr)\}, \end{aligned} \tag{6.151}$$

where C is the normalization constant, and we have introduced the following parameter

$$\exp(i2\xi) = \frac{-(j + 1/2) - i\eta m_0c^2/E}{\nu - i\eta}. \tag{6.152}$$

The normalization constant C is determined by the asymptotical form of the wave function at $r \to \infty$. Finally, for the normalized wave function we get

$$\begin{aligned} f(r) &= 4\sqrt{\frac{E + m_0c^2}{2E}}\frac{|\Gamma(\nu + 1 + i\eta)|}{\Gamma(2\nu + 1)}\frac{(2kr)^{\nu}}{r}\exp\left(\frac{\pi\eta}{2}\right)\times \\ &\quad \times \operatorname{Im}\{\exp[i(kr + \xi)]F(\nu - i\eta, 2\nu + 1, -i2kr)\}, \end{aligned} \tag{6.153}$$

$$g(r) = 4\sqrt{\frac{m_0c^2 - E}{2E}} \frac{|\Gamma(\nu + 1 + i\eta)|}{\Gamma(2\nu + 1)} \frac{(2kr)^\nu}{r} \exp(\pi\eta/2) \times \\ \times \operatorname{Re}\{\exp[i(kr + \xi)] F(\nu - i\eta, 2\nu + 1, -i2kr)\}. \quad (6.154)$$

The asymptotical form of the functions (6.153), (6.154) at $r \to \infty$ is

$$\begin{aligned} f_{kl}(r) &= \frac{\sqrt{2}}{r}\sqrt{\frac{E + m_0c^2}{E}} \sin\left(kr + \delta_k + \eta\ln(2kr) - \frac{\pi l}{2}\right), \\ g_{kl}(r) &= \frac{\sqrt{2}}{r}\sqrt{\frac{E - m_0c^2}{E}} \cos\left(kr + \delta_k + \eta\ln(2kr) - \frac{\pi l}{2}\right), \end{aligned} \quad (6.155)$$

where

$$\exp(2i\delta_k) = \frac{-(j + 1/2) - i\eta m_0c^2/E}{\nu - i\eta} \frac{\Gamma(\nu + 1 - i\eta)}{\Gamma(\nu + 1 + i\eta)} \exp[i\pi(l - \nu)]. \quad (6.156)$$

PART II

THEORY OF LAMB SHIFT

Chapter 7

THEORY OF SPIN-1/2 PARTICLES INTERACTING WITH ELECTROMAGNETIC FIELD

The hydrogenic spectrum, calculated on the basis of the Dirac theory [4, 55, 56], was in good agreement with the experimental data of that time. Indeed, the fine structure splitting, $\alpha^4 m_0 c^2/32$, is in good agreement with the experimental data and this value, which is only the first order correction to the Bohr formula, coincides with the correction, calculated earlier by Sommerfeld [53] with the help of quantization rules. As we have mentioned, the next step in the development of the theory of atomic spectra was stimulated by the experimental observation of the Lamb shift [57]. The researches, directed towards the explanation of the Lamb shift, triggered the development of the quantum field theory. The application of the powerful technique of the quantum field theory enables to calculate precisely the hyperfine structure of the hydrogenic spectra. The radiative correction theory is based on the account of the virtual processes of the charged particle interaction with the electromagnetic vacuum and vacuum of the electron-positron pairs. The main technique here is the invariant theory of perturbations. The smallness parameter of the perturbation theory is the fine structure constant. Indeed, the mean potential energy of electron in the hydrogen atom is equal to $U_{\mathrm{at}} = e^2/a_{\mathrm{B}}$, where a_{B} is the Bohr radius. The ratio of U_{at} to the electron rest energy is $U_{\mathrm{at}}/(m_0 c^2) = \alpha^2$. However, in the case of the hydrogenlike ions, the characteristic constant of interaction is the parameter $Z\alpha$, which increases with the increase of the ion charge Z. Hence, the perturbation series becomes less convergent with the increase of Z. Therefore, the development of the non-perturbative approach is of the great interest.

By comparing the quantum mechanics and quantum field theory approaches, we can see that the principle difference between them consists in the following. In the frames of the Schrödinger and Dirac theories the charge density is $\rho(\mathbf{r},t) = e\Psi^{+}(\mathbf{r},t)\Psi(\mathbf{r},t)$. Hence, the charge conservation law fixes both the integral charge and integral number of particles, because

$$\int \Psi^{+}(\mathbf{r},t)\Psi(\mathbf{r},t)\,dV = \sum_{i} |A_i|^2,$$

where the index i numerates the linear independent solutions of the quantum mechanical equation, and A_i are the amplitudes of these solutions.

On the other hand, the quantum field theory, including into consideration the virtual processes, removes any restrictions for a number of particles involved in the interaction. Hence, to account for the many-particle processes we should reconstruct the normalization condition in a way allowing the variation of the total number of particles. Notice here, that the particle and antiparticle has the opposite sign of charge. It gives us some hint, how we can construct the wave function normalization condition in order to satisfy the charge conservation law, on the one hand, and to remove the restrictions on the integral number of particles, on the other hand. It is evident that such kind normalization condition could not be incorporated into the theory, which is based on the differential equation of the first order with respect to time derivative, because the continuity equation should have the relativistic invariant form. The theory, which met the all above mentioned requirements, was proposed recently [58]. The calculated spectrum [59] is in reasonably good agreement with the experimentally measured spectra of hydrogen and deuterium. Here, we give an overview of the theory and demonstrate its application to the theory of hydrogen atom, geonium atom, and problem of the electric dipole moment of spin-1/2 particles.

7.1 Action principle

Let the action for the spin-1/2 particle interacting with the electromagnetic field be

$$S = \frac{1}{16\pi}\int F_{\mu\nu}F_{\nu\mu}\,dV dt -$$
$$-\frac{1}{2m_0}\int\left[\left(i\hbar\nabla_\mu\bar{\Psi} - \frac{q_0}{c}A_\mu\bar{\Psi}\right)\left(-i\hbar\nabla_\mu\Psi - \frac{q_0}{c}A_\mu\Psi\right) + m_0^2c^2\bar{\Psi}\Psi\right] dV\,dt +$$
$$+\frac{i\mu_0}{2}\int \bar{\Psi}\gamma_\mu\gamma_\nu F_{\nu\mu}\Psi\,dV\,dt, \quad (7.1)$$

where $F_{\mu\nu} = \nabla_\mu A_\nu - \nabla_\nu A_\mu$ is the electromagnetic field tensor; $\nabla_\mu = \partial/\partial x_\mu$, $x_\mu = (\mathbf{r}, ict)$; $A_\mu = (\mathbf{A}, i\varphi)$ is the four-potential of the electromagnetic field; γ_μ are the four by four matrices; m_0, q_0 and μ_0 is mass, charge, and magneton (i.e. magnitude of the magnetic moment), respectively. The wave function Ψ of the equation (7.1) is the bispinor

$$\Psi(\mathbf{r},t) = \begin{pmatrix} \varphi \\ \chi \end{pmatrix}, \tag{7.2}$$

where φ and χ are the three-dimensional spinors. The Dirac adjoint wave function is

$$\bar{\Psi} = \Psi^{+}\gamma_4.$$

It is seen that the main difference of the action (7.1) and action for the Dirac equation is in the following: Firstly, the action (7.1) is the quadratic form of the four-momentum operator. Secondly, it enables to introduce the three independent material constants characterizing the particle properties — mass m_0, charge q_0, and magneton μ_0.

The Euler-Lagrange equation, when S is varied with respect to $\bar{\Psi}$, is

$$\frac{1}{m_0}\left[\left(-i\hbar\frac{\partial}{\partial x_\mu} - \frac{q_0}{c}A_\mu\right)^2 + m_0^2c^2\right]\Psi = i\mu_0\,\gamma_\mu\gamma_\nu F_{\nu\mu}\Psi, \tag{7.3}$$

The variation of S with respect to A_μ results in the following equation

$$\nabla_\nu^2 A_\mu - \nabla_\mu\nabla_\nu A_\nu = -\frac{4\pi}{c}j_\mu, \tag{7.4}$$

where

$$j_\mu = \frac{q_0}{m_0}\left[\frac{i\hbar}{2}\left(\frac{\partial\bar{\Psi}}{\partial x_\mu}\Psi - \bar{\Psi}\frac{\partial\Psi}{\partial x_\mu}\right) - \frac{q_0}{c}\bar{\Psi}A_\mu\Psi\right] - \\ - \frac{ic\mu_0}{2}\frac{\partial}{\partial x_\nu}\left(\bar{\Psi}\left(\gamma_\mu\gamma_\nu - \gamma_\nu\gamma_\mu\right)\Psi\right). \tag{7.5}$$

It is seen that the current density four vector satisfies the continuity equation

$$\nabla_\mu j_\mu = 0.$$

The spatial component of the current density four vector is the current density

$$\mathbf{j}(\mathbf{r},t) = \frac{q_0}{m_0}\left[\frac{i\hbar}{2}\left(\nabla\bar{\Psi}\cdot\Psi - \bar{\Psi}\cdot\nabla\Psi\right) - \frac{q_0}{c}\bar{\Psi}\mathbf{A}\Psi\right] + \\ + c\mu_0\,\mathrm{curl}\left(\bar{\Psi}\boldsymbol{\Sigma}\Psi\right) - i\mu_0\frac{\partial}{\partial t}\left(\bar{\Psi}\boldsymbol{\alpha}\Psi\right), \tag{7.6}$$

and the time component is the charge density

$$\rho(\mathbf{r},t) = \frac{q_0}{m_0 c}\left[-\frac{i\hbar}{2c}\left(\frac{\partial\bar{\Psi}}{\partial t}\Psi - \bar{\Psi}\frac{\partial\Psi}{\partial t}\right) - \frac{q_0}{c}\bar{\Psi}\varphi\Psi\right] + i\mu_0\nabla\left(\bar{\Psi}\boldsymbol{\alpha}\Psi\right). \tag{7.7}$$

7.2 Connections with the Dirac equation

Let us use the following identity

$$\gamma_\mu P_\mu \gamma_\nu P_\nu = P_\mu^2 + \frac{1}{2}\gamma_\mu \gamma_\nu \left(P_\mu P_\nu - P_\nu P_\mu\right) = P_\mu^2 - \frac{iq\hbar}{2c}\gamma_\mu \gamma_\nu F_{\nu\mu},$$

where

$$P_\mu = p_\mu - \frac{q_0}{c} A_\mu = -i\hbar \frac{\partial}{\partial x_\mu} - \frac{q_0}{c} A_\mu .$$

With the help of this identity the equation (7.3) can be rewritten in the following form

$$\left(\gamma_\mu P_\mu + i m_0 c\right)\left(\gamma_\mu P_\mu - i m_0 c\right)\Psi = i m_0 \left(\mu_0 - \frac{q_0 \hbar}{2 m_0 c}\right)\gamma_\mu \gamma_\nu F_{\nu\mu}\Psi . \quad (7.8)$$

Let us assume now that the magneton, introduced in the equations (7.1), is equal to the Bohr magneton

$$\mu_0 = \mu_B = \frac{|e|\,\hbar}{2 m_0 c}. \quad (7.9)$$

In this case the equation (7.3) becomes

$$\left(\gamma_\mu P_\mu + i m_0 c\right)\left(\gamma_\mu P_\mu - i m_0 c\right)\Psi = 0. \quad (7.10)$$

Thus, we can see, that in the case, when the magneton is equal to the Bohr magneton, any solution of the Dirac equation

$$\left(\gamma_\mu \left(-i\hbar \frac{\partial}{\partial x_\mu} - \frac{q_0}{c} A_\mu\right) - i m_0 c\right)\Psi = 0 \quad (7.11)$$

is, at the same time, the solution of the equation (7.3). However, the opposite is not true, because the number of the linear independent solutions of the second order differential equation is twice larger.

If the assumption (7.9) is applied to the equation (7.5), the current density four vector takes the form

$$j_\mu = \frac{q_0}{2 m_0 c}\left\{\bar{\Psi}\gamma_\mu \cdot \left[\left(\gamma_\nu \left(-i\hbar \frac{\partial}{\partial x_\nu} - \frac{q_0}{c} A_\nu\right) - i m_0 c\right)\Psi\right] - \right.$$
$$\left. - \left[\bar{\Psi}\left(\gamma_\nu \left(-i\hbar \frac{\partial}{\partial x_\nu} + \frac{q_0}{c} A_\nu\right) - i m_0 c\right)\right] \cdot \gamma_\mu \Psi\right\}. \quad (7.12)$$

The equation (7.11) for the Dirac adjoint wave function reads

$$\bar{\Psi}\left(\gamma_\mu \left(-i\hbar \frac{\partial}{\partial x_\mu} + \frac{q_0}{c} A_\mu\right) + i m_0 c\right) = 0. \quad (7.13)$$

If we apply the equations (7.11) and (7.13) to the right-hand-side of the equation (7.12), then the equation for the current density four vector becomes

$$j_\mu^{(D)} = icq_0 \bar{\Psi} \gamma_\mu \Psi. \tag{7.14}$$

Thus, if a particle wave function obeys the Dirac equation (7.11), then the equation for the current density four vector (7.5) takes the form of the current density four vector in the Dirac theory:

$$j_\mu^{(D)} = (\mathbf{j},\, ic\rho) = \left(cq_0 \Psi^+ \boldsymbol{\alpha} \Psi,\, icq_0 \Psi^+ \Psi\right).$$

Thus, we can seen that in the case of $\mu_0 = \mu_{\mathrm{B}}$ (i.e. when the magneton is equal to the Bohr magneton), any solution of the Dirac equation is also a solution of the equation (7.3).

7.3 Symmetry properties with respect to orthogonal transformations

Let us study the symmetry properties of the equation (7.3) with respect to the orthogonal transformations

$$x'_\mu = a_{\mu\nu} x_\nu + a_\mu, \tag{7.15}$$

where the matrix $a_{\mu\nu}$ obeys the condition

$$a_{\mu\nu} a_{\mu\lambda} = \delta_{\nu\lambda}. \tag{7.16}$$

We have explored symmetries in the previous chapters, therefore we can exclude some specific cases. Firstly, there is no necessity to study the transformation properties of the equation (7.3) for the case of free particle, because the left-hand-side of this equation has evidently relativistic form. Hence, the transformation properties of the equation (7.3) for the free particle are completely determined by the transformation properties of the bispinor wave function Ψ. We have discussed them in the previous chapter. Secondly, there is no necessity to discuss the transformation properties with respect to the translation in spacetime, because they are the same for all quantum-mechanical equations, and we have discussed them in the previous chapters. Hence, we can assume $a_\mu = 0$ and consider further the following transformations

$$x'_\mu = a_{\mu\nu} x_\nu. \tag{7.17}$$

If the transformations (7.17) are applied to the equation (7.3) we get

$$\frac{1}{m_0} \left(P_\mu^2 + m_0^2 c^2\right) \Psi' \left(x'\right) = i\mu_0 \gamma_\mu \gamma_\nu a_{\mu\rho} a_{\nu\lambda} F_{\lambda\rho} \Psi' \left(x'\right). \tag{7.18}$$

As we have mentioned in the previous chapters, the invariance of the equation with respect to the orthogonal transformations means that there is a matrix S, defined by

$$\Psi'\left(x'\right) = S\Psi\left(x\right), \tag{7.19}$$

which transforms the equation (7.18) into the equation (7.3). By substituting the transformation (7.19) into (7.18) and then multiplying both sides of the obtained equation by S^{-1} from the left, we get the equation, which coincides with the equation (7.3), if the following condition holds:

$$S^{-1}\gamma_\mu\gamma_\nu S a_{\mu\rho} a_{\nu\lambda} = \gamma_\rho\gamma_\lambda. \tag{7.20}$$

By taking into account the orthonormality of matrices $a_{\mu\nu}$ (see (7.16)), we can rewrite the equation (7.20) in the form

$$S^{-1}\gamma_\mu\gamma_\nu S = a_{\mu\rho} a_{\nu\lambda}\gamma_\rho\gamma_\lambda. \tag{7.21}$$

In spite of the fact, that the obtained equations (7.20), (7.21) differ from the corresponding equations of the Dirac theory (see (6.24), (6.25)), the transformation properties of the equation (7.3) are quite predictable. Indeed, both equations (7.3) and (7.11) are the relativistic invariant equations. Hence, their transformation properties with respect to the three- and four-dimensional rotations will coincide, because these properties are determined by the symmetry properties of the wave function, i.e. the internal symmetry of the particle.

The equivalence of the symmetry properties of equations (7.3) and (7.11) with respect to the space inversion is not so evident. Indeed, the left-hand-side of the equation (7.3) is invariant with respect to the space inversion, but the right-hand-side of this equation includes the electromagnetic field tensor $F_{\mu\nu}$. Some components of the electromagnetic field tensor are projections of the polar vector, some of them are projections of the axial vector. However, the right-hand-side of the equation (7.3) is proportional to the product of the two antisymmetric tensors, therefore there is some reason to assume that the symmetry properties of the equations (7.3) and (7.11) with respect to the space inversion may be similar.

The time reversal and charge conjugation transformations require the separate consideration, because these transformations include the complex conjugation. In the frames of the Dirac theory formalism, the time reversal transformation is applied to show the symmetry of the free particle equations. In some sense, the time reversal plays an auxiliary role in comparison with the charge conjugation transformation. Indeed, the time reversal transformation changes sign only of the time derivative. The time-reversed equation can be returned to the initial form only with the help of the complex conjugation. So, the positive and negative energy solutions of the differential equations of the first order with respect to time derivative are, in general sense, the solutions of the different equations. As we have discussed in the previous chapter, the positive and negative energy solutions are associated with the particles and antiparticles, respectively. Hence, the particles and antiparticles are not completely equivalent in the frames of theories, based on the differential equations of the first order with respect to time derivative. Contrary, the particles and antiparticles are certainly equivalent in the frames of theories based on the equations of second order with respect to the time derivative. However, the real symmetry of particles and antiparticles can be revealed only with exploring of their interaction with the electromagnetic field, because the equations describing the interaction depend explicitly on the particle charge. At the charge conjugation transformation we change sign of the charge and make complex conjugation simultaneously, as a result the equation (7.11) as well as the left-hand-side of the equation (7.8) do not vary under this transformation. Hence, the transformation properties of the equation (7.3) with respect to charge conjugation are determined primarily by the transformation properties of the right-hand-side of the equation (7.8).

7.3.1 Space inversion

In the case of the space inversion transformation, the matrix $a_{\mu\nu}$ is

$$a_{\mu\nu} = \begin{pmatrix} -1 & 0 & 0 & 0 \\ 0 & -1 & 0 & 0 \\ 0 & 0 & -1 & 0 \\ 0 & 0 & 0 & 1 \end{pmatrix}.$$

By multiplying both sides of equation (7.20) by matrix S from the left, we get

$$\gamma_i\gamma_j S = S\gamma_i\gamma_j, \quad \gamma_i\gamma_4 S = -S\gamma_i\gamma_4, \quad \gamma_4^2 S = S\gamma_4^2, \tag{7.22}$$

where $i, j = 1, 2, 3$. It can be easily seen that the matrix

$$S_P = \lambda_P\gamma_4, \tag{7.23}$$

is the solution of the equations (7.22). In complete analogy with the case of the Dirac equation, there is a freedom in the choice of the value of constant λ_P. If we assume that the double space inversion transformation is the identical transformation, then we get $\lambda_P = \pm 1$. However, if we assume that the double space inversion transformation is equivalent to the rotation by the angle 2π, then we get $\lambda_P = \pm i$. In both cases $|\lambda_P| = 1$. Thus, the space inversion transformation of the wave function of the equation (7.3) is realized by the matrix γ_4.

7.3.2 Three-dimensional rotations

When we explore the continuous transformations, it is convenient to consider initially the infinitesimally small transformations

$$a_{\mu\nu} = \delta_{\mu\nu} + \varepsilon_{\mu\nu}, \tag{7.24}$$

where $\varepsilon_{\mu\nu}$ is the infinitesimally small tensor of the second rank. As far as the matrix $a_{\mu\nu}$ obeys the condition (7.16), then the tensor $\varepsilon_{\mu\nu}$ is completely antisymmetric tensor

$$\varepsilon_{\mu\nu} = -\varepsilon_{\nu\mu}. \tag{7.25}$$

At the infinitesimally small transformations (7.24), the transformation matrix S differs from the identity matrix by a small component proportional to the tensor $\varepsilon_{\mu\nu}$:

$$S_{\mu\nu} = \delta_{\mu\nu} + \frac{1}{2}C^{\alpha\beta}_{\mu\nu}\varepsilon_{\alpha\beta}. \tag{7.26}$$

By substituting the equation (7.26) into the equation (7.21), we get

$$\frac{1}{2}\left(\gamma_\mu\gamma_\nu C^{\alpha\beta} - C^{\alpha\beta}\gamma_\mu\gamma_\nu\right)\varepsilon_{\alpha\beta} = \gamma_\mu\gamma_\beta\varepsilon_{\nu\beta} + \gamma_\beta\gamma_\nu\varepsilon_{\mu\beta}$$

or

$$\left[\frac{1}{2}\left(\gamma_\mu\gamma_\nu C^{\alpha\beta} - C^{\alpha\beta}\gamma_\mu\gamma_\nu\right) - \delta_{\alpha\nu}\gamma_\mu\gamma_\beta - \delta_{\alpha\mu}\gamma_\beta\gamma_\nu\right]\varepsilon_{\alpha\beta} = 0. \tag{7.27}$$

It is seen from the structure of the last equation, that its solution should have the form $C^{\alpha\beta} = \lambda\gamma_\alpha\gamma_\beta$, where λ is the constant. By substituting this expression into the equation (7.27), we get

$$\begin{aligned}
&[(\lambda-1)(\gamma_\mu\gamma_\beta\delta_{\alpha\nu} + \gamma_\beta\gamma_\nu\delta_{\alpha\mu}) - \lambda(\gamma_\mu\gamma_\alpha\delta_{\beta\nu} + \gamma_\alpha\gamma_\nu\delta_{\beta\mu})]\varepsilon_{\alpha\beta} = \\
&= (\lambda-1)(\gamma_\mu\gamma_\beta\delta_{\alpha\nu} + \gamma_\beta\gamma_\nu\delta_{\alpha\mu})\varepsilon_{\alpha\beta} - \lambda(\gamma_\mu\gamma_\beta\delta_{\alpha\nu} + \gamma_\beta\gamma_\nu\delta_{\alpha\mu})\varepsilon_{\beta\alpha} = \\
&= (2\lambda-1)(\gamma_\mu\gamma_\beta\delta_{\alpha\nu} + \gamma_\beta\gamma_\nu\delta_{\alpha\mu})\varepsilon_{\alpha\beta} = 0.
\end{aligned}$$

Thus, we get finally

$$C^{\alpha\beta} = \frac{1}{2}\gamma_\alpha\gamma_\beta. \tag{7.28}$$

The three-dimensional rotations touche only the spatial components of the four-vectors. Therefore, by taking into account the following properties of the matrices γ_1, γ_2, γ_3:

$$\gamma_i\gamma_j = ie_{ijk}\Sigma_k, \tag{7.29}$$

for the matrix of the three-dimensional rotations we get

$$S_R(\delta\boldsymbol{\theta}) = I + \frac{i}{2}\delta\theta\mathbf{n}\boldsymbol{\Sigma}, \tag{7.30}$$

where $\mathbf{n}$ is the unit vector of the rotation axis, and the matrix $\boldsymbol{\Sigma}$ is defined by the equation (6.33). Thus, in both cases of the equation (7.3) and the Dirac equation, the three-dimensional rotations are realized by the same matrix. Therefore, it can be easily understood, that the transformation matrix for the finite angle of rotation will coincide with that given by equation (6.36):

$$S_R = \exp\left(\frac{i}{2}\boldsymbol{\theta}\boldsymbol{\Sigma}\right) = \cos\frac{\theta}{2} + i\left(\mathbf{n}\boldsymbol{\Sigma}\right)\sin\frac{\theta}{2}. \tag{7.31}$$

7.3.3 Lorentz transformation

The Lorentz transformation, or the four-dimensional rotation, is also continuous transformation. Hence, to find the matrix of this transformation we can again use the equations (7.26) and (7.28). The matrix $\varepsilon_{\mu\nu}$ of the Lorentz transformation to the new reference frame, moving along the x axis with the velocity δv, is

$$\varepsilon_{\mu\nu} = \begin{pmatrix} 0 & 0 & 0 & \delta\varphi \\ 0 & 0 & 0 & 0 \\ 0 & 0 & 0 & 0 \\ -\delta\varphi & 0 & 0 & 0 \end{pmatrix}$$

where $\delta\varphi = i\delta v/c$. By taking into account the equalities

$$\gamma_1\gamma_4 = -\gamma_4\gamma_1 = i\alpha_1,$$

we get

$$S_x(\delta v) = I - \frac{1}{2}\frac{\delta v}{c}\alpha_1. \tag{7.32}$$

By combining the similar transformations touching other spatial axes, we get the following vectorial equation

$$S_L(\delta\mathbf{v}) = I - \frac{1}{2}\frac{\delta\mathbf{v}}{c}\boldsymbol{\alpha}. \tag{7.33}$$

Thus, we can see again, that in both cases of the equation (7.3) and Dirac equation, the Lorentz transformation is realized by the same matrix. Hence, at the arbitrary finite velocity $\mathbf{v} = \mathbf{n}v$, the matrix of the Lorentz transformation is

$$S_L(\mathbf{v}) = \cosh\left(\frac{1}{2}\tanh^{-1}\frac{v}{c}\right) - (\mathbf{n}\boldsymbol{\alpha})\sinh\left(\frac{1}{2}\tanh^{-1}\frac{v}{c}\right). \tag{7.34}$$

7.3.4 Time reversal

As already mentioned, the time reversal transformation establishes the connection between the positive and negative energy solutions. Therefore we should compare the equation for bispinor wave function $\Psi(\mathbf{r},t)$ and Dirac adjoint wave function $\bar{\Psi}(\mathbf{r},t)$. It is convenient to rewrite the equation (7.3) in the following identical form

$$\left[\left(-i\hbar\nabla - \frac{q_0}{c}\mathbf{A}\right)^2 - \frac{1}{c^2}\left(i\hbar\frac{\partial}{\partial t} - q_0\varphi\right)^2 + \right. \\ \left. + m_0^2c^2 + 2m_0\mu_0\left(i\boldsymbol{\alpha}\mathbf{E} - \boldsymbol{\Sigma}\mathbf{B}\right)\right]\Psi = 0. \quad (7.35)$$

The equation for the Dirac adjoint wave function $\bar{\Psi}(\mathbf{r},t)$ is

$$\bar{\Psi}(\mathbf{r},t)\left[\left(i\hbar\nabla - \frac{q_0}{c}\mathbf{A}\right)^2 - \frac{1}{c^2}\left(i\hbar\frac{\partial}{\partial t} + q_0\varphi\right)^2 + \right. \\ \left. + m_0^2c^2 + 2m_0\mu_0\left(i\boldsymbol{\alpha}\mathbf{E} - \boldsymbol{\Sigma}\mathbf{B}\right)\right] = 0. \quad (7.36)$$

The time reversal transformation is usually applied to the free-particle equation. However, the free-particle equation (7.3) is evidently invariant with respect to time reversal. Therefore, it is useful to generalize the transformation and to consider the equation for a particle interacting with the electromagnetic field. Let us use the following notations

$$p_\mu(\mathbf{r},t) = \left(-i\hbar\nabla, -\frac{\hbar}{c}\frac{\partial}{\partial t}\right) = -\left(i\hbar\nabla, \frac{\hbar}{c}\frac{\partial}{\partial t}\right) = -p_\mu^*(\mathbf{r},-t).$$

It is seen, that the same equalities are valid for the generalized four-momentum

$$P_\mu(\mathbf{r},t) = \left(-i\hbar\nabla - \frac{q_0}{c}\mathbf{A}(\mathbf{r},t), -\frac{\hbar}{c}\frac{\partial}{\partial t} - i\frac{q_0}{c}\varphi(\mathbf{r},t)\right) = -P_\mu^*(\mathbf{r},-t),$$

when

$$\mathbf{A}(\mathbf{r},-t) = -\mathbf{A}(\mathbf{r},t), \quad \varphi(\mathbf{r},-t) = \varphi(\mathbf{r},t). \quad (7.37)$$

Hence, the four-momentum p_μ and generalized four-momentum P_μ are transformed in the same way, if the four-potential of electromagnetic field is transformed under time reversal according to (7.37). As we have mentioned in subsection 3.2.5, the classical particle makes the time reversal motion in the case when the following conditions hold

$$\mathbf{E}(\mathbf{r},-t) = \mathbf{E}(\mathbf{r},t), \quad \mathbf{B}(\mathbf{r},-t) = -\mathbf{B}(\mathbf{r},t). \quad (7.38)$$

It is seen that the conditions (7.38) are identical to the conditions (7.37).

By using the equations (7.37), we can transform the equation (7.36) to the following form

$$\left[\left(-i\hbar\nabla - \frac{q}{c}\mathbf{A}\right)^2 - \frac{1}{c^2}\left(i\hbar\frac{\partial}{\partial t} - q\varphi\right)^2 + \right.$$
$$\left. + m_0^2c^2 + 2m_0\mu\left(i\tilde{\boldsymbol{\alpha}}\mathbf{E} + \tilde{\boldsymbol{\Sigma}}\mathbf{B}\right)\right]\tilde{\bar{\Psi}}(\mathbf{r}, -t) = 0, \quad (7.39)$$

where tilde matrix is the transposed matrix. Thus, the transformation matrix, defined by

$$\tilde{\bar{\Psi}}(-t) = S_T\Psi(t), \quad (7.40)$$

should satisfy the following equations

$$S_T^{-1}\tilde{\boldsymbol{\alpha}}S_T = \boldsymbol{\alpha}, \quad S_T^{-1}\tilde{\boldsymbol{\Sigma}}S_T = -\boldsymbol{\Sigma}. \quad (7.41)$$

The solution of the equations (7.41) is

$$S_T = \lambda_T\gamma_4\alpha_3\alpha_1 = \lambda_T\gamma_4\gamma_3\gamma_1, \quad (7.42)$$

where, as above, we have used the standard representation of the matrices γ_μ. By taking into account that the double time reversal transformation is identical transformation, we get $|\lambda_T|^2 = 1$.

7.3.5 Charge conjugation

The charge conjugation transformation establishes the connection between the solutions of the equation (7.3) and equation, obtained from it, with the help of the following replacement: $q_0 \to -q_0$ and $\mu_0 \to -\mu_0$. Thus, the wave function of the charge conjugated particle is a solution of the following equation

$$\left[\left(i\hbar\nabla - \frac{q_0}{c}\mathbf{A}\right)^2 - \frac{1}{c^2}\left(i\hbar\frac{\partial}{\partial t} + q_0\varphi\right)^2 + \right.$$
$$\left. + m_0^2c^2 - 2m_0\mu_0\left(i\boldsymbol{\alpha}\mathbf{E} - \boldsymbol{\Sigma}\mathbf{B}\right)\right]\Psi_C = 0. \quad (7.43)$$

We should find the matrix transforming the transposed equation (7.36) into the equation (7.43), i.e.

$$\Psi_C = S_C\tilde{\bar{\Psi}},$$

It is easily seen from the comparison of equations (7.36) and (7.43), that the transformation matrix should satisfy the following equations

$$S_C\tilde{\boldsymbol{\alpha}}S_C^{-1} = -\boldsymbol{\alpha}, \quad S_C\tilde{\boldsymbol{\Sigma}}S_C^{-1} = -\boldsymbol{\Sigma}. \quad (7.44)$$

In the standard representation, there are the following relationships between the transposed and direct matrices: $\tilde{\alpha}_{1,3} = \alpha_{1,3}$, $\tilde{\alpha}_2 = -\alpha_2$, $\tilde{\Sigma}_{1,3} = \Sigma_{1,3}$, and $\tilde{\Sigma}_2 = -\Sigma_2$. With the help of these relationships, the solution of the equations (7.44) can be easily found. It is

$$S_C = \lambda_C \alpha_2 = i\lambda_C \gamma_4 \gamma_2, \tag{7.45}$$

where $|\lambda_C|^2 = 1$.

7.3.6 *CPT* invariance

The combined transformations of the time reversal, space inversion, and charge conjugation can be written in the following way:

a) T-transformation

$$S_T \Psi(\mathbf{r}, t) = \tilde{\bar{\Psi}}(\mathbf{r}, -t),$$

b) PT-transformation

$$S_P S_T \Psi(\mathbf{r}, t) = \tilde{\bar{\Psi}}(-\mathbf{r}, -t),$$

c) CPT-transformation

$$S_C S_P S_T \Psi(\mathbf{r}, t) = \Psi_C(-\mathbf{r}, -t).$$

Hence, the combined transformation is

$$S_C S_P S_T \Psi(\mathbf{r}, t) = -i\lambda_C \lambda_P \lambda_T \gamma_1 \gamma_2 \gamma_3 \gamma_4 \Psi(\mathbf{r}, t) = \Psi_C(-\mathbf{r}, -t). \tag{7.46}$$

It is seen, that the combined CPT-transformation is realized by the matrix

$$\gamma_5 = \gamma_1 \gamma_2 \gamma_3 \gamma_4 = -\begin{pmatrix} 0 & I \\ I & 0 \end{pmatrix}, \tag{7.47}$$

this matrix coincides with the matrix realizing the CPT-transformation of the Dirac wave function. The difference between the transformations (7.46) and (6.58) can only be in the choice of the coefficients in these equations. In the equation (7.46) the coefficient is equal to $-i\lambda_C \lambda_P \lambda_T$. The exact value of this coefficient depends on the internal symmetry of a particle. The internal symmetry of a particle describing by the equation (7.3) will be discussed below.

As it is seen from the equation (7.46), the CPT-invariance provides a precise correspondence between the particle motion and reversed in time and space motion of antiparticle. The matrix of the CPT-transformation, γ_5, satisfies the anticommutation relations

$$\gamma_5 \gamma_\mu + \gamma_\mu \gamma_5 = 0, \tag{7.48}$$

where $\mu = 1, 2, 3, 4$. As a result, it does not commute with the Hamiltonian of the Dirac equation. It is not surprised, because the particle and antiparticle obey the different equations in the frame of the Dirac theory. Indeed, the Dirac equation for the wave function Ψ_D is

$$(\gamma_\mu P_\mu - im_0 c)\,\Psi_D = 0.$$

Operating on this equation with matrix γ_5, we get

$$-\left(\gamma_\mu P_\mu + im_0 c\right)\left(\gamma_5 \Psi_D\right) = 0.$$

However, the equation

$$(\gamma_\mu P_\mu + im_0 c)\,\Psi_C = 0$$

can, in principle, have the solutions, which do not coincide with the solution given by $-\gamma_5 \Psi_D(\mathbf{r}, t)$.

Contrary, the matrix of the CPT-transformation, γ_5, commutes with the Hamiltonian of the equation (7.3). We shall see in the next chapters, that it is this difference between the Dirac equation and equation (7.3), which results in the crucial difference between the solutions of these two equations.

7.4 Wave function normalization condition

The wave function normalization condition is unambiguously determined by the continuity equation

$$\frac{\partial \rho}{\partial t} + \operatorname{div} \mathbf{j} = 0, \tag{7.49}$$

where the current $\mathbf{j}$ and charge ρ density are defined by the equations (7.6) and (7.7), respectively. Integrating the equation (7.49) over the volume V, we get

$$\frac{d}{dt}\int_V \rho(\mathbf{r}, t)\, dV = -\oint_S \mathbf{j}(\mathbf{r}, t)\, d\mathbf{S},$$

where S is the boundary surface of the volume V.

If the initial and final states of the particle interacting with the external fields satisfy the boundary condition

$$\Psi(\mathbf{r})|_{r\to\infty} = 0, \tag{7.50}$$

then the equation (7.49) yields

$$\int_V \rho(\mathbf{r}, t) = \text{const}, \tag{7.51}$$

where the charge density is integrated over the infinite volume ($V \to \infty$).

There are at least the two reasons, indicating that the space integral of the charge density is not definitely positive defined value. Firstly, the time derivative $\partial\Psi/\partial t$ may be both positive and negative, and, in principle, it can change sign in the process of evolution of the particle state. Secondly, the product $\bar{\Psi}\Psi$ for the case of the bispinor wave function $\Psi = \begin{pmatrix} \varphi \\ \chi \end{pmatrix}$ is

$$\bar{\Psi}(\mathbf{r}, t)\,\Psi(\mathbf{r}, t) = \varphi^{+}(\mathbf{r}, t)\,\varphi(\mathbf{r}, t) - \chi^{+}(\mathbf{r}, t)\,\chi(\mathbf{r}, t). \tag{7.52}$$

Therefore, it is seen, that the space integral of the function (7.52) is not definitely positive defined value.

Exploring the Klein–Gordon–Fock equation, we have mentioned that the problem of the positivity condition, $\int \rho(\mathbf{r}, t)\, dV > 0$, can be eliminated by a proper choice of the sign of a particle charge. But, as we have mentioned above, the Hamiltonian of the equation (7.3) commutes with the operator γ_5. Hence, if the wave function Ψ is a solution of the equation (7.3), then the wave function $\Psi' = \gamma_5\Psi$ is also a solution of this equation. However, it is seen from the equation (7.52), that $\bar{\Psi}'\Psi' = -\bar{\Psi}\Psi$. We shall see later, that the sign of $\int \rho(\mathbf{r}, t)\, dV$ is the fundamental characteristic of the solutions of equation (7.3), which provides the invariant definition of the particle and antiparticle states.

As we have discussed above, it is assumed, in the frames of the quantum field theory, that the positive energy solutions

$$\Psi_p(\mathbf{r}, t) = \Psi(\mathbf{r}) \exp\left(-i\frac{Et}{\hbar}\right), \tag{7.53}$$

corresponds to particles, and the negative energy solutions

$$\Psi_a(\mathbf{r}, t) = \Psi(\mathbf{r}) \exp\left(i\frac{Et}{\hbar}\right) \tag{7.54}$$

corresponds to antiparticles. However, in the frames of the Dirac theory, it is impossible to differ the particle state (7.53) from the antiparticle state (7.54), because the Dirac equation is the first order differential equation with respect to the time derivative. Hence, in order to define unambiguously the initial state we need only in the initial value of the wave function. Contrary, the charge density $\rho(\mathbf{r}, t)$, defined by the equation (7.6), depends not only on the initial value of the wave function $\Psi(\mathbf{r}, 0)$, but on the value of the time derivative $\partial\Psi(\mathbf{r}, 0)/\partial t$, too. Therefore, in the frames of the theory based on the equation (7.3), the initial value of the charge density will be different for particle and antiparticle even in the case, when the particle and antiparticle wave functions differ only in the sign of energy.

The wave function of particle, which is in the stationary state of energy E_n, is

$$\Psi(\mathbf{r},t) = \Psi_n(\mathbf{r}) \exp\left(-i\frac{E_n t}{\hbar}\right). \tag{7.55}$$

If the state of a particle is the bound state, then the wave function satisfies the boundary condition (7.50). In this case the equation (7.51) reads

$$\frac{q_0}{m_0 c^2} \int\limits_V \bar{\Psi}_n(\mathbf{r}) \left(E_n - q_0 \varphi(\mathbf{r})\right) \Psi_n(\mathbf{r})\, dV = \text{const}, \tag{7.56}$$

where $V \to \infty$.

Thus the normalization condition can be written in the following form

$$\frac{1}{q_0} \int \rho(\mathbf{r})\, dV = \frac{1}{m_0 c^2} \int \bar{\Psi}_n(\mathbf{r}) \left(E_n - q\varphi(\mathbf{r})\right) \Psi_n(\mathbf{r})\, dV = \pm 1. \tag{7.57}$$

It is seen that the bound state normalization condition (7.57) means that the charge $\int \rho\, dV$ is equal to $+q_0$ or $-q_0$, where q_0 is the elementary electric charge appearing in the equation for action (7.1). We can always assume, that the condition

$$\int \rho(\mathbf{r})\, dV = q_0 \tag{7.58}$$

corresponds to particle, and the condition

$$\int \rho(\mathbf{r})\, dV = -q_0 \tag{7.59}$$

corresponds to antiparticle.

It should be noted that the normalization conditions (7.58) and (7.59) do not impose any restrictions. As already mentioned, the Hamiltonian of the equation (7.3) commutes with the operator γ_5, it means that the particle and antiparticle possess the equivalent properties in the frames of the theory based on the equation (7.3).

7.5 Plane waves

The momentum operator commutes with the free-particle Hamiltonian of the equation (7.3), therefore the free-particle wave function reads

$$\Psi_{E,p}(\mathbf{r},t) = u_{\mathbf{p}}^{(E)} \exp\left[-i\frac{E_p t - \mathbf{p}\mathbf{r}}{\hbar}\right]. \tag{7.60}$$

Substituting this wave function into the equation (7.3), we get

$$E_p = \pm\Gamma_p = \pm\sqrt{m_0^2 c^4 + p^2 c^2}. \tag{7.61}$$

Hence, the linear independent solutions can be taken in the form

$$\Psi^{(1)}_{\pm\mathbf{p}} = \begin{pmatrix} \varphi_+ \\ 0 \end{pmatrix} \exp\left[-i\frac{\Gamma_p t \mp \mathbf{pr}}{\hbar}\right], \quad \Psi^{(2)}_{\pm\mathbf{p}} = \begin{pmatrix} 0 \\ \chi_+ \end{pmatrix} \exp\left[-i\frac{\Gamma_p t \mp \mathbf{pr}}{\hbar}\right],$$
$$\Psi^{(3)}_{\pm\mathbf{p}} = \begin{pmatrix} 0 \\ \chi_- \end{pmatrix} \exp\left[i\frac{\Gamma_p t \mp \mathbf{pr}}{\hbar}\right], \quad \Psi^{(4)}_{\pm\mathbf{p}} = \begin{pmatrix} \varphi_- \\ 0 \end{pmatrix} \exp\left[i\frac{\Gamma_p t \mp \mathbf{pr}}{\hbar}\right]. \tag{7.62}$$

As we have mentioned above, the dimension of the phase space of the equation (7.3) is doubled with respect to that of the Dirac equation. Indeed, the equation (7.3) is the second order differential equation both in time and space, and the wave function is the bispinor. Hence, there are the eight linear independent solutions of the equation (7.3) in general case. To label the free particle states we can use the energy, momentum, and charge. The charge of particle in the states (7.62) is defined by

$$\begin{aligned} \frac{q_1}{q_0} &= \frac{\Gamma_p}{m_0c^2}\int \varphi_+^+\varphi_+\,dV > 0, \quad \frac{q_2}{q_0} = -\frac{\Gamma_p}{m_0c^2}\int \chi_+^+\chi_+\,dV < 0, \\ \frac{q_3}{q_0} &= \frac{\Gamma_p}{m_0c^2}\int \chi_-^+\chi_-\,dV > 0, \quad \frac{q_4}{q_0} = -\frac{\Gamma_p}{m_0c^2}\int \varphi_-^+\varphi_-\,dV < 0. \end{aligned} \tag{7.63}$$

The four states, $\Psi^{(1,3)}_{\pm\mathbf{p}}$, correspond to the positively charged particle, and the four states, $\Psi^{(2,4)}_{\pm\mathbf{p}}$, correspond to the negatively charged particle.

Thus, at a given value of energy Γ_p, the eight linear independent free-particle solutions of the equation (7.3) may be classified according to the values of the following three binary quantum numbers: energy $E_p/\Gamma_p = \pm 1$, momentum $\pm\mathbf{p}/|\mathbf{p}| = \pm 1$, and charge $q/q_0 = \pm 1$.

The normalization condition of the free-particle wave functions is completely similar to that of the Schrödinger equation:

$$\frac{(E_{\mathbf{p}} + E_{\mathbf{p}'})}{2m_0c^2}\int \bar{\Psi}_{\mathbf{p}}\Psi_{\mathbf{p}'}\,dV = \pm(2\pi\hbar)^3\,\delta\left(\mathbf{p}-\mathbf{p}'\right) \tag{7.64}$$

In this case, the charge of the particle is defined by the sign in the right-hand-side of the last equation, $q = \pm q_0$. The values of the energy E_p/Γ_p, momentum projection $\mathbf{e}^{(+)}\mathbf{p}$, where $\mathbf{e}^{(+)} = +\mathbf{p}/p$, and charge q/q_0 for the eight linear independent states of free particle is shown in Table 7.1.

It is seen from the equations (7.62) that the positive energy solutions, $E_p = \Gamma_p$, for particle $\Psi^{(1)}_{\pm\mathbf{p}}$ and antiparticle $\Psi^{(2)}_{\pm\mathbf{p}}$ are orthogonal. The negative energy solutions, $E_p = -\Gamma_p$, for particle $\Psi^{(3)}_{\pm\mathbf{p}}$ and antiparticle $\Psi^{(4)}_{\pm\mathbf{p}}$ are also orthogonal. The solutions $\Psi^{(i)}_{\pm\mathbf{p}}$ are orthogonal due to

Table 7.1. Classification of linear independent free-particle solutions

	$\Psi_{+\mathbf{p}}^{(1)}$	$\Psi_{+\mathbf{p}}^{(2)}$	$\Psi_{+\mathbf{p}}^{(3)}$	$\Psi_{+\mathbf{p}}^{(4)}$	$\Psi_{-\mathbf{p}}^{(1)}$	$\Psi_{-\mathbf{p}}^{(2)}$	$\Psi_{-\mathbf{p}}^{(3)}$	$\Psi_{-\mathbf{p}}^{(4)}$
E_p/Γ_p	+1	+1	−1	−1	+1	+1	−1	−1
$\mathbf{e}^{(+)}\mathbf{p}/p$	+1	+1	+1	+1	−1	−1	−1	−1
q/q_0	+1	−1	+1	−1	+1	−1	+1	−1

normalization condition (7.64). Thus, the general solution for the free spin-1/2 particle has the following form

$$\Psi(\mathbf{r},t) = \sum_{\mathbf{p}} \sum_{\lambda_E=\pm 1} \left(A_{\mathbf{p},\lambda_E} u_{\lambda_E} + B_{\mathbf{p},\lambda_E} v_{\lambda_E}\right) \exp\left(-i\lambda_E \frac{\Gamma_p t - \mathbf{pr}}{\hbar}\right), \tag{7.65}$$

where $\lambda_E = E_p/\Gamma_p = \pm 1$, and the bispinors u and v are

$$\begin{aligned} u_{\lambda_E=+1} &= \begin{pmatrix} \varphi_+ \\ 0 \end{pmatrix}, \quad u_{\lambda_E=-1} = \begin{pmatrix} 0 \\ \chi_- \end{pmatrix}, \\ v_{\lambda_E=+1} &= \begin{pmatrix} 0 \\ \chi_+ \end{pmatrix}, \quad v_{\lambda_E=-1} = \begin{pmatrix} \varphi_- \\ 0 \end{pmatrix}. \end{aligned} \tag{7.66}$$

The general solution depends on the two pairs of the arbitrary three-dimensional spinors $\varphi_\pm$ and $\chi_\pm$. These spinors can be chosen in the following way. The Hamiltonian of the equation (7.3) commutes with the operators of momentum, angular momentum, spin, and helicity. If the direction of the momentum coincides with the direction of the z axis of the spin state reference frame, then the eigenfunctions of the equation

$$\sigma_z w^{(\sigma)} = \sigma w^{(\sigma)},$$

can be taken as the basis spinors. They are

$$w^{(\sigma=+1)} = \begin{pmatrix} 1 \\ 0 \end{pmatrix}, \quad w^{(\sigma=-1)} = \begin{pmatrix} 0 \\ 1 \end{pmatrix}. \tag{7.67}$$

In general case, when the particle moves in the direction determined by the angles θ and φ in the spin state reference frame, the basis spinors are the eigenfunctions of the equation

$$(\boldsymbol{\sigma}\mathbf{n})\, w^{(\sigma)} = \sigma w^{(\sigma)},$$

which are

$$w_{\mathbf{p}}^{(\sigma=+1)} = \begin{pmatrix} \exp\left(-i\frac{\varphi}{2}\right)\cos\frac{\theta}{2} \\ \exp\left(i\frac{\varphi}{2}\right)\sin\frac{\theta}{2} \end{pmatrix}, \quad w_{\mathbf{p}}^{(\sigma=-1)} = \begin{pmatrix} -\exp\left(-i\frac{\varphi}{2}\right)\sin\frac{\theta}{2} \\ \exp\left(i\frac{\varphi}{2}\right)\cos\frac{\theta}{2} \end{pmatrix}, \tag{7.68}$$

In the latter case, the direction of the momentum $-\mathbf{p}$ is determined by the angles $\theta' = \pi - \theta$ and $\varphi' = \varphi + \pi$, hence

$$w_{-\mathbf{p}}^{(\sigma=+1)} = iw_{\mathbf{p}}^{(\sigma=-1)}, \quad w_{-\mathbf{p}}^{(\sigma=-1)} = iw_{\mathbf{p}}^{(\sigma=+1)}. \tag{7.69}$$

At $\theta = 0$ and $\varphi = 0$, the equations (7.68) and (7.67) coincide, therefore, in general case, we can assume

$$\varphi_+ = w_{\mathbf{p}}^{(\pm 1)}, \quad \varphi_- = w_{\mathbf{p}}^{(\mp 1)}. \tag{7.70}$$

7.5.1 Particle-antiparticle transformation

Let us compare the CPT-transformation of the Dirac equation and equation (7.3). As it was shown in the previous chapter, the CPT-transformation of the Dirac equation is realized by the matrix $\gamma_5 = \gamma_1\gamma_2\gamma_3\gamma_4$ (see (6.58), (6.59)). The matrix γ_5 anticommutes with matrices γ_μ

$$\gamma_5\gamma_\mu + \gamma_\mu\gamma_5 = 0, \tag{7.71}$$

where $\mu = 1, 2, 3, 4$. As a result, the matrix γ_5 does not commute with the Hamiltonian of the Dirac equation. Indeed, let the wave function Ψ_p obey the Dirac equation

$$\left(\gamma_\mu p_\mu - im_0 c\right)\Psi_p = 0. \tag{7.72}$$

The current density four-vector of this equation is

$$j_\mu^{(p)} = iq_0 c\bar{\Psi}_p\gamma_\mu\Psi_p. \tag{7.73}$$

Let us introduce the wave function

$$\Psi_a\left(\mathbf{r}, t\right) = \gamma_5\Psi_p\left(\mathbf{r}, t\right). \tag{7.74}$$

The Dirac adjoint wave function is

$$\bar{\Psi}_a = -\bar{\Psi}_p\gamma_5. \tag{7.75}$$

By taking into account the commutation relations (7.71), we can see that the wave function Ψ_a obeys the equation

$$\left(\gamma_\mu p_\mu + im_0 c\right)\Psi_a = 0. \tag{7.76}$$

The current density four-vector, corresponding to the last equation, is

$$j_\mu^{(a)} = iq_0 c\bar{\Psi}_a\gamma_\mu\Psi_a. \tag{7.77}$$

The current density four-vectors (7.73) and (7.77) coincide completely. Indeed, with the help of equations (7.74) and (7.75), we get

$$j_\mu^{(a)} = iq_0 c\bar{\Psi}_a\gamma_\mu\Psi_a = -iq_0 c\bar{\Psi}_p\gamma_5\gamma_\mu\gamma_5\Psi_p = j_\mu^{(p)}.$$

On the other hand, the solution of the equation (7.76) can be written as follows $\Psi_a(\mathbf{r},t) = \lambda\Psi_p(-\mathbf{r},-t)$, where λ is the constant. Hence, the equation (7.76) describes the time-reversed and space-inverted particle motion, remaining invariable the current density four-vector. If, simultaneously with the transformation (7.74), we change the sign of charge $q_0 \to -q_0$, then the current density four-vector will change sign too.

As we have seen above, the CPT-transformation of the wave function of the equation (7.3) is also realized by the matrix γ_5 (see (7.46), (7.47)). However, the operator γ_5 commutes with the Hamiltonian of the equation (7.3). As a result, if the wave function Ψ_p is the solution of the equation (7.3), then the wave function Ψ_a is also the solution of the equation (7.3). With the help of matrix γ_5, the positive energy solutions (7.62) of the equation (7.3) can be written as follows

$$\Psi^{(2)}_{\pm\mathbf{p}} = \gamma_5\Psi^{(1)}_{\pm\mathbf{p}}. \tag{7.78}$$

The similar relation holds for the negative energy solutions (7.62).

We have already shown, that the free-particle wave functions Ψ_p and $\Psi_a = \gamma_5\Psi_p$ correspond to the oppositely charged particles. By substituting the wave functions (7.74) and (7.75) into the equation for the current density four-vector (7.5), we get

$$j^{(a)}_\mu = -j^{(p)}_\mu. \tag{7.79}$$

Thus, the wave functions Ψ_p and Ψ_a correspond to the spin-1/2 particles, that have the opposite charges and opposite magnetic moments, i.e. they correspond to the particle and antiparticle. So, we can see the principle difference between the physical meaning, attributed to the wave functions Ψ_p and $\Psi_a = \gamma_5\Psi_p$ in the frames of the Dirac theory and theory based on the equation (7.3).

7.5.2 Space inversion, three-dimensional rotation, Lorentz transformation, and time reversal

The relativistic parity operator is defined by

$$\hat{P} = \gamma_4 P_3, \tag{7.80}$$

where P_3 is the three-dimensional space inversion operator acting as follows: $P_3 f(\mathbf{r}) = f(-\mathbf{r})$.

The wave functions of the even $\Psi^{(+)}_p$ and odd $\Psi^{(-)}_p$ states are the eigenfunctions of the following equation

$$\hat{P}\Psi^{(\pm)}_p(\mathbf{r},t) = \gamma_4\Psi^{(\pm)}_p(-\mathbf{r},t) = \pm\Psi^{(\pm)}_p(\mathbf{r},t).$$

If $\Psi^{(\pm)}_p$ is the particle wave function, then the antiparticle wave function is $\Psi_a^{(\pm)} = \gamma_5 \Psi_p^{(\pm)}$. By applying the operator $\hat{P}$ to the wave function $\Psi_a^{(\pm)}$ we get

$$\hat{P}\big(\gamma_5 \Psi_p^{(\pm)}(\mathbf{r},t)\big) = \gamma_4 \gamma_5 \Psi_p^{(\pm)}(-\mathbf{r},t) = -\gamma_5 \hat{P} \Psi_p^{(\pm)}(\mathbf{r},t) = \mp \gamma_5 \Psi_p^{(\pm)}(\mathbf{r},t). \tag{7.81}$$

Thus, the particle and antiparticle wave functions have the opposite parity.

The operators of the three-dimensional rotation S_R and Lorentz transformation S_L are defined by

$$S_R = \cos\frac{\theta}{2} + i\,(\mathbf{n}\boldsymbol{\Sigma})\sin\frac{\theta}{2},$$

$$S_L = \cosh\left(\frac{1}{2}\tanh^{-1}\frac{v}{c}\right) - (\mathbf{n}\boldsymbol{\alpha})\sinh\left(\frac{1}{2}\tanh^{-1}\frac{v}{c}\right).$$

As far as $\Sigma_i = -\frac{i}{2} e_{ijk}\gamma_j\gamma_k$ and $\boldsymbol{\alpha} = i\gamma_4\boldsymbol{\gamma}$, then the operator γ_5 commutes with S_R and S_L. Hence, the particle and antiparticle have the same transformation properties with respect to the three-dimensional rotations and Lorentz transformation.

The time reversal transformation is

$$\hat{T}\Psi(\mathbf{r},t) = \lambda_T^* \gamma_1\gamma_3 \Psi^*(\mathbf{r},-t). \tag{7.82}$$

By applying the transformation (7.82) to the wave function

$$\Psi^{(1)}_{\mathbf{p},\sigma=\pm1} = \begin{pmatrix} w_{\mathbf{p}}^{(\sigma=\pm1)} \\ 0 \end{pmatrix} \exp\left(-i\frac{\Gamma t - \mathbf{pr}}{\hbar}\right), \tag{7.83}$$

we get

$$\hat{T}\Psi^{(1)}_{\mathbf{p},\sigma=\pm1}(\mathbf{r},t) = \mp i\lambda_T^* \Psi^{(1)}_{-\mathbf{p},\sigma=\pm1}(\mathbf{r},t). \tag{7.84}$$

It is seen that the wave function (7.84) describes the particle motion, which is time-reversed with respect to motion described by the wave function (7.83), because the particle momentum changes its sign. It can be easily understood that the transformation properties of the antiparticle wave function $\Psi^{(2)}_{\mathbf{p},\sigma}$ are similar to (7.84), because, in the standard representation, we have $\gamma_5^* = \gamma_5$, and matrix γ_5 commutes with the products of matrices γ_μ and γ_ν.

7.5.3 Charge conjugation

The charge conjugation transformation is

$$\hat{C}\Psi(\mathbf{r},t) = -i\lambda_C \gamma_2 \Psi^*(\mathbf{r},t). \tag{7.85}$$

By applying the transformation (7.85) to the wave function (7.83), we get

$$\hat{C}\Psi^{(1)}_{\mathbf{p},\sigma=\pm1}(\mathbf{r},t) = \pm i\lambda_C\Psi^{(3)}_{\mathbf{p},\sigma=\mp1}(\mathbf{r},t), \tag{7.86}$$

where

$$\Psi^{(3)}_{\mathbf{p},\sigma}(\mathbf{r},t) = \begin{pmatrix} 0 \\ w_{\mathbf{p}}^{(\sigma)} \end{pmatrix} \exp\left(i\frac{\Gamma t - \mathbf{pr}}{\hbar}\right).$$

As far as the charge conjugation operator(7.85) is the linear operator with respect to matrices γ_μ (the matrix γ_2, in the standard representation), then the particle Ψ_p and antiparticle Ψ_a wave functions are transformed with the opposite signs.

Let us compare the parity of the particle and charged conjugated particle. The wave function of the charged conjugated particle $\Psi_C(\mathbf{r},t) = -i\lambda_C\gamma_2\Psi^*(\mathbf{r},t)$ is transformed under space inversion in the following way

$$\hat{P}\Psi_C(\mathbf{r},t) = i\lambda_C\gamma_2\hat{P}\Psi^*(\mathbf{r},t) = \mp\Psi_C(\mathbf{r},t),$$

therefore the particle and charge conjugated particle have the opposite parity. Hence, the positive energy particle solution is transformed, under charge conjugation, into the negative energy antiparticle solution. For the case of a free particle, it is directly seen from the equations (7.62).

Thus, the general positive energy solution for the free particle is

$$\Psi(\mathbf{r},t) = \sum_{\mathbf{p}}\sum_{\sigma=\pm1}\left(A_{\mathbf{p}\sigma}u_{\mathbf{p}\sigma} + B_{\mathbf{p}\sigma}\gamma_5 u_{\mathbf{p}\sigma}\right)\exp\left(-i\frac{\Gamma_p t - \mathbf{pr}}{\hbar}\right), \tag{7.87}$$

where

$$u_{\mathbf{p}\sigma} = \sqrt{\frac{m_0c^2}{\Gamma_p}}\begin{pmatrix} w_{\mathbf{p}}^{(\sigma)} \\ 0 \end{pmatrix}. \tag{7.88}$$

The charged conjugated solution is the negative energy solution

$$\Psi_C(\mathbf{r},t) = i\lambda_C\sum_{\mathbf{p}}\sum_{\sigma=\pm1}\sigma\left(A^*_{\mathbf{p}\sigma}\gamma_5 u_{\mathbf{p}\sigma} - B^*_{\mathbf{p}\sigma}u_{\mathbf{p}\sigma}\right)\exp\left(i\frac{\Gamma_p t - \mathbf{pr}}{\hbar}\right). \tag{7.89}$$

By comparing the equations (7.87) and (7.89), we can see that the positive and negative energy solutions are really the degenerated solutions. We shall see below, that the degeneracy is appropriate not only to the case of free particle, but to the case of particle motion in the external fields too. Therefore, we can really take into account the positive energy solutions only. Notice, that the physical sense of the charge conjugation transformation may be completely understood only if we consider the interactions of the particles. We shall see later, that this symmetry means that the change of sign of all particles in an isolated system does not affect on the dynamics of the system evolution.

By substituting the wave function (7.87) into the normalization condition (7.64), we get

$$\sum_{\mathbf{p}}\sum_{\sigma=\pm1}\left(|A_{\mathbf{p}\sigma}|^2-|B_{\mathbf{p}\sigma}|^2\right)=1. \tag{7.90}$$

It is seen, that, in contrast to the Schrödinger theory and Dirac theory, the normalization condition (7.51), applied to the wave function of the equation (7.3), means only the conservation of the charge, but it does not demand the conservation of the integral number of particles and antiparticles.

7.6 Spherical waves

7.6.1 Spherical spinors

The free-particle Hamiltonian of the equation (7.3) commutes with the total angular momentum operator

$$\hbar\mathbf{j}=\hbar\left(\mathbf{l}+\mathbf{s}\right)=[\mathbf{r}\,\mathbf{p}]+\frac{\hbar}{2}\mathbf{\Sigma} \tag{7.91}$$

and operator of its projection j_z, therefore the angular part of the free-particle wave function can be expressed in terms of the spherical spinors of the total angular momentum j and its projection m. As we have mentioned in the previous chapter, according to the rules of the angular-momentum addition, at a given value of the total angular momentum j, total angular-momentum z component m, and parity $(-1)^{j\mp1/2}$, there are the two linear independent spherical spinors corresponding to the two possible values of the orbital angular momentum $l=j\mp1/2$. These spinors are

$$\begin{aligned}\Omega^{(1)}_{j,l,m}\left(\theta,\varphi\right)&=\begin{pmatrix}\sqrt{\dfrac{j+m}{2j}}Y_{l,m-1/2}\left(\theta,\varphi\right)\\ \sqrt{\dfrac{j-m}{2j}}Y_{l,m+1/2}\left(\theta,\varphi\right)\end{pmatrix},\\ \Omega^{(2)}_{j,l+1,m}\left(\theta,\varphi\right)&=\begin{pmatrix}-\sqrt{\dfrac{j-m+1}{2j+2}}Y_{l+1,m-1/2}\left(\theta,\varphi\right)\\ \sqrt{\dfrac{j+m+1}{2j+2}}Y_{l+1,m+1/2}\left(\theta,\varphi\right)\end{pmatrix}.\end{aligned} \tag{7.92}$$

We have mentioned in the previous chapter, that the spinors (7.92) are orthonormalized

$$\int\Omega^*_{jlm}\Omega_{j'l'm'}\sin\theta\,d\theta\,d\varphi=\delta_{jj'}\delta_{ll'}\delta_{mm'}.$$

There are the following useful relationships between these spinors

$$\Omega^{(2)}_{j,l+1,m} = -i\sigma_r \Omega^{(1)}_{j,l,m}, \quad \Omega^{(1)}_{j,l,m} = i\sigma_r \Omega^{(2)}_{j,l+1,m}, \tag{7.93}$$

where $\sigma_r = \mathbf{e}_r \boldsymbol{\sigma}$.

For further applications, it is helpful to express the free-particle solution in terms of the spinors (7.92). Accounting the previous discussion, the general solution of the positive energy E, total angular momentum j, and its projection m reads

$$\Psi_{kjm}(\mathbf{r}) = \sum_{n=1,2} A^{(n)}_{jm} u^{(n)}_{jm} f^{(n)}_{kj}(r) + \gamma_5 \sum_{n=1,2} B^{(n)}_{jm} u^{(n)}_{jm} f^{(n)}_{kj}(r), \tag{7.94}$$

where

$$k^2 = \frac{E^2 - m^2 c^4}{\hbar^2 c^2} \tag{7.95}$$

and

$$u^{(1)}_{jm} = \begin{pmatrix} \Omega^{(1)}_{j,l,m} \\ 0 \end{pmatrix}, \quad u^{(2)}_{jm} = \begin{pmatrix} \Omega^{(2)}_{j,l+1,m} \\ 0 \end{pmatrix}. \tag{7.96}$$

The four linear independent solutions (7.94) have the following parity

$$\hat{P} u^{(1)}_{jm} = (-1)^l u^{(1)}_{jm}, \quad \hat{P} u^{(2)}_{jm} = (-1)^{l+1} u^{(2)}_{jm},$$

$$\hat{P} \gamma_5 u^{(1)}_{jm} = (-1)^{l+1} \gamma_5 u^{(1)}_{jm}, \quad \hat{P} \gamma_5 u^{(2)}_{jm} = (-1)^l \gamma_5 u^{(2)}_{jm},$$

where $\hat{P}$ is the parity operator defined by the equation (7.80). Thus, if we deal with the problem of a particle motion in the external field, then the solutions possessing the definite parity are the following superpositions of the eigenfunctions: $u^{(1)}_{jm} \pm \gamma_5 u^{(2)}_{jm}$ and $u^{(2)}_{jm} \pm \gamma_5 u^{(1)}_{jm}$.

By substituting the wave function (7.94) into the equation (7.3), we get for the radial wave functions $f^{(n)}$ the following equations

$$\begin{aligned} \frac{d^2 f^{(1)}}{dr^2} + \frac{2}{r}\frac{df^{(1)}}{dr} - \frac{l(l+1)}{r^2} f^{(1)} + k^2 f^{(1)} = 0, \\ \frac{d^2 f^{(2)}}{dr^2} + \frac{2}{r}\frac{df^{(2)}}{dr} - \frac{(l+1)(l+2)}{r^2} f^{(2)} + k^2 f^{(2)} = 0. \end{aligned} \tag{7.97}$$

The solutions of the equations (7.97) are the spherical Bessel function

$$j_l(x) = \sqrt{\frac{\pi}{2x}} J_{l+1/2}(x),$$

therefore, the general solution (7.94) reads

$$\Psi_{kjm}(\mathbf{r}) = \Big(A_l u_{jm}^{(1)} j_l(kr) + B_{l+1} u_{jm}^{(2)} j_{l+1}(kr)\Big) + \\ + +\gamma_5 \Big(C_l u_{jm}^{(1)} j_l(kr) + D_{l+1} u_{jm}^{(2)} j_{l+1}(kr)\Big). \quad (7.98)$$

With the help of the obtained free-particle solutions, we can easily construct the general solution for the particle moving in the spherically symmetric external field. Indeed, the general free-particle solutions of the parity $(-1)^l$ or $(-1)^{l+1}$ are, respectively:

$$\Psi_{kjm}^{(l)} = \begin{pmatrix} A_l \Omega_{j,l,m}^{(1)} j_l(kr) \\ -D_{l+1} \Omega_{j,l+1,m}^{(2)} j_{l+1}(kr) \end{pmatrix}, \quad \Psi_{kjm}^{(l+1)} = \begin{pmatrix} B_{l+1} \Omega_{j,l+1,m}^{(2)} j_{l+1}(kr) \\ -C_l \Omega_{j,l,m}^{(1)} j_l(kr) \end{pmatrix}.$$

It should be noted that, in the case of particle motion in the external electric field, the equations for the radial wave functions of the upper and lower spinors of the bispinor wave function form the coupled set of equations. Hence, instead of the two independent equations (7.97) we shall have the coupled set of equations, the general solution of which has the following form

$$f^{(n=1,2)}(r) = \sum_i A_i^{(n=1,2)} F_i(r),$$

where the index i numerates the linear independent solutions of the coupled set of equations. Thus, the general solution of the parity $(-1)^l$ for the particle moving in the external field has the following form

$$\Psi_{kjm}^{(l)} = \sum_i \begin{pmatrix} A_i^{(1)} \Omega_{j,l,m}^{(1)} \\ -A_i^{(2)} \Omega_{j,l+1,m}^{(2)} \end{pmatrix} F_i(r).$$

7.6.2 Plane wave expansion in spherical harmonics series

In the study of the scattering problems, it is usually assumed that the incident particle is in the plane wave state. Therefore, it is helpful to express the plane waves in terms of the spherical waves. Let the incident particle be in the following plane wave state

$$\Psi_{k\sigma}(\mathbf{r}, t) = \begin{pmatrix} u_\sigma \\ 0 \end{pmatrix} \exp(ikz), \quad (7.99)$$

where the spinors u_σ are the eigenfunctions of the equation

$$\sigma_z u_\sigma = \sigma u_\sigma,$$

which are

$$u_{+1} = \begin{pmatrix} 1 \\ 0 \end{pmatrix}, \quad u_{-1} = \begin{pmatrix} 0 \\ 1 \end{pmatrix}. \tag{7.100}$$

In the case of free motion, the particle and antiparticle solutions are not coupled, therefore, without loss of generality, we can take the wave function in the form $\Psi = \begin{pmatrix} \varphi \\ 0 \end{pmatrix}$, where the spinors φ_{kjm}, in accordance with the equation (7.98), are

$$\varphi_{kjm}(\mathbf{r}) = A_l \Omega^{(1)}_{j,l,m} j_l(kr) + B_l \Omega^{(2)}_{j,l+1,m} j_{l+1}(kr). \tag{7.101}$$

The projection of the total angular momentum in the states (7.99) is equal to $m = \pm 1/2$. Hence, we get

$$\begin{aligned} \Omega^{(1)}_{j,l,m=1/2} &= \frac{1}{\sqrt{4\pi}} \begin{pmatrix} i^l \sqrt{l+1}\, P_l \\ -\dfrac{i^l}{\sqrt{l+1}} P_l^{(1)} \exp(i\varphi) \end{pmatrix}, \\ \Omega^{(2)}_{j,l+1,m=1/2} &= \frac{1}{\sqrt{4\pi}} \begin{pmatrix} -i^{l+1} \sqrt{l+1} P_{l+1} \\ -\dfrac{i^{l+1}}{\sqrt{l+1}} P_{l+1}^{(1)} \exp(i\varphi) \end{pmatrix}, \end{aligned} \tag{7.102}$$

and

$$\begin{aligned} \Omega^{(1)}_{j,l,m=-1/2} &= \frac{1}{\sqrt{4\pi}} \begin{pmatrix} \dfrac{i^l}{\sqrt{l+1}} P_l^{(1)} \exp(-i\varphi) \\ i^l \sqrt{l+1} P_l \end{pmatrix}, \\ \Omega^{(2)}_{j,l+1,m=-1/2} &= \frac{1}{\sqrt{4\pi}} \begin{pmatrix} -\dfrac{i^{l+1}}{\sqrt{l+1}} P_{l+1}^{(1)} \exp(-i\varphi) \\ i^{l+1} \sqrt{l+1} P_{l+1} \end{pmatrix}, \end{aligned} \tag{7.103}$$

where $P_l^{(m)}(\cos\theta)$ is the associated Legendre polynomial.

Let us use the well known expansion

$$\exp(ikz) = \sum_{l=0}^{\infty} (2l+1)\, i^l P_l(\cos\theta)\, j_l(kr). \tag{7.104}$$

By substituting the equations (7.102) into the equation (7.101) and summing over l, we get, with the help of (7.104), the following results:

$$\varphi_{k,m=1/2}(\mathbf{r}) = \begin{pmatrix} 1 \\ 0 \end{pmatrix} \exp(ikz)$$

when the coefficients A_l and B_l in the equation (7.101) are equal to

$$A_l = -B_l = \sqrt{4\pi (l+1)}, \tag{7.105}$$

and

$$\varphi_{j,m=-1/2}(\mathbf{r}) = \begin{pmatrix} 0 \\ 1 \end{pmatrix} \exp(ikz)$$

when the coefficients A_l and B_l in the equation (7.101) are equal to

$$A_l = B_l = \sqrt{4\pi (l+1)}. \tag{7.106}$$

Thus, the wave function (7.101) is transformed into the wave function (7.99) under appropriate choice of the coefficients A_l and B_l in the equation (7.101).

7.6.3 Convergent and divergent spherical waves

According to the definition of the orbital angular momentum operator $\mathbf{l}$, its radial projection $l_r = \mathbf{l}\mathbf{e}_r = 0$ is identically equal to zero. Hence, the radial projection of the spin Σ_r conserves when particle moves in the spherically symmetric external field. Therefore, it is useful to find the eigenfunctions of the operator σ_r. These eigenfunctions can be directly obtained from the relationships (7.93). Indeed, we can easily get

$$\sigma_r \Omega^{(\pm)}_{jlm} = (\pm 1)\, \Omega^{(\pm)}_{jlm}$$

where

$$\Omega^{(\pm)}_{jlm} = \frac{1}{\sqrt{2}} \left(\Omega^{(1)}_{j,l,m} \pm i \Omega^{(2)}_{j,l+1,m} \right). \tag{7.107}$$

The spinors $\Omega^{(\pm)}$ are orthonormalized

$$\int \Omega^{(\pm)+}_{jlm} \Omega^{(\pm)}_{jlm}\, do = 1, \quad \int \Omega^{(\pm)+}_{jlm} \Omega^{(\mp)}_{jlm}\, do = 0.$$

By substituting the equation (7.107) into the equation (7.101), we get

$$\varphi_{kjm}(\mathbf{r}) = \frac{1}{\sqrt{2}} \Omega^{(+)}_{jlm} \left(A_l j_l(kr) - i B_l j_{l+1}(kr) \right) + \\ + \frac{1}{\sqrt{2}} \Omega^{(-)}_{jlm} \left(A_l j_l(kr) + i B_l j_{l+1}(kr) \right).$$

Particularly, at $m = \pm 1/2$, with the help of equations (7.105) and (7.106), we can easily obtain the following asymptotical wave functions

$$\varphi_{k,m=\pm 1/2}(\mathbf{r}) = \sqrt{2\pi} \sum_{l=0}^{\infty} \left[\sqrt{l+1}\, (\mp i)^{l+1}\, \Omega^{(+)}_{j,m=\pm 1/2} \frac{\exp(\pm ikr)}{kr} + \right. \\ \left. + \sqrt{l+1}\, (\pm i)^{l+1}\, \Omega^{(-)}_{j,m=\pm 1/2} \frac{\exp(\mp ikr)}{kr} \right]. \tag{7.108}$$

Thus, at $m = 1/2$, the wave function is the sum of the divergent spherical wave with the spin radial projection $\sigma_r = +1$ and convergent spherical wave with the spin radial projection $\sigma_r = -1$. At $m = -1/2$, the spin radial projections of the divergent and convergent waves change their signs.

With the help of the recurrence relations for the associated Legendre polynomials, the spinors $\Omega^{(\pm)}$ can be transformed to the following form

$$\Omega^{(+)}_{j,m=1/2} = i^l \sqrt{\frac{l+1}{8\pi}} \left(P_l + P_{l+1}\right) \begin{pmatrix} 1 \\ \exp\left(i\varphi\right) \tan\dfrac{\theta}{2} \end{pmatrix},$$

$$\Omega^{(-)}_{j,m=1/2} = i^l \sqrt{\frac{l+1}{8\pi}} \left(P_l - P_{l+1}\right) \begin{pmatrix} 1 \\ -\exp\left(i\varphi\right) \cot\dfrac{\theta}{2} \end{pmatrix},$$

$$\Omega^{(+)}_{j,m=-1/2} = i^l \sqrt{\frac{l+1}{8\pi}} \left(P_l - P_{l+1}\right) \begin{pmatrix} \exp\left(-i\varphi\right) \cot\dfrac{\theta}{2} \\ 1 \end{pmatrix},$$

$$\Omega^{(-)}_{j,m=-1/2} = i^l \sqrt{\frac{l+1}{8\pi}} \left(P_l + P_{l+1}\right) \begin{pmatrix} -\exp\left(-i\varphi\right) \tan\dfrac{\theta}{2} \\ 1 \end{pmatrix}.$$

Thus, in the case $m = \pm 1/2$, the spinors $\Omega^{(\pm)}$ are the products of the Legendre polynomials and spinors $w^{(\sigma)}$, which are the eigenfunctions of the equation $\sigma_r w^{(\sigma)} = \sigma w^{(\sigma)}$:

$$w^{(\sigma=+1)} = \begin{pmatrix} \exp\left(-i\dfrac{\varphi}{2}\right) \cos\dfrac{\theta}{2} \\ \exp\left(i\dfrac{\varphi}{2}\right) \sin\dfrac{\theta}{2} \end{pmatrix}, \quad w^{(\sigma=-1)} = \begin{pmatrix} -\exp\left(-i\dfrac{\varphi}{2}\right) \sin\dfrac{\theta}{2} \\ \exp\left(i\dfrac{\varphi}{2}\right) \cos\dfrac{\theta}{2} \end{pmatrix}.$$

With the help of the last equations, the asymptotical wave function (7.108) can be transformed to the following form

$$\varphi_{k,m=1/2}\left(\mathbf{r}\right)\Big|_{r\to\infty} = -\frac{i}{\sqrt{2}} \frac{\sum\limits_{l=0}^{\infty} \left(2l+1\right) P_l\left(\cos\theta\right)}{\sqrt{1+\cos\theta}} w^{(+1)} \frac{\exp\left(ikr + i\varphi/2\right)}{kr} +$$

$$+ \frac{i}{\sqrt{2}} \frac{\sum\limits_{l=0}^{\infty} \left(-1\right)^l \left(2l+1\right) P_l\left(\cos\theta\right)}{\sqrt{1-\cos\theta}} w^{(-1)} \frac{\exp\left(-ikr + i\varphi/2\right)}{kr}. \quad (7.109)$$

To interpret the equation (7.109), it is helpful to use the following equations

$$\begin{aligned}\sum_{n=0}^{\infty} P_n(\cos\theta) &= \frac{1}{\sqrt{2(1-\cos\theta)}},\\ \sum_{n=0}^{\infty} (-1)^n P_n(\cos\theta) &= \frac{1}{\sqrt{2(1+\cos\theta)}},\\ \sum_{n=0}^{\infty} nP_n(\cos\theta) &= -\frac{1}{2\sqrt{2(1-\cos\theta)}},\\ \sum_{n=0}^{\infty} (-1)^n nP_n(\cos\theta) &= -\frac{1}{2\sqrt{2(1+\cos\theta)}}.\end{aligned} \tag{7.110}$$

The equations (7.110) generate the following equations

$$\sum_{n=0}^{\infty} (2n+1) P_n(\cos\theta) = 2\delta(1-\cos\theta),$$

$$\sum_{n=0}^{\infty} (-1)^n (2n+1) P_n(\cos\theta) = 2\delta(1+\cos\theta).$$

Thus, the asymptotical wave function (7.109) at $z \to -\infty$, i.e. $\theta = \pi$, is

$$\varphi_{k,m=1/2}\big|_{z\to-\infty} = -\begin{pmatrix}1\\0\end{pmatrix} \frac{\exp(-ikr)}{ikr}\,\delta(1+\cos\theta).$$

By taking into account the following transformation

$$\exp(ikz) = \exp(ikr\cos\theta)|_{\theta=\pi} = \exp(-ikr),$$

we can see that the divergent wave in the equation (7.109) is the incident wave. Indeed, at $z = -\infty$, this wave is the plane wave propagating into the positive direction of the z axis. At $z \to \infty$, the asymptotical wave function (7.109) takes the form

$$\varphi_{k,m=1/2}\big|_{z\to\infty} = \begin{pmatrix}1\\0\end{pmatrix} \frac{\exp(ikr)}{ikr}\,\delta(1-\cos\theta),$$

i.e. the angular spectrum of the transmitted wave (at $z = +\infty$) coincides with the angular spectrum of the incident wave. In the presence of the external field the angular spectrum of the scattered wave will differ from the delta function. The obtained equation for the asymptotical wave function (7.109) is of interest for the study the scattering processes. It enables us to exclude the incident plane wave from the general continuous spectrum solution of the problem on particle motion in the external field.

Chapter 8

PARTICLE MOTION IN STATIC EXTERNAL FIELDS

In previous chapter, the general principles of the relativistic second order differential equation, describing the spin-1/2 particle, have been mainly applied to the free particle states. We have seen that there is a number of specific features of this equation, that give us some grounds to assume that, in this case, particle and antiparticle behave themselves in a way different of that prescribed by the Dirac equation. But the real specificity of particle behavior can be understood only in the study of their interaction with the external fields. In this chapter, we will consider the basic problems on particle motion in the external fields: the electron motions in Coulomb field and uniform magnetic field, and the neutron interaction with the static magnetic fields of the different spatial profile.

8.1 Integrals of motion

The very important category of the physical variables is the conservative variables. Let L be the operator of some physical variable. The mean value of the operator L in the state, described by the arbitrary wave function $\Psi(\mathbf{r},t)$, is defined by

$$\langle L\rangle = \int \bar{\Psi}(\mathbf{r},t)\, L\Psi(\mathbf{r},t)\, dV. \tag{8.1}$$

The physical variable corresponding to operator L is conservative, if the mean value (8.1) does not vary in time. The time derivative of the mean value is

$$\frac{d}{dt}\int \bar{\Psi}L\Psi\, dV = \int \bar{\Psi}\frac{\partial L}{\partial t}\Psi\, dV + \int \frac{\partial\bar{\Psi}}{\partial t}L\Psi\, dV + \int \bar{\Psi}L\frac{\partial\Psi}{\partial t}\, dV. \tag{8.2}$$

Let us rewrite the equation (7.3) in the following form

$$\frac{1}{2m_0c^2}\left[\left(i\hbar\frac{\partial}{\partial t}\right)^2 - Ui\hbar\frac{\partial}{\partial t} - i\hbar\frac{\partial}{\partial t}U\right]\Psi = H\Psi, \tag{8.3}$$

where

$$H = \frac{1}{2m_0c^2}\left[c^2\left(\mathbf{p} - \frac{q_0}{c}\mathbf{A}\right)^2 + m_0^2c^4 - U^2 + 2m_0c^2\mu_0(i\boldsymbol{\alpha}\mathbf{E} - \boldsymbol{\Sigma}\mathbf{B})\right] \tag{8.4}$$

and $U(\mathbf{r},t) = q_0\varphi(\mathbf{r},t)$. Multiplying the equation (8.3) by $\bar{\Psi}L$ from the left, then, by multiplying the equation for the Dirac adjoint function $\bar{\Psi}$ by $L\Psi$ from the right, and, finally, subtracting the obtained equations, we get

$$\frac{\partial}{\partial t}\left[\left(\frac{\partial\bar{\Psi}}{\partial t} - \frac{i}{\hbar}U\bar{\Psi}\right)L\Psi - \bar{\Psi}L\left(\frac{\partial\Psi}{\partial t} + \frac{i}{\hbar}U\Psi\right)\right] =$$
$$= -\frac{1}{\hbar^2}\left((\bar{\Psi}\tilde{H})L\Psi - \bar{\Psi}L(H\Psi)\right),$$

where $\tilde{H} = \gamma_4^{-1}H^+\gamma_4$, notice, that this operator acts on the wave function $\bar{\Psi}$. Integrating both sides of the last equation over the whole space and transposing the action of the operator $\tilde{H}$ to the function Ψ, we get

$$\frac{d}{dt}\int\left[\left(\frac{\partial\bar{\Psi}}{\partial t} - \frac{i}{\hbar}\bar{\Psi}U\right)L\Psi - \bar{\Psi}L\left(\frac{\partial\Psi}{\partial t} + \frac{i}{\hbar}U\Psi\right)\right]dV =$$
$$= -\frac{1}{\hbar^2}\int\bar{\Psi}\left((HL - LH)\Psi\right)dV. \tag{8.5}$$

The wave function Ψ is an arbitrary wave function. Hence, if the operator L commutes with the Hamiltonian (8.4)

$$[H, L] = HL - LH = 0, \tag{8.6}$$

we get

$$\frac{\partial\Psi}{\partial t} = -\frac{i}{\hbar}U\Psi, \quad \frac{\partial\bar{\Psi}}{\partial t} = \frac{i}{\hbar}\bar{\Psi}U.$$

Substituting the last equations into the equation (8.2), we can see, that if the operator L does not depend explicitly on time and commutes with the Hamiltonian, then the mean value of this operator does not depend on time:

$$\frac{d}{dt}\int\bar{\Psi}L\Psi\,dV = \frac{i}{\hbar}\int\bar{\Psi}(UL - LU)\Psi\,dV = 0.$$

Notice, if L commutes with H, then it commutes with U. Thus, the operator L corresponds to the conservative variable, if it does not explicitly depend on time and commutes with the Hamiltonian (8.4).

The Hamiltonian H commutes with itself, therefore the energy of a particle interacting with the static external fields is the conservative variable, or integral of motion. The wave function of the equation (7.3), for the case of particle interacting with the static external fields, is

$$\Psi_n(\mathbf{r},t) = \Psi_n(\mathbf{r})\exp\left(-iE_n t/\hbar\right),$$

where the spatial part of the wave function is determined by the solution of the eigenvalue problem

$$H(E_n)\Psi_n(\mathbf{r}) = 0, \tag{8.7}$$

where

$$H(E_n) = \frac{1}{2m_0}\left[\left(-i\hbar\nabla - \frac{q_0}{c}\mathbf{A}\right)^2 + m_0^2c^2 - \frac{1}{c^2}\left(E_n - U(\mathbf{r})\right)^2\right] + \\ + i\mu_0\boldsymbol{\alpha}\mathbf{E} - \mu_0\boldsymbol{\Sigma}\mathbf{B}, \tag{8.8}$$

The boundary conditions for the eigenvalue problem (8.7) are the same as in all previous chapters. The eigenfunctions of the equation (8.7) are orthonormalized. Indeed, if the wave function $\Psi_m(\mathbf{r})$ satisfies the equation (8.7), then the Dirac adjoint wave function satisfies the equation

$$\bar{\Psi}_m(\mathbf{r})\left[\frac{1}{2m_0}\left(\left(\mathbf{p} + \frac{q_0}{c}\mathbf{A}\right)^2 + m_0^2c^2 - \frac{1}{c^2}\left(E_m - U(\mathbf{r})\right)^2\right) + \\ + i\mu_0\boldsymbol{\alpha}\mathbf{E} - \mu_0\boldsymbol{\Sigma}\mathbf{B}\right] = 0. \tag{8.9}$$

Multiplying the equation (8.9) by Ψ_n from the right, integrating the obtained equation over the whole space and transposing the action of Hamiltonian from the wave function $\bar{\Psi}_m$ to Ψ_n, we finally get

$$\left(E_n - E_m\right)\frac{1}{m_0c^2}\int \bar{\Psi}_m\left(\frac{E_n + E_m}{2} - U(\mathbf{r})\right)\Psi_n\, dV = 0. \tag{8.10}$$

Thus, the wave functions of the non-degenerated states are orthonormalized by the condition

$$\frac{1}{m_0c^2}\int \bar{\Psi}_m(\mathbf{r})\left(\frac{E_n + E_m}{2} - U(\mathbf{r})\right)\Psi_n(\mathbf{r})\, dV = \pm\delta_{nm}. \tag{8.11}$$

At $n = m$ the last equation coincides with the normalization condition (7.56).

8.1.1 Free particle

The solution of the equation (8.7) for the case of the free particle has been already obtained in the previous chapter. The free-particle Hamiltonian (8.8) is

$$H_0(E) = \frac{\mathbf{p}^2}{2m_0} - \frac{E^2 - m_0^2c^4}{2m_0c^2}. \tag{8.12}$$

The Hamiltonian (8.12) commutes with the operators of parity, momentum $\mathbf{p} = -i\hbar\nabla$, angular momentum $\hbar\mathbf{l} = [\mathbf{r}\mathbf{p}]$, spin $\hbar\mathbf{s} = \hbar\mathbf{\Sigma}/2$, and, hence, total angular momentum

$$\hbar\mathbf{j} = \hbar\mathbf{l} + \hbar\mathbf{s} = [\mathbf{r}\mathbf{p}] + \frac{\hbar}{2}\mathbf{\Sigma}. \tag{8.13}$$

Thus, the free-particle states are the eigenfunctions of all these operators. However, we have seen that there are the eight linear independent solutions at a given energy eigenvalue and momentum direction $\mathbf{p} = \pm\mathbf{n}p$. Hence, to label the eight linear independent solutions we need only in the three binary quantum numbers. In the previous chapter we have chosen the following quantum numbers: $\boldsymbol{\lambda} = (\lambda_E, \lambda_{\mathbf{p}}, \lambda_q)$, where $\lambda_E = E/|E|$, $\lambda_{\mathbf{p}} = (\mathbf{n}\mathbf{p})/|\mathbf{p}|$, and $\lambda_q = q/q_0$. But, the sign of the energy is unambiguously related with the charge conjugation transformation, therefore we can use the quantum number $\lambda_C = q_0/|q_0|$ instead of quantum number λ_E. Instead of quantum number $\lambda_{\mathbf{p}}$, we can use the spin quantum number $\sigma = \pm 1$, which determines the spin state of the particle at given momentum $\mathbf{p}$.

As we have seen in the previous chapter, the quantum number λ_q indicates whether a given state corresponds to particle or antiparticle solution. This quantum number is closely related with the eigenvalues of the operator γ_5. The operator γ_5 commutes with the Hamiltonian (8.8). In the standard representation of the matrices γ_μ, the eigenvalue problem for the operator γ_5:

$$\gamma_5 u_{\pm 1} = (\pm 1)\, u_{\pm 1}, \tag{8.14}$$

has the following solutions

$$u_{+1} = \frac{1}{\sqrt{2}}\begin{pmatrix}\varphi \\ -\varphi\end{pmatrix}, \quad u_{-1} = \frac{1}{\sqrt{2}}\begin{pmatrix}\varphi \\ \varphi\end{pmatrix}, \tag{8.15}$$

where φ is the arbitrary spinor satisfying the normalization condition $\varphi^+\varphi = 1$. The wave functions of the particle u_p and antiparticle u_a are the superpositions of the wave functions (8.15):

$$u_p = \frac{1}{\sqrt{2}}(u_{+1} + u_{-1}) = \begin{pmatrix}\varphi \\ 0\end{pmatrix}, \quad u_a = \frac{1}{\sqrt{2}}(u_{+1} - u_{-1}) = \begin{pmatrix}0 \\ -\varphi\end{pmatrix} = \gamma_5 u_p. \tag{8.16}$$

It can be easily seen, that, in standard representation of matrices γ_μ, the wave functions u_p and u_a are the eigenfunctions of the operator γ_4. Indeed, the eigenvalue problem for the operator γ_4,

$$\gamma_4 v_{\pm 1} = (\pm 1)\, v_{\pm 1},$$

has the following solutions

$$v_{+1} = \begin{pmatrix} I \\ 0 \end{pmatrix}, \quad v_{-1} = \begin{pmatrix} 0 \\ I \end{pmatrix}.$$

The operator γ_4 commutes with the free-particle Hamiltonian (8.12) and does not commute with the Hamiltonian (8.8) at $\mathbf{E} \neq 0$. Hence, a state of fermion is the pure state of particle or antiparticle only in the absence of the external electric field. If the amplitude of the external electric field is non-zero, then the particle state is a superposition of the pure particle and antiparticle states.

Notice, that, to simplify the reading, in this book we use the standard representation of the matrices γ_μ, given by (6.12). However, we can choose these matrices as follows

$$\boldsymbol{\gamma}' = i \begin{pmatrix} 0 & -\boldsymbol{\sigma} \\ \boldsymbol{\sigma} & 0 \end{pmatrix}, \quad \gamma_4' = - \begin{pmatrix} 0 & 1 \\ 1 & 0 \end{pmatrix}, \quad \gamma_5' = \begin{pmatrix} 1 & 0 \\ 0 & -1 \end{pmatrix}. \tag{8.17}$$

In this case the notation γ_5 is particularly appropriate, because the given above commutation relations for matrices γ_μ and γ_5 show that the matrices $\gamma_1, \gamma_2, \gamma_3, \gamma_4, \gamma_5$ provide the Clifford algebra in the five spacetime dimensions. It is seen that, in this case, the bispinors (8.16) are the eigenfunctions of the operator γ_5.

8.1.2 Particle motion in centro-symmetric fields

When we analyze the particle motion in the external fields of the spherical or cylindrical symmetry, it is convenient to use the curvilinear coordinates instead of Cartesian ones. The relationships between the components of the matrix $\boldsymbol{\sigma}$ in the curvilinear and Cartesian coordinates are determined by the general equations of the vectorial analysis. The projections of $\boldsymbol{\sigma}$ in the spherical coordinates are

$$\begin{aligned} \sigma_r &= \sigma_+ \sin\theta \exp(-i\varphi) + \sigma_- \sin\theta \exp(i\varphi) + \sigma_z \cos\theta, \\ \sigma_\theta &= \sigma_+ \cos\theta \exp(-i\varphi) + \sigma_- \cos\theta \exp(i\varphi) - \sigma_z \sin\theta, \\ \sigma_\varphi &= -i\sigma_+ \exp(-i\varphi) + i\sigma_- \exp(i\varphi), \end{aligned} \tag{8.18}$$

where $\sigma_\pm = (\sigma_x \pm i\sigma_y)/2$. The commutation relations for the matrices (8.18) are

$$[\sigma_\alpha, \sigma_\beta] = 2ie_{\alpha\beta\gamma}\sigma_\gamma, \tag{8.19}$$

where $\alpha, \beta, \gamma = r, \theta, \varphi$ and $e_{r\theta\varphi} = 1$.

In spherical coordinates, the operator of the orbital angular momentum has the following form

$$\mathbf{l} = -i\left[\mathbf{r}\nabla\right] = \mathbf{e}_\theta \frac{i}{\sin\theta}\frac{\partial}{\partial\varphi} - \mathbf{e}_\varphi i \frac{\partial}{\partial\theta}. \tag{8.20}$$

The operator of the square of the orbital angular momentum is

$$\mathbf{l}^2 = l_x^2 + l_y^2 + l_z^2 = -\left[\frac{1}{\sin^2\theta}\frac{\partial^2}{\partial\varphi^2} + \frac{1}{\sin\theta}\frac{\partial}{\partial\theta}\left(\sin\theta\frac{\partial}{\partial\theta}\right)\right].$$

The operator $\mathbf{l}^2$ coincides with the angular part of the Laplace operator

$$\Delta = \frac{1}{r^2}\frac{\partial}{\partial r}\left(r^2\frac{\partial}{\partial r}\right) - \frac{1}{r^2}\mathbf{l}^2. \tag{8.21}$$

When a particle moves in the spherically symmetric external field, i.e. $\varphi(\mathbf{r}) = \varphi(r)$, the Hamiltonian (8.8) takes the form

$$H^{(s)}(E_n) = \frac{1}{2m_0c^2}\left[-\hbar^2c^2\Delta + m_0^2c^4 - (E_n - U(r))^2\right] + i\mu_0\alpha_r E_r(r). \tag{8.22}$$

It follows from the equations (8.20) and (8.21), that the angular momentum operator $\mathbf{l}$ commutes with terms in square brackets of the Hamiltonian (8.22). It is convenient to use the equations (8.18) in order to find the commutation relations of the operator $\mathbf{l}$ and the last term in the Hamiltonian (8.22). According to (8.18), we get

$$\frac{\partial\sigma_r}{\partial\varphi} = \sigma_\varphi\sin\theta, \qquad \frac{\partial\sigma_r}{\partial\theta} = \sigma_\theta.$$

Hence,

$$[l_\theta, \alpha_r E_r(r)] = iE_r\frac{1}{\sin\theta}\frac{\partial\alpha_r}{\partial\varphi} = iE_r\alpha_\varphi = i\left[\boldsymbol{\alpha}\mathbf{E}\right]_\theta,$$

$$[l_\varphi, \alpha_r E_r(r)] = -iE_r\frac{\partial\alpha_r}{\partial\theta} = -iE_r\alpha_\theta = i\left[\boldsymbol{\alpha}\mathbf{E}\right]_\varphi.$$

For the commutator of $\mathbf{l}$ and $H^{(s)}$ we get finally

$$\left[\mathbf{l}, H^{(s)}\right] = -\mu_0[\boldsymbol{\alpha}\mathbf{E}]. \tag{8.23}$$

Thus, in contrast to the spinless particle, the orbital angular momentum ceases to be the integral of motion for the problem of the particle motion in the spherically symmetric external field.

The commutation relation, for the spin operator and Hamiltonian (8.8), is

$$[\boldsymbol{\Sigma}, H(E_n)] = 2\mu_0([\boldsymbol{\alpha}\mathbf{E}] + i\left[\boldsymbol{\Sigma}\mathbf{B}\right]). \tag{8.24}$$

Hence, in the spherically symmetric external field ($U(\mathbf{r}) = U(r)$ and $\mathbf{A} = 0$) the total angular momentum operator $\mathbf{j}$ is the integral of motion. Indeed,

$$[\mathbf{j}, H^{(s)}] = [\mathbf{l}, H^{(s)}] + \frac{1}{2}[\mathbf{\Sigma}, H^{(s)}] = -\mu_0[\boldsymbol{\alpha}\mathbf{E}] + \frac{1}{2}2\mu_0[\boldsymbol{\alpha}\mathbf{E}] = 0. \quad (8.25)$$

According to the definition of the orbital angular momentum, its radial projection $l_r = (\mathbf{e}_r\mathbf{l})$ is equal to zero. Hence, the conservation of the total angular momentum results in the conservation of the spin radial projection

$$[\Sigma_r, H^{(s)}] = 0. \quad (8.26)$$

Thus, the total angular momentum and radial projection of the spin are the conservative variables in the case of particle motion in the spherically symmetric external field. The orbital angular momentum and other spin projections are not the conservative variables.

The relativistic parity is also the integral of motion for the spherically symmetric external field. Indeed,

$$\hat{P}\boldsymbol{\alpha}\mathbf{E} = \gamma_4 P_3 \boldsymbol{\alpha}\mathbf{E} = -\gamma_4\boldsymbol{\alpha}\mathbf{E}P_3 = \boldsymbol{\alpha}\mathbf{E}\gamma_4 P_3 = \boldsymbol{\alpha}\mathbf{E}\hat{P}.$$

8.1.3 Cylindrically symmetric external fields

Let us consider the particle motion in the axially symmetric static fields: $\varphi = \varphi(\rho, z)$ and $\mathbf{A} = \mathbf{e}_\varphi A(\rho, z)$. In this case we have

$$\mathbf{E} = -\mathbf{e}_\rho \frac{\partial\varphi}{\partial\rho} - \mathbf{e}_z\frac{\partial\varphi}{\partial z}, \quad \mathbf{B} = -\mathbf{e}_\rho\frac{\partial A}{\partial z} + \mathbf{e}_z \frac{1}{\rho}\frac{\partial(\rho A)}{\partial\rho}, \quad (8.27)$$

and the Hamiltonian (8.8) reads

$$H^{(c)} = \frac{1}{2m_0c^2}\left[-\hbar^2c^2\Delta + 2i\hbar c q_0 \frac{A(\rho,z)}{\rho}\frac{\partial}{\partial\varphi} + \right.$$
$$\left. + m_0^2c^4 - (E_n - U(\rho,z))^2 + q_0^2A^2(\rho,z)\right] + \mu_0(i\boldsymbol{\alpha}\mathbf{E} - \mathbf{\Sigma}\mathbf{B}). \quad (8.28)$$

The projections of the vectorial spin operator $\boldsymbol{\sigma}$ in the cylindrical coordinates are given by

$$\begin{aligned} \sigma_\rho &= \sigma_+ \exp(-i\varphi) + \sigma_- \exp(i\varphi), \\ \sigma_\varphi &= -i\sigma_+ \exp(-i\varphi) + i\sigma_- \exp(i\varphi), \\ \sigma_z &= \sigma_z. \end{aligned} \quad (8.29)$$

The commutation relations for the operators σ_ρ, σ_φ, σ_z are given by

$$[\sigma_\alpha, \sigma_\beta] = 2ie_{\alpha\beta\gamma}\sigma_\gamma, \quad (8.30)$$

where $\alpha, \beta, \gamma = \rho, \varphi, z$ and $e_{\rho\varphi z} = 1$.

In cylindrical coordinates, the operator of the orbital angular momentum has the following form

$$\mathbf{l} = \mathbf{e}_\rho\Big(i\frac{z}{\rho}\frac{\partial}{\partial\varphi}\Big) + \mathbf{e}_\varphi\Big[-i\Big(z\frac{\partial}{\partial\rho} - \rho\frac{\partial}{\partial z}\Big)\Big] + \mathbf{e}_z\Big(-i\frac{\partial}{\partial\varphi}\Big). \tag{8.31}$$

It can be easily seen that the axial projection of the orbital angular momentum, $l_z = -i(\partial/\partial\varphi)$, commutes with the terms in the square brackets of the Hamiltonian (8.28). It is convenient to use the equations (8.29) in order to find the commutator of the operators l_z and the last term in the Hamiltonian (8.28). It follows from the equations (8.29), that

$$\frac{\partial\sigma_\rho}{\partial\varphi} = \sigma_\varphi, \quad \frac{\partial\sigma_\varphi}{\partial\varphi} = -\sigma_\rho.$$

With the help of the last equations we get

$$\big[l_z, H^{(c)}\big] = -\mu_0([\boldsymbol{\alpha}\mathbf{E}] + i\,[\boldsymbol{\Sigma}\mathbf{B}])_z\,. \tag{8.32}$$

Thus, in contrast to the case of the spinless particle, the axial projection of the orbital angular momentum ceases to be the integral of motion in the case of the axially symmetric external fields. This is due to the interaction of the orbital angular momentum and spin.

The commutation relation of the operator Σ_z and Hamiltonian (8.8) follows from the general equation (8.24):

$$[\Sigma_z, H] = 2\mu_0([\boldsymbol{\alpha}\mathbf{E}] + i\,[\boldsymbol{\Sigma}\mathbf{B}])_z\,. \tag{8.33}$$

It is seen from the equations (8.32) and (8.33) that the axial projection of the total angular momentum is the conservative variable: $\big[j_z, H^{(c)}\big] = 0$. Thus, the axial projection of the total angular momentum is only conservative variable in the case of the particle motion in the external fields of the cylindrical symmetry. In general case, the axial projections of the orbital angular momentum and spin are not the conservative variables. However, it follows from the equation (8.33), that the axial projection of the spin is conservative variable, when a particle moves in the homogeneous magnetic field, $\mathbf{B} = \mathbf{e}_z B$, or in the superposition of the parallel homogeneous magnetic and electric fields, $\mathbf{B} = \mathbf{e}_z B$, $\mathbf{E} = \mathbf{e}_z E$.

The relativistic parity is the integral of motion, when a particle moves in the cylindrically symmetric external fields of the following type: $\varphi(\rho, |z|)$ and $A(\rho, |z|)$. Indeed, it is evident that in this case the parity operator commutes with the terms in the square brackets of the Hamiltonian (8.28). It also commutes with the last term of the Hamiltonian (8.28), because the electric field is antisymmetric in this case, $\mathbf{E}(-\mathbf{r}) = -\mathbf{E}(\mathbf{r})$, and $\gamma_4\boldsymbol{\alpha} = -\boldsymbol{\alpha}\gamma_4$, while the magnetic field is symmetric, $\mathbf{B}(-\mathbf{r}) = \mathbf{B}(\mathbf{r})$, and $\gamma_4\boldsymbol{\Sigma} = \boldsymbol{\Sigma}\gamma_4$.

8.2 Electron motion in Coulomb field

8.2.1 General solution

Let a particle move in the attracting Coulomb field. In this case

$$U(\mathbf{r}) = e\varphi(\mathbf{r}) = -\frac{Ze^2}{r}, \tag{8.34}$$

and the equation (7.3) becomes

$$\left[\Delta - \frac{m_0^2c^4 - E^2}{\hbar^2c^2} + \frac{2EZ\alpha}{\hbar c}\frac{1}{r} + \frac{Z^2\alpha^2}{r^2}\right]\Psi = -i\frac{\mu_0}{\mu_{\mathrm{B}}}\frac{Z\alpha}{r^2}\alpha_r\Psi \tag{8.35}$$

where $\alpha = e^2/\hbar c$ is the fine structure constant, $\alpha_r = \boldsymbol{\alpha}\mathbf{e}_r = \begin{pmatrix} 0 & \sigma_r \\ \sigma_r & 0 \end{pmatrix}$, here $\mathbf{e}_r$ is the radial unit vector of the spherical coordinates. In the equation (8.35) we have assumed $q_0 = -|e|$ and taken into account that the electron magneton is negative, hence the constant μ_0 is here the magnitude of the magneton.

As far as the total angular momentum and parity are the integrals of motion in the Coulomb field, then the angular part of the wave function is given by the spinors (7.92). Therefore the wave function of the equation (8.35) has the following form

$$\Psi(\mathbf{r}) = \begin{pmatrix} \Omega^{(1)}_{jlm} f(r) \\ \Omega^{(2)}_{jl'm} g(r) \end{pmatrix}$$

where $l = j \pm 1/2$, $l' = 2j - l$, and spinors $\Omega^{(1,2)}_{jlm}$ are:

$$\Omega^{(1)}_{jlm} = \begin{pmatrix} \sqrt{\frac{j+m}{2j}}Y_{l,m-1/2} \\ \sqrt{\frac{j-m}{2j}}Y_{l,m+1/2} \end{pmatrix}, \quad \Omega^{(2)}_{jlm} = \begin{pmatrix} -\sqrt{\frac{j-m+1}{2j+2}}Y_{l,m-1/2} \\ \sqrt{\frac{j+m+1}{2j+2}}Y_{l,m+1/2} \end{pmatrix}. \tag{8.36}$$

Let us start with the case of $j = l + 1/2$. The equations for the radial wave functions $f(r)$ and $g(r)$ can be easily obtained, if we use the following relationship: $\Omega^{(2)}_{jl+1m} = -i\sigma_r\Omega^{(1)}_{jlm}$. With the help of this relationship, we get

$$\left[\frac{d^2}{dr^2} + \frac{2}{r}\frac{d}{dr} + \frac{Z^2\alpha^2 - l(l+1)}{r^2} + \frac{2EZ\alpha}{\hbar c}\frac{1}{r} - \kappa^2\right] f = -\frac{Z_1\alpha}{r^2}g, \tag{8.37}$$

$$\left[\frac{d^2}{dr^2} + \frac{2}{r}\frac{d}{dr} + \frac{Z^2\alpha^2 - (l+1)(l+2)}{r^2} + \frac{2EZ\alpha}{\hbar c}\frac{1}{r} - \kappa^2\right] g = \frac{Z_1\alpha}{r^2}f, \tag{8.38}$$

where Z_1 and κ are defined by

$$Z_1 = Z\frac{\mu_0}{\mu_{\mathrm{B}}}, \quad \kappa^2 = \frac{m_0^2 c^4 - E^2}{\hbar^2 c^2}.$$

In order to find the solutions of the equations (8.37), (8.38), it is convenient to use the solution of the following equation

$$\frac{d^2 f}{dx^2} + \frac{2}{x}\frac{df}{dx} - \left(a - \frac{b}{x} - \frac{c}{x^2}\right) f(x) = 0.$$

The solution of the last equation, which is not divergent at $r = 0$, is

$$f(x) = \exp\left(-\sqrt{a}x + \frac{\sqrt{1-4c}-1}{2}\ln x\right) \times$$
$$\times F\left(\frac{1+\sqrt{1-4c}}{2} - \frac{b}{2\sqrt{a}},\ 1+\sqrt{1-4c},\ 2\sqrt{a}x\right), \quad (8.39)$$

where $F(p, q, z)$ is the confluent hypergeometric function. We have omitted the second linear independent solution of this equation, which is proportional to $z^{1-q}F(p-q+1, 2-q, z)$, because it is convergent at $z \to 0$. It is convenient to introduce the following function

$$G(\xi, \nu, r) = \exp(-\kappa r)\, r^{\nu-1} F\left(\nu - \frac{EZ\alpha}{\hbar c \kappa},\ 2\nu,\ 2\kappa r\right), \quad (8.40)$$

where

$$\xi = \nu - \frac{EZ\alpha}{\hbar c\kappa} \quad \text{and} \quad \nu = \frac{1+\sqrt{1-4\gamma}}{2}.$$

So, let the solutions of the equations (8.37), (8.38) be

$$f(r) = f_0 G(\xi, \nu, r), \quad g(r) = g_0 G(\xi, \nu, r), \quad (8.41)$$

where f_0 and g_0 are the constants. By substituting the equations (8.41) into the equations (8.37), (8.38), we get the following algebraic set of equations for the coefficients f_0 and g_0:

$$\left(Z^2\alpha^2 - l(l+1) - \gamma\right) f_0 + Z_1 \alpha g_0 = 0,$$

$$Z_1 \alpha f_0 - \left(Z^2\alpha^2 - (l+1)(l+2) - \gamma\right) g_0 = 0.$$

The condition of existence of the non-trivial solutions of the last equations enables us to determine the unknown parameter γ in the solutions (8.41). Employing this condition, we have

$$\gamma_{1,2} = Z^2\alpha^2 - (j+1/2)^2 \pm \sqrt{(j+1/2)^2 - (Z_1\alpha)^2}. \quad (8.42)$$

8.2.2 Discrete spectrum

The radial wave functions (8.41) satisfy the boundary conditions at $r \to \infty$, when the parameter ξ is the non-positive integer

$$\nu - \frac{EZ\alpha}{\hbar c\kappa} = -n, \tag{8.43}$$

where n is the non-negative integer. By solving the last equation with respect to E, we obtain the following formula for the energy spectrum

$$E_{nj}^{(i)} = \frac{m_0 c^2 (\nu_i + n)}{\sqrt{(\nu_i + n)^2 + Z^2\alpha^2}}, \tag{8.44}$$

where the index i enumerates the roots of the equation (8.42), and

$$\nu_{1,2} = \frac{1 + \sqrt{1 - 4\gamma_{1,2}}}{2}.$$

The wave functions, associated with ν_1 and ν_2, are

$$\begin{aligned} \Psi_{njm}^{(1)}(\mathbf{r}) &= C_1 \begin{pmatrix} \Omega_{jlm}^{(1)} \\ -\zeta\Omega_{jl+1m}^{(2)} \end{pmatrix} G(-n, \nu_1, r), \\ \Psi_{njm}^{(2)}(\mathbf{r}) &= C_2 \begin{pmatrix} -\zeta\Omega_{jlm}^{(1)} \\ \Omega_{jl+1m}^{(2)} \end{pmatrix} G(-n, \nu_2, r), \end{aligned} \tag{8.45}$$

where $C_{1,2}$ are the normalization constants, and

$$\zeta = \frac{Z_1\alpha}{j + 1/2 + \sqrt{(j+1/2)^2 - (Z_1\alpha)^2}}.$$

As we have mentioned above, there is no need to consider separately the case of $j = l - 1/2$, because the corresponding solutions follow from the above obtained with the help of transposition of the upper and lower spinors in the equations (8.45):

$$\begin{aligned} \Psi_{njm}^{(3)}(\mathbf{r}) &= C_3 \begin{pmatrix} -\zeta\Omega_{jl+1m}^{(2)} \\ \Omega_{jlm}^{(1)} \end{pmatrix} G(-n, \nu_1, r), \\ \Psi_{njm}^{(4)}(\mathbf{r}) &= C_4 \begin{pmatrix} \Omega_{jl+1m}^{(2)} \\ -\zeta\Omega_{jlm}^{(1)} \end{pmatrix} G(-n, \nu_2, r). \end{aligned} \tag{8.46}$$

According to the discussion given in the previous chapter, one of the solutions (8.45) (or (8.46)) corresponds to the particle, and another corresponds to antiparticle. Indeed, the particle charge in the states described by the equations (8.45) is

$$\frac{1}{q_0}\int \rho_{1,2}\left(\mathbf{r}\right)dV = \\ = \pm\frac{1}{m_0c^2}\left(1-\zeta^2\right)\int\left(E_{nj}^{(1,2)}+\frac{Ze^2}{r}\right)\left|G\left(-n,v_{1,2},r\right)\right|^2 r^2\,dr. \quad (8.47)$$

By taking into account that the constant ζ is proportional to the fine structure constant α, we can see that the solutions $\Psi^{(1)}$ and $\Psi^{(2)}$ correspond to the particle and antiparticle, respectively.

Similarly, the solutions $\Psi^{(3)}$ and $\Psi^{(4)}$ correspond to the antiparticle and particle. It is also seen that the sign of the constant q_0 in the equation (7.3) does not determine the sign of the particle charge. In complete analogy with the case of the free particle, the sign of particle charge is determined by the integral $\int \rho\, dV$.

If we apply the charge conjugation transformation $q_0 \to -q_0$ and $\mu_0 \to -\mu_0$ to the equation (8.35), then the negative energy solutions will only satisfy the condition (8.43). However, the simultaneous change of sign of energy E and potential energy $U(r) = q_0\varphi(r)$ results in the reversion of sign of the right-hand-side of the equation (8.47). Notice, that the wave functions of the charge conjugated states can be also obtained by the action of the operator $\hat{C}$ on the wave functions (8.45) and (8.46). Thus, the degeneracy of energy spectrum with respect to its sign is the demonstration of the charge symmetry, which means that the change of sign of all charges in the isolated system does not result in the change of the physical state of a system.

The analysis given above shows that the positive energy solutions, corresponding to electron, are

$$\begin{aligned}
\Psi_{njm}^{(l=j-1/2)}(\mathbf{r}) &= C_1\begin{pmatrix}\Omega_{jlm}^{(1)}\\ -\zeta\Omega_{jl+1m}^{(2)}\end{pmatrix}G\left(-n,\nu_+,r\right),\\
\Psi_{njm}^{(l=j+1/2)}(\mathbf{r}) &= C_2\begin{pmatrix}\Omega_{jlm}^{(2)}\\ -\zeta\Omega_{jl-1m}^{(1)}\end{pmatrix}G\left(-n,\nu_-,r\right),
\end{aligned} \quad (8.48)$$

the energies of these states are defined by

$$E_{nj}^{(l=j\mp1/2)} = \frac{m_0c^2(n+\nu_\pm)}{\sqrt{(n+\nu_\pm)^2+Z^2\alpha^2}}, \quad (8.49)$$

where $\nu_+ = \nu_1$, $\nu_- = \nu_2$. The normalization constants of the wave functions (8.48) can be easily calculated. They are

$$C_{njl} = \frac{(2\kappa_{njl})^{\nu}}{\sqrt{n!\,\Gamma(n+2\nu)}}\sqrt{\frac{m_0c/\hbar}{(1-\zeta_j^2)}}\left[\frac{\sqrt{m_0c^2 - E_{njl}^2}}{E_{njl}(n+\nu) + Z\alpha\sqrt{m_0c^2 - E_{njl}^2}}\right]^{1/2}.$$

At a given value of j, the two solutions (8.48) have the opposite parity.

It is natural to compare the spectrum (8.49) with the spectrum (6.132), that was obtained from the solution of the Dirac equation. The principle difference between these two spectra consists in the fact, that the spectrum (6.132) depends on the total angular momentum j and does not depend on the mutual orientation of the orbital angular momentum $\mathbf{l}$ and spin $\mathbf{s}$. Contrary, the spectrum (8.49) depends on the mutual orientation of the orbital angular momentum and spin. It means that the energy of the states with the same value of j and different values of l (or spin s) is different. Particularly, the Lamb shift is appeared in the spectrum of hydrogen atom. Indeed, for the $2s_{1/2}$ and $2p_{1/2}$ states we have

$$\gamma(2s_{1/2}) = (Z\alpha)^2 - 1 + \sqrt{1-(Z_1\alpha)^2},$$

$$\gamma(2p_{1/2}) = (Z\alpha)^2 - 1 - \sqrt{1-(Z_1\alpha)^2}.$$

Of course, the fine structure of the hydrogenic spectra still remains. Indeed,

$$\gamma(2p_{3/2}) = (Z\alpha)^2 - 4 + \sqrt{4-(Z_1\alpha)^2}.$$

Thus, the equation (8.49) shows the presence of the shift between the $2s_{1/2}$ and $2p_{1/2}$ states.

It should be noted finally, that at $\mu_0 = \mu_{\mathrm{B}}$, i.e. when the electron magneton is equal to the Bohr magneton, the shift of the $2s_{1/2}$ and $2p_{1/2}$ states disappears and the energy spectrum (8.49) transforms into the energy spectrum (6.132)

$$E_{nj} = \frac{m_0c^2\left(n_r + \sqrt{k^2 - Z^2\alpha^2}\right)}{\sqrt{\left(n_r + \sqrt{k^2 - Z^2\alpha^2}\right)^2 + Z^2\alpha^2}}, \tag{8.50}$$

where n_r is the radial quantum number, and $k = l, -(l+1)$.

8.2.3 Continuous spectrum

In the case, when $E > m_0c^2$, the parameter κ becomes the pure imaginary number

$$\kappa = \frac{\sqrt{m_0^2c^4 - E^2}}{\hbar c} = -i\frac{\sqrt{E^2 - m_0^2c^4}}{\hbar c} = -ik. \tag{8.51}$$

The radial wave function (8.40) takes in this case the following form

$$G(\xi,\nu,r) = \exp(ikr)(2kr)^{\nu-1}F(\nu - i\eta, 2\nu, -i2kr), \tag{8.52}$$

where

$$\eta = \frac{EZ\alpha}{\hbar ck}. \tag{8.53}$$

The asymptotical form of the wave function (8.52) at $r \to \infty$ is

$$\begin{aligned} G(\xi,\nu,r) &= \exp\left(-\frac{\pi\eta}{2}\right)\frac{\Gamma(2\nu)}{|\Gamma(\nu+i\eta)|}\frac{1}{2kr}\times \\ &\times\left\{\exp\left[i\left(kr + \eta\ln 2kr - \frac{\pi\nu}{2} - \arg\Gamma(\nu+i\eta)\right)\right] + \right. \\ &\left. + \exp\left[-i\left(kr + \eta\ln 2kr - \frac{\pi\nu}{2} - \arg\Gamma(\nu+i\eta)\right)\right]\right\}. \end{aligned} \tag{8.54}$$

The states of the continuous spectrum are infinitely degenerated with respect to the total angular momentum and its projection.

8.3 Geonium atom

8.3.1 Electron motion in homogeneous magnetic field

Let us consider the problem on electron motion in the homogeneous magnetic field, $\mathbf{B} = \mathbf{e}_z B_0$. In this case the vector potential is $\mathbf{A} = \mathbf{e}_\varphi B_0\rho/2$ and the equation (7.3) takes the form

$$\left[\Delta + \frac{E^2 - m_0^2c^4}{\hbar^2c^2} + \frac{i|q|B_0}{\hbar c}\frac{\partial}{\partial\varphi} - \left(\frac{qB_0}{2\hbar c}\right)^2\rho^2\right]\Psi(\mathbf{r}) = \frac{2m_0\mu_0B_0}{\hbar^2}\Sigma_z\Psi(\mathbf{r}). \tag{8.55}$$

where μ_0 is again the magnitude of the electron magneton. As we have mentioned above, the axial projection j_z of the total angular momentum,

$$\mathbf{j} = -i\hbar[\mathbf{r}\nabla] + \frac{\hbar}{2}\mathbf{\Sigma},$$

is the conservative variable, when a particle moves in the homogeneous magnetic field. However, the Hamiltonian (8.55) commutes, separately, with axial projection of the orbital angular momentum

$$l_z = -i\hbar[\mathbf{r}\nabla]_z = -i\hbar\frac{\partial}{\partial\varphi},$$

and the axial projection of spin

$$s_z = \frac{\hbar}{2}\Sigma_z.$$

Hence, without loss of generality, we can assume that one of the spinors of the bispinor wave function is equal to zero. Let the lower spinor be equal zero and we can take the upper spinor in the form

$$\varphi(\mathbf{r}) = Cf(\rho)\exp(im\varphi + ik_z z)\, u_\sigma,$$

where the spinors u_σ are the eigenfunctions of the operator σ_z:

$$\sigma_z u_\sigma = \sigma u_\sigma,$$

they have the form

$$u_{+1} = \begin{pmatrix} 1 \\ 0 \end{pmatrix}, \quad u_{-1} = \begin{pmatrix} 0 \\ 1 \end{pmatrix}.$$

The substitution of the above wave function into the equation (8.55) results in the following equation for the radial wave function $f(\rho)$:

$$\frac{d^2 f}{d\rho^2} + \frac{1}{\rho}\frac{df}{d\rho} + \left(\beta_{m\sigma} - \frac{m^2}{\rho^2} - \left(\frac{qB_0}{2\hbar c}\right)^2 \rho^2\right) f = 0, \tag{8.56}$$

where

$$\beta_{m\sigma} = \left(\frac{E^2 - m_0^2 c^4}{\hbar^2 c^2} - k_z^2 - \frac{|q|\, B_0}{\hbar c}\left(m + \frac{\mu_0}{\mu_{\mathrm{B}}}\sigma\right)\right).$$

The general solution of the equation (8.56) is

$$f_{m\sigma}(\rho) = C_1\left(\kappa\rho^2\right)^{m/2}\exp\left(-\frac{\kappa\rho^2}{2}\right) F\left(\frac{1+m}{2} - \frac{\beta_{m\sigma}}{4\kappa},\ 1+m,\ \kappa\rho^2\right) +$$
$$+ C_2\left(\kappa\rho^2\right)^{-m/2}\exp\left(-\frac{\kappa\rho^2}{2}\right) F\left(\frac{1-m}{2} - \frac{\beta_{m\sigma}}{4\kappa},\ 1-m,\ \kappa\rho^2\right), \tag{8.57}$$

where

$$\kappa = \frac{|q|\, B_0}{2\hbar c} = \frac{m_0\omega_H}{2\hbar}$$

and

$$\omega_H = \frac{|q|\, B_0}{m_0 c}. \tag{8.58}$$

It is seen that the substitution $m \to -m$ transforms the first term in the equation (8.57) into the second one. Hence, let us consider the function

$$f_{m\sigma}(\rho) = C\left(\kappa\rho^2\right)^{m/2}\exp\left(-\frac{\kappa\rho^2}{2}\right) F\left(\frac{1+m}{2} - \frac{\beta_{m\sigma}}{4\kappa},\ 1+m,\ \kappa\rho^2\right).$$

The function $F(p, q, z)$ becomes polynomial, and, hence, satisfies the boundary condition at $\rho \to \infty$, when the following condition holds

$$\frac{1+m}{2} - \frac{\beta_{m\sigma}}{4\kappa} = -n, \tag{8.59}$$

where n is the non-negative integer. The normalized wave function is

$$\varphi_{nm\sigma}(\mathbf{r}) = \sqrt{\frac{\kappa}{\pi L}\frac{n!}{(n+m)!}}\left(\kappa\rho^2\right)^{m/2}\exp\left(-\frac{\kappa\rho^2}{2}\right)\times$$
$$\times L_n^{(m)}\left(\kappa\rho^2\right)\exp\left(im\varphi + ik_z z\right)u_\sigma, \quad (8.60)$$

where L is the length of the region available for electron motion in the direction of the applied magnetic field, $L_n^{(m)}(z)$ is the generalized Laguerre polynomial. The solution (8.60) is not divergent at $r = 0$, when the following condition holds $m \geqslant -n$.

It should be noted that in the case, when the condition (8.59) holds, the two solutions (8.57) coincide, because

$$L_{n+m}^{(-m)}(z) = (-z)^m \frac{n!}{(n+m)!} L_n^{(m)}(z).$$

8.3.2 Energy spectrum

The condition (8.59) yields the following equation for the energy spectrum

$$E_{nm\sigma}^2 = m_0^2c^4 + \hbar^2c^2k_z^2 + 2m_0c^2\left[\hbar\omega_H\left(n+m+\frac{1}{2}\right) + \mu_0 B_0\sigma\right]. \quad (8.61)$$

The solution (8.60) satisfies the boundary conditions at $\rho = 0$ and $\rho \to \infty$, when $-n \leq m$. Hence, by taking into account the wave function symmetric form given by (8.57), we should assume that in the equation (8.61) the quantum number m lies inside the interval $-n \leq m \leq n$.

Notice, that at $\mu_0 = \mu_\mathrm{B}$, i.e. when the electron magneton is equal to the Bohr magneton, the equation (8.61) becomes

$$E_{nm\sigma}^2 = m_0^2c^4 + c^2\hbar^2k_z^2 + 2m_0c^2\hbar\omega_H\left(n+m+\frac{1+\sigma}{2}\right). \quad (8.62)$$

Introducing the eigenvalues of the axial projection of the total angular momentum

$$j_z\Psi_M = \hbar M\Psi_M = \hbar\left(m + \frac{\sigma}{2}\right)\Psi_M,$$

the spectrum (8.61) can be rewritten in the form

$$E_{nM} = \sqrt{m_0^2c^4 + \hbar^2c^2k_z^2 + 2m_0c^2\hbar\omega_H\left(n+M+1/2\right)}. \quad (8.63)$$

Thus, we can see that, at $\mu_0 = \mu_\mathrm{B}$, the energy spectrum becomes degenerated with respect to the sum of the quantum numbers, $n + M$. This is the characteristic feature of the spectra obtained from the solution of the Pauli and Dirac equations. In the case, when $\hbar\omega_H \ll m_0 c^2$, the equation (8.63) is simplified and takes the form

$$\Delta E_{nM} = E_{nM} - m_0 c^2 \approx \hbar\omega_H \left(n + M + \frac{1}{2}\right) + \frac{\hbar^2 k_z^2}{2m_0}.$$

It follows from the last equation, that the states with the smallest projection of the total angular momentum $M = -n - 1/2$, at given j and $k_z = 0$, have the zero energy, $\Delta E_{n,-n-1/2} = 0$.

In general case, when $\mu_0 \neq \mu_\mathrm{B}$, the hyperfine structure is appeared in the spectrum of electron moving in uniform magnetic field

$$\Delta E_{nM\sigma} = \hbar\omega_H \left(n + M + 1/2\right) + \left(\mu_0 - \mu_\mathrm{B}\right) B_0 \sigma. \tag{8.64}$$

The energy level (n, M), characterized by the quantum numbers n and M, splits into the two sublevels (n, M, σ) with the energy difference of $2(\mu_0 - \mu_\mathrm{B})\, B_0$ between them. It is this splitting that was observed experimentally by Dehmelt and co-authors [15].

8.3.3 Induced magnetic field

The wave functions (8.60) enable us to calculate the current density produced by electron in the state with the quantum numbers (n, m, σ). Substituting the wave function into the equation (7.6), we can easily get

$$\begin{aligned}\mathbf{j}_{nm\sigma} = &-\mathbf{e}_\varphi 4c\mu_0 \frac{\kappa^{3/2}}{\pi L} \frac{n!}{(n + M - 1/2)!} \exp\left(-\kappa\rho^2\right) \left(\kappa\rho^2\right)^M \times \\ &\times L_n^{(M+1/2)}\left(\kappa\rho^2\right) L_n^{(M-1/2)}\left(\kappa\rho^2\right) + \\ &+ \mathbf{e}_\varphi 2c\left(\mu_0 - \mu_\mathrm{B}\right) f_{nm\sigma}^2(\rho) \left(\kappa\rho + \frac{m}{\rho}\right) - \mathbf{e}_z \frac{|q|\,\hbar k_z}{m_0} f_{nm\sigma}^2(\rho),\end{aligned} \tag{8.65}$$

where $L_n^{(m)}(z)$ is the generalized Laguerre polynomial. It is seen from the last equation, that, in the state with the largest negative projection of the total angular momentum, i.e. $M = -n - 1/2$, and $k_z = 0$, the current density is non-zero and proportional to $\mu_0 - \mu_\mathrm{B}$. Contrary, if we shall use the wave functions obtained from the solution of the Pauli or Dirac equations, the current density is exactly equal to zero in this specific state (see section 4.2).

To calculate the magnetic field, induced by the current density (8.65), we can use the Maxwell equation

$$\operatorname{curl} \mathbf{B} = \frac{4\pi}{c}\mathbf{j}. \tag{8.66}$$

Substituting the current density (8.65) into the equation (8.66) we get

$$\mathbf{B} = -\mathbf{e}_z B_0 \frac{1}{m_0 c^2} \frac{q^2}{L} \left[\sigma \frac{\mu_0}{\mu_\mathrm{B}} \frac{n!}{(n+m)!} \left(L_n^{(m)} \left(\kappa\rho^2\right) \right)^2 \left(\kappa\rho^2\right)^m \exp\left(-\kappa\rho^2\right) + \right.$$
$$\left. + 2 \frac{n!}{(n+m)!} \int\limits_{\sqrt{\kappa}\rho}^{\infty} \left(L_n^{(m)} \left(x^2\right) \right)^2 x^{2m} \exp\left(-x^2\right) \left(x + \frac{m}{x} \right) dx \right]. \quad (8.67)$$

A simple interpretation can be given to the equation (8.67). Indeed, it is well known that the capacity of a cylindrical capacitor is proportional to

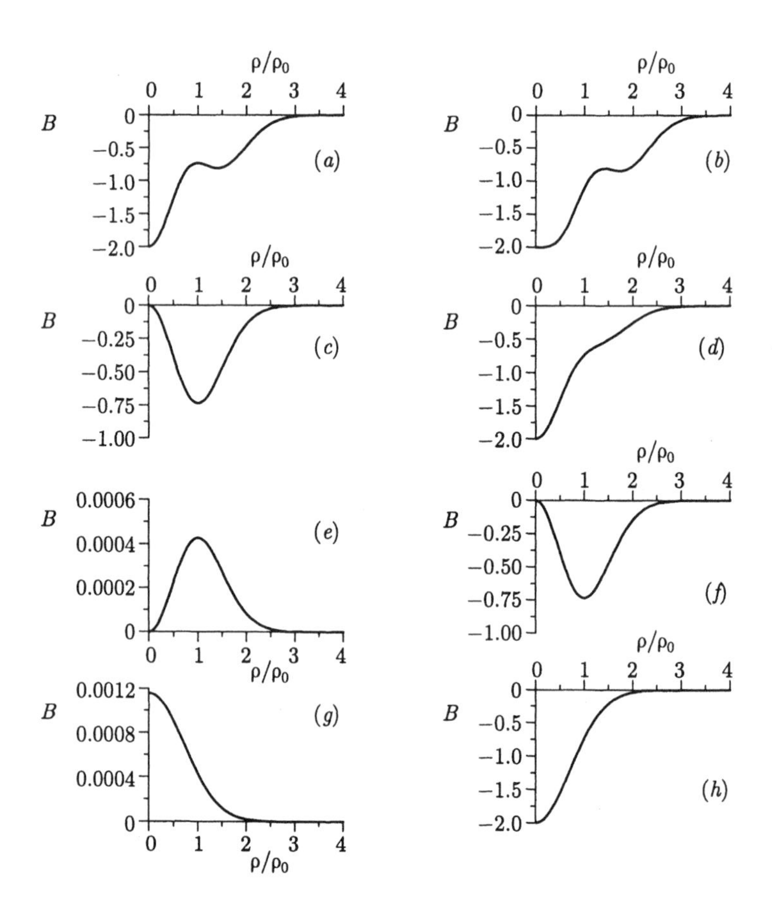

Figure 8.1. The spatial profile of the induced magnetic field for the states of electron: $n=1$, $m=1$, $s=-1$ (a); $n=1$, $m=1$, $s=1$ (b); $n=1$, $m=0$, $s=-1$ (c); $n=1$, $m=0$, $s=1$ (d); $n=1$, $m=-1$, $s=-1$ (e); $n=1$, $m=-1$, $s=1$ (f); $n=0$, $m=0$, $s=-1$ (g); $n=0$, $m=0$, $s=1$ (h)

its length. By taking this into account, one can guess that the variable

$$W_E(\rho) = \frac{q^2 I_{nM}(\rho)}{L}$$

determines the energy density of the electrostatic field produced by the electron in the state (n, M). Thus, we can see that the strength of the magnetic field, induced by the electron motion in the external magnetic field B_0, is proportional to the ratio of the energy of electrostatic field, produced by the electron in the given quantum state, to the electron rest energy. It is seen that the equation (8.67) consists of two terms, one of them is proportional to the magneton μ_0 and another is proportional to the Bohr magneton μ_B. The detailed interpretation of these two terms will be given in the next chapter.

The induced magnetic field as a function of the distance ρ is shown in the Fig. 8.1. It is seen that the induced magnetic field is opposite to the applied external field $\mathbf{B}_0$ in all states of electron, excepting the states of the smallest energy $(M = -n - 1/2,\ \sigma = -1)$. In the state of the smallest energy the induced field is parallel to the applied external field. Indeed, at $m = -n$ the equation (8.67) becomes

$$\mathbf{B} = -\mathbf{e}_z B_0 \frac{1}{m_0 c^2} \frac{q^2}{L} \left(\frac{\mu_0}{\mu_B} \sigma + 1 \right) \frac{1}{n!} \left(\kappa \rho^2 \right)^n \exp\left(-\kappa \rho^2 \right),$$

as far as $\mu_0 > \mu_B$, then at $\sigma = -1$ the right-hand-side of the last equation stands positive. Thus, the response of an ensemble of free electrons may be both diamagnetic and paramagnetic. It is this feature that shows the qualitative difference between the Dirac equation and equation (7.3), because in the frames of the Dirac theory the response is always diamagnetic.

8.4 Neutron motion in static magnetic field

The specific feature of the equation (7.3) is that it includes the magneton as an independent material constant, as a result the particle, of zero charge and non-zero magneton, interacts with the electric and magnetic fields. In this section we consider the problems of the neutron motion in the static magnetic field. The problems on the neutron motion in the electric field will be considered in the next chapter, because the successive interpretation of the phenomena appearing in the neutron motion in the electric field is possible only with the application of the concept of the electric polarization vector. This concept will be successively introduced in the next chapter.

8.4.1 Neutron reflection by magnetic field

Let us consider the neutron motion in static uniform magnetic field. The magnetic field is really uniform in the finite region of space, therefore it is more realistic to consider the problem, when an initially free neutron enters into the region of the non-zero magnetic field. Therefore, it is natural to assume that the strength of the magnetic field is determined by

$$\mathbf{B}(z) = \frac{\mathbf{B}_0}{1+\exp(-\beta z)}, \tag{8.68}$$

i.e. the strength of the field changes gradually from the zero value at $z \to -\infty$ to the value B_0 at $z \to \infty$. The characteristic spatial width of the varying field region is about $l = 1/\beta$.

In this case, the equation (7.3) is

$$\left(\Delta + \kappa^2\right)\Psi(\mathbf{r}) = -\frac{2m_0\mu_0}{\hbar^2}(\mathbf{\Sigma B})\Psi(\mathbf{r}), \tag{8.69}$$

where

$$\kappa^2 = \frac{E^2 - m_0^2c^4}{\hbar^2c^2}.$$

As already mentioned, the equations for spinors of the bispinor wave function are independent , when a particle moves in the static magnetic field. Therefore, the neutron wave function can be taken in the form

$$\Psi(\mathbf{r}) = \begin{pmatrix} \varphi(\mathbf{r}) \\ 0 \end{pmatrix}. \tag{8.70}$$

The Hamiltonian of the equation (8.69) commutes with the operator of the spin projection on the direction of the magnetic field $(\mathbf{n}_{\mathrm{B}}\mathbf{\Sigma})$, where $\mathbf{n}_{\mathrm{B}} = \mathbf{B}_0/B_0$, therefore it is convenient to take the three-dimensional spinor of the wave function (8.70) in the form

$$\varphi(\mathbf{r}) = u_\sigma f_\sigma(\mathbf{r}), \tag{8.71}$$

where the spinors u_σ are the eigenfunctions of the equation

$$(\mathbf{n}_{\mathrm{B}}\boldsymbol{\sigma})u_\sigma = \sigma u_\sigma. \tag{8.72}$$

The operator of momentum projection on the plane perpendicular to the z axis commutes with the Hamiltonian of the equation (8.69). Hence, this momentum projection is the conservative variable, and we can assume $\Psi(\mathbf{r}) = \Psi(z)$.

By substituting the equations (8.68) and (8.71) into the equation (8.69), we get

$$\left(\frac{d^2}{dz^2} + \kappa^2\right) f_\sigma(z) = \frac{2m_0|\mu_0|\,B_0}{\hbar^2}\frac{\sigma}{1+\exp(-\beta z)} f_\sigma(z),$$

where we have taken into account that the neutron magneton is negative. By introducing the new variable

$$\xi = -\exp(-\beta z),$$

we can transform the last equation to the hypergeometric type equation

$$\xi^2(1-\xi)\, f''_\sigma + \xi\,(1-\xi)\, f'_\sigma + \kappa^2(1-\xi)\, f_\sigma - a\sigma f_\sigma = 0, \qquad (8.73)$$

where

$$k^2 = \frac{\kappa^2}{\beta^2}, \qquad a = \frac{2m_0|\mu_0|\, B_0}{\hbar^2\beta^2}.$$

The general solution of the equation (8.73) is

$$f_\sigma(\xi) = C_1(-\xi)^{-i\nu_\sigma} F(-ik - i\nu_\sigma, ik - i\nu_\sigma, 1 - 2i\nu_\sigma, \xi) + \\ + C_2(-\xi)^{i\nu_\sigma} F(-ik + i\nu_\sigma, ik + i\nu_\sigma, 1 + 2i\nu_\sigma, \xi), \qquad (8.74)$$

where

$$\nu_\sigma = \sqrt{k^2 - a\sigma}. \qquad (8.75)$$

At $z \to \infty$ the solution (8.74) has the following asymptotic form

$$f_\sigma(z)|_{z\to\infty} = C_1 \exp(i\nu_\sigma\beta z) + C_2 \exp(-i\nu_\sigma\beta z).$$

Hence, only the first term in equation (8.74) satisfies the required boundary condition at $z \to \infty$. At $z \to -\infty$, the asymptotical form of this solution is

$$f_\sigma(z)|_{z\to-\infty} = \exp(-ik\beta z)\frac{\Gamma(1 - i2\nu_\sigma)\,\Gamma(i2k)}{\Gamma(ik - i\nu_\sigma)\,\Gamma(1 + ik - i\nu_\sigma)} + \\ + \exp(ik\beta z)\frac{\Gamma(1 - i2\nu_\sigma)\,\Gamma(-i2k)}{\Gamma(-ik - i\nu_\sigma)\,\Gamma(1 - ik - i\nu_\sigma)}.$$

By normalizing this solution to the unit current of the incident particles

$$f_\sigma(z) = \begin{cases} \exp(i\kappa z) + r_\sigma \exp(-i\kappa z), & z \to -\infty \\ t_\sigma \exp(i\nu_\sigma\beta z), & z \to \infty \end{cases} \qquad (8.76)$$

for the coefficients r_σ and t_σ we get

$$r_\sigma = \frac{\Gamma(-ik - i\nu_\sigma)\,\Gamma(1 - ik - i\nu_\sigma)\,\Gamma(i2k)}{\Gamma(ik - i\nu_\sigma)\,\Gamma(1 + ik - i\nu_\sigma)\,\Gamma(-i2k)},$$

$$t_\sigma = \frac{\Gamma(-ik - i\nu_\sigma)\,\Gamma(1 - ik - i\nu_\sigma)}{\Gamma(1 - i2\nu_\sigma)\,\Gamma(-i2k)}.$$

The continuity equation, applied to the wave functions of the continuous spectrum, results in the current conservation law

$$\int\limits_{z\to\infty} \mathbf{j}\, d\mathbf{S} + \int\limits_{z\to-\infty} \mathbf{j}\, d\mathbf{S} = 0.$$

Hence, the energy reflection R_σ and transmission T_σ coefficients are defined by

$$R_\sigma = \frac{\mathbf{e}_{-z}\left(\nabla\bar{\Psi}_r \cdot \Psi_r - \bar{\Psi}_r \cdot \nabla\Psi_r\right)\big|_{z\to-\infty}}{\mathbf{e}_z\left(\nabla\bar{\Psi}_0 \cdot \Psi_0 - \bar{\Psi}_0 \cdot \nabla\Psi_0\right)} = |r_\sigma|^2,$$

$$T_\sigma = \frac{\mathbf{e}_z\left(\nabla\bar{\Psi}_t \cdot \Psi_t - \bar{\Psi}_t \cdot \nabla\Psi_t\right)\big|_{z\to\infty}}{\mathbf{e}_z\left(\nabla\bar{\Psi}_0 \cdot \Psi_0 - \bar{\Psi}_0 \cdot \nabla\Psi_0\right)} = \frac{\mathrm{Re}\,(\nu_\sigma)}{k}\,|t_\sigma|^2.$$

By substituting here the obtained wave function, we get

$$R_\sigma = \left(\frac{\sinh\left(\pi\left(k - \mathrm{Re}\,(\nu_\sigma)\right)\right)}{\sinh\left(\pi\left(k + \mathrm{Re}\,(\nu_\sigma)\right)\right)}\right)^2, \quad T_\sigma = \frac{\sinh\left(2\pi\mathrm{Re}\,(\nu_\sigma)\right)\sinh\left(2\pi k\right)}{\sinh^2\!\left(\pi\left(k + \mathrm{Re}\,(\nu_\sigma)\right)\right)}. \tag{8.77}$$

It is seen, that

$$R_\sigma + T_\sigma = 1.$$

Notice, that according to the definition (8.75) we have $\mathrm{Re}\,(\nu_+) = 0$ at $k < a$ and $\mathrm{Re}\,(\nu_+) > 0$ at $k > a$, while $\mathrm{Re}\,(\nu_-) > 0$ at any k. Thus, the total reflection of neutrons with the polarization $\sigma = +1$ occurs at $k < a$. It is quite natural, because the energy of the neutron polarized along the magnetic field (and, hence, the magnetic moment directed oppositely magnetic field) increases with the increase of the magnetic field strength. The reflection coefficient for the incident neutrons of polarization $\sigma = -1$ is notably non-zero, only at $k \to 0$. The reflection $R_\pm$ and transmission $T_\pm$ coefficients as a function of k are shown at Fig. 8.2.

As we have mentioned in the Chapter 1, the main idea of experiments on search of the electric dipole moment of neutron is based on the comparison of the neutron spin precession frequency in the parallel and antiparallel magnetic and electric fields. Therefore, for further discussion it is convenient to calculate the spin precession frequency in the uniform magnetic field. It is seen from the Fig. 8.2 that the reflection coefficient sharply drops at $k \approx a$ and transmission coefficient tends to unity with the increase of k. If the spin of the incident neutron is polarized along the magnetic field, then the direction of the spin does not vary with neutron propagation through the magnetic field, because the spin projection Σ_z is integral of motion. If the incident neutron has the non-zero spin

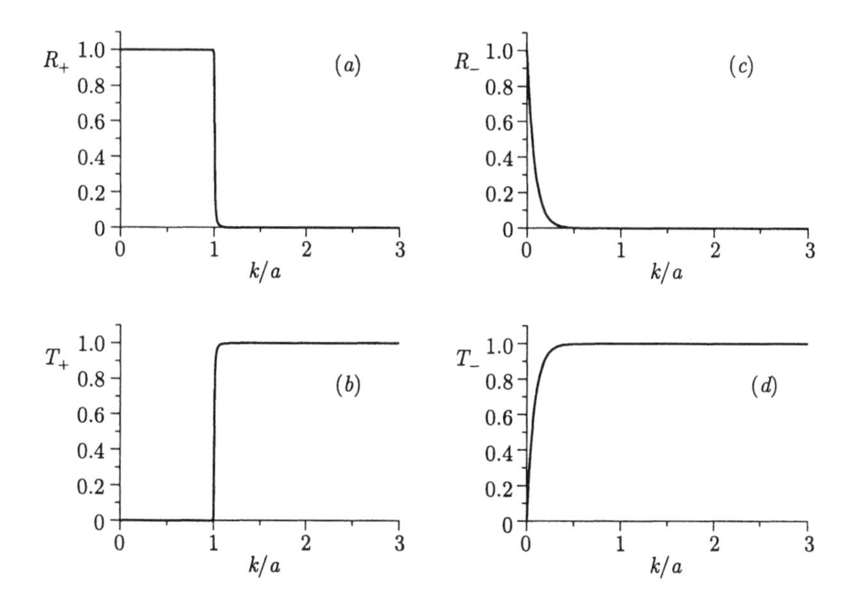

Figure 8.2. The reflection R_σ and transmission T_σ coefficients as functions of wavenumber k for spin projection: $\sigma = 1$ (a), (b); $\sigma = -1$ (c), (d)

projection on the plane perpendicular to the magnetic field, then the spin precession around the direction of the magnetic field occurs. Let the incident neutron be polarized along the x axis, then the spin projection vary with the distance traveled in the magnetic field in the following way

$$\varphi^+ \Sigma_x \varphi = \cos\left(\left(\nu_+ - \nu_-\right) z\right), \quad \varphi^+ \Sigma_y \varphi = \sin\left(\left(\nu_+ - \nu_-\right) z\right), \tag{8.78}$$

where $\nu_\pm$ is defined by the equation (8.75).

8.4.2 Neutron scattering by localized magnetic field

In the experiments on the neutron scattering by the gas of the polarized atoms the magnetic field is strongly localized in the volume of the atomic size. To model this process we can assume that the spatial profile of the magnetic field is

$$\mathbf{B}(z) = \frac{\mathbf{B}_0}{\cosh^2(\beta z)}. \tag{8.79}$$

In this case, the equation (7.3) takes the form

$$\left(\frac{d^2}{dz^2} + k^2\right) f_\sigma(z) = \frac{2m_0|\mu_0|\, B_0}{\hbar^2 \beta^2} \frac{\sigma}{\cosh^2(z)} f_\sigma(z). \tag{8.80}$$

In accordance with the discussion given in the previous subsection, we have taken the wave function in the form of (8.70)–(8.72). We have introduced in the equation (8.80) the dimensionless coordinate $z' = \beta z$ (in (8.80) and later, the primes are omitted), therefore the equation (8.80) depends on the parameter $k = \kappa/\beta$, where the pararmeter κ is the same as in previous subsection.

By introducing the new unknown function

$$f(z) = (\cosh z)^{-ik} g(z)$$

and new variable

$$\eta = \frac{1}{1+\exp(2z)},$$

we can transform the equation (8.80) to the hypergeometric type equation

$$\eta(1-\eta) g''_\sigma(\eta) + (1+ik)(1-2\eta) g'_\sigma(\eta) - ik(1+ik) g_\sigma(\eta) = a\sigma g_\sigma(\eta), \tag{8.81}$$

where

$$a = \frac{2m_0|\mu_0| B_0}{\hbar^2\beta^2}. \tag{8.82}$$

The general solution of the equation (8.81) is

$$g_\sigma(\eta) = C_1 F(1+ik+s_\sigma, ik-s_\sigma, 1+ik, \eta) + \\ + C_2 \eta^{-ik} F(1+s_\sigma, -s_\sigma, 1-ik, \eta),$$

where

$$s_\sigma = \frac{1}{2}\left(\sqrt{1-4a\sigma} - 1\right). \tag{8.83}$$

The asymptotical form of the solution $f(z)$ at $z \to \infty$ is

$$f(z)|_{z\to\infty} = C_1 \exp(-ikz) + C_2 \exp(ikz).$$

Hence, the term proportional to the coefficient C_2 is only satisfied to the required boundary condition at $z \to \infty$. Thus, the solution reads

$$f_\sigma(z) = \exp(ikz) F\left(1+s_\sigma, -s_\sigma, 1-ik, \frac{1}{1+\exp(2z)}\right). \tag{8.84}$$

At $z \to -\infty$ this solution takes the following asymptotical form

$$f(z)|_{z\to-\infty} = \frac{\Gamma(1-ik)\Gamma(-ik)}{\Gamma(-ik-s)\Gamma(1-ik+s)} \exp(ikz) + \\ + \frac{\Gamma(1-ik)\Gamma(ik)}{\Gamma(1+s)\Gamma(-s)} \exp(-ikz). \tag{8.85}$$

Thus, by normalizing the function $f(z)$ to the unit current of the incident particles (see (8.76)) and taking into account the definition (8.83) of the reflection and transmission coefficients, we get

$$R_\sigma = |r_\sigma|^2 = \frac{\sin(\pi s_\sigma)\sin(\pi s_\sigma^*)}{\sinh^2(\pi k) + \sin(\pi s_\sigma)\sin(\pi s_\sigma^*)},$$
$$T_\sigma = |t_\sigma|^2 = \frac{\sinh^2(\pi k)}{\sinh^2(\pi k) + \sin(\pi s_\sigma)\sin(\pi s_\sigma^*)}, \tag{8.86}$$

where we have taken into account, that at $a > 1/4$ and $\sigma = +1$ the parameter s_σ is the complex number. Indeed, in this case, the parameter s_+ becomes

$$s_{\sigma=+1} = \frac{1}{2}\left(-1 + i\sqrt{4a-1}\right).$$

It follows directly from the equations (8.86), that

$$R_\sigma + T_\sigma = 1.$$

It is also seen from the equations (8.86), that the boundary of the reflection region is determined by the condition $k_0 = |s|$. It is seen from the equation (8.82), that the parameter a is equal to the ratio of the energy of magnetic dipole interaction, $|\mu_0| B_0$ to the energy $\hbar^2\beta^2/(2m_0)$ that determines the kinetic energy of a particle localized in the region of the non-zero magnetic field. At $a \ll 1/4$, the boundary energy of reflection is given by

$$\Delta E_0 = E_0 - m_0 c^2 \approx \mu_0 B_0 \cdot a.$$

Thus, at small values of the parameter a the boundary energy is the product of the magnetic dipole interaction energy and the parameter a. The non-zero difference of the reflection coefficients for the two states of the incident neutron polarization, $\sigma = \pm 1$, occurs in the region $\Delta E_+ - - \Delta E_- = 2|\mu_0|\, B_0 \cdot a$. At $a \gg 1/4$ we get $\Delta E_+ \approx |\mu_0|\, B_0$. Fig. 8.3 shows the difference of the reflection coefficients $\Delta R = R_+ - R_-$ as a function of the wavenumber k at following values of the parameter a: 1/40 (a), 1/20 (b), 1/8 (c), 1/4 (d), 1 (e), 2 (f), 4 (g), 8 (h). It is seen that the increase in the strength of the magnetic field results in broadening of the region, where the neutrons of polarization $\sigma = +1$ are reflected from the barrier and neutrons of polarization $\sigma = -1$ pass through it. It is seen from the equations (8.86) that the reflection coefficient for the neutrons of polarization $\sigma = -1$ becomes zero when the parameter s is equal to an integer. In this case we get $\Delta R = R_+$. Particularly, the curve f in Fig. 8.3 shows the case of $s_- = 1$.

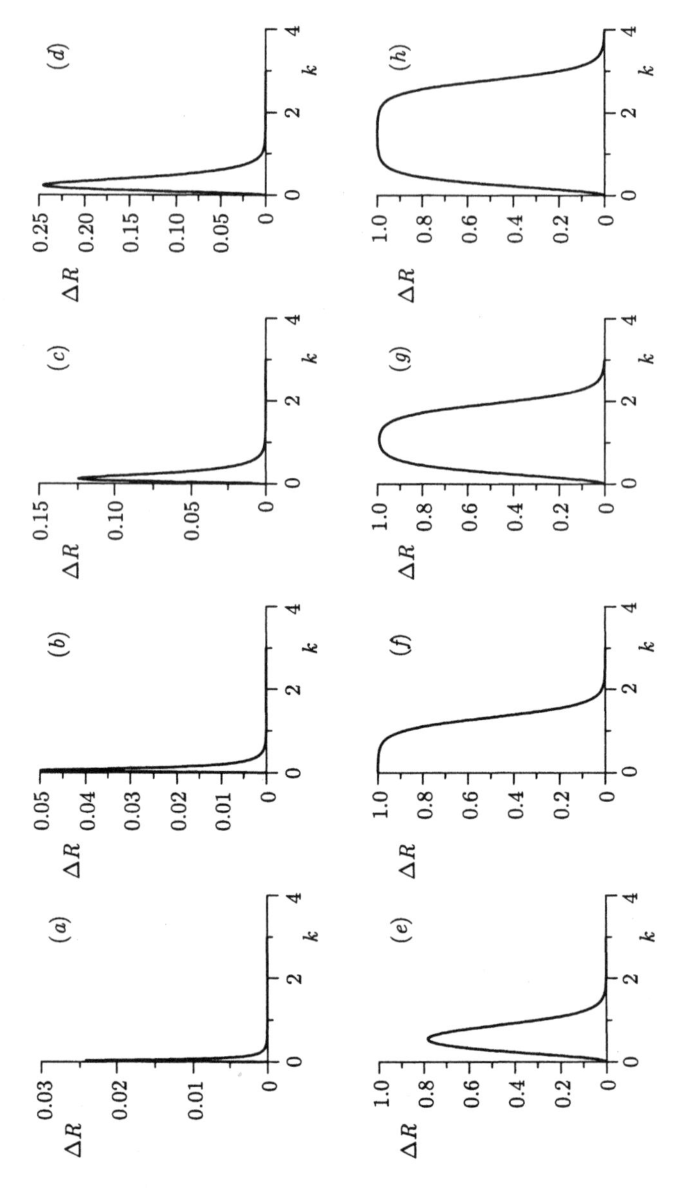

Figure 8.3. The difference of the reflection coefficients $\Delta R = R_+ - R_-$ as a function of the wavenumber k at following values of the parameter a: 1/40 (a), 1/20 (b), 1/8 (c), 1/4 (d), 1 (e), 2 (f), 4 (g), 8 (h)

8.4.3 The bound states of neutron in magnetic field

As we have discussed above, the energy of the magnetic dipole interaction is positive for the neutrons of polarization $\sigma = +1$, as a result these neutrons are reflected by the magnetic field, when their kinetic energy is small. Contrary, the energy of the magnetic dipole interaction for neutrons of polarization $\sigma = -1$ is negative, hence, these neutrons can form the bound states inside the magnetic field barrier. The energy of the bound state is negative, therefore

$$k = \frac{\sqrt{E^2 - m_0^2 c^4}}{\hbar c \beta} = i \frac{\sqrt{m_0^2 c^4 - E^2}}{\hbar c \beta} = i\varepsilon.$$

With the help of the hypergeometric function transformation

$$F(a, b, c, z) = (1 - z)^{c-a-b} F(c - a, c - b, c, z),$$

the wave function (8.84) can be written as follows

$$f_\sigma(z) = \left(\frac{1}{\exp(z) + \exp(-z)}\right)^\varepsilon F\left(\varepsilon - s_\sigma, 1 + \varepsilon + s_\sigma, 1 + \varepsilon, \frac{1}{1 + \exp(2z)}\right). \tag{8.87}$$

The function (8.87) tends to zero at $z \to \infty$. In order this function to be zero at $z \to -\infty$, the following condition should hold

$$\varepsilon - s_\sigma = -n, \tag{8.88}$$

where n is the non-negative integer. The condition $\varepsilon > 0$ means that

$$n < s_\sigma.$$

As far as s_+ is negative at $a > 0$, then the last condition holds true only for $\sigma = -1$. Thus, the neutrons polarized opposite to the magnetic field can form the bound states in the magnetic field, because their magnetic moment is parallel to the magnetic field and, hence, the energy of the magnetic dipole interaction is negative. The condition (8.88) yields the energy spectrum of the bound states

$$E_n = \sqrt{m_0^2 c^4 - \hbar^2 c^2 \beta^2 (s_- - n)^2}. \tag{8.89}$$

Hence

$$\Delta E_n = m_0 c^2 - E_n \approx \frac{\hbar^2 \beta^2}{2 m_0} (s_- - n)^2.$$

There is a finite number of the bound states at a given magnitude of the magnetic field strength.

Chapter 9

ORIGIN OF LAMB SHIFT

In the previous chapter we have applied the equation (7.3) to the analysis of a number of problems on particle motion in the static external fields. The analysis has shown that the equation (7.3) results in a number of specific features in the particle behavior that are qualitatively different from the predictions of other theories. Indeed, the spectrum of the electron moving in the Coulomb field shows the presence of the splitting of $2s_{1/2}$ and $2p_{1/2}$ states. The numerical estimations of the magnitude of this splitting will be given later, but it is essentially more important to understand what is the origin of the splitting. The splitting occurs also for the levels of electron in the uniform magnetic field, and the magnetic field of response demonstrates a number of the unusual features. The analysis of the problem on the neutron motion in the magnetic field has shown that the neutron can be reflected by the magnetic field, and it can form the bound states in the localized magnetic field. The amplitude of reflection significantly depends on the incident neutron polarization. This is also agree qualitatively with the experimental data discussed in the Chapter 1.

Before we start with the numerical estimations and comparison between the theory and experiment, it is quite useful, at least qualitatively, to answer the question: why, in spite of the close connection between the equation (7.3) and the Dirac equation, so drastic difference arises between the behavior of the particles obeying these two equations. By comparing these two equations we have not touched yet the equations for electromagnetic field, that follow from the action (7.1). The equation (7.4) for the four-potential of the electromagnetic field A_{μ} differs from that in the Dirac theory due to the difference in the current density four vector, the time and spatial components of which are

$$\rho(\mathbf{r},t)=\frac{q_0}{m_0c}\left[-\frac{i\hbar}{2c}\left(\frac{\partial\bar{\Psi}}{\partial t}\Psi-\bar{\Psi}\frac{\partial\Psi}{\partial t}\right)-\frac{q_0}{c}\bar{\Psi}\varphi\Psi\right]+i\mu_0\nabla\left(\bar{\Psi}\boldsymbol{\alpha}\Psi\right), \quad (9.1)$$

$$\mathbf{j}(\mathbf{r},t)=\frac{q_0}{m_0}\left[\frac{i\hbar}{2}\left(\nabla\bar{\Psi}\cdot\Psi-\bar{\Psi}\cdot\nabla\Psi\right)-\frac{q_0}{c}\bar{\Psi}\mathbf{A}\Psi\right]+ \\ +c\mu_0\,\mathrm{curl}\left(\bar{\Psi}\boldsymbol{\Sigma}\Psi\right)-i\mu_0\frac{\partial}{\partial t}\left(\bar{\Psi}\boldsymbol{\alpha}\Psi\right). \quad (9.2)$$

9.1 Static fields

Let us write the equations for the strength of the electric and magnetic fields produced by the particle in the state $\Psi(\mathbf{r},t)=\Psi_n(\mathbf{r})\times \times\exp(-iE_nt/\hbar)$. In the next chapter we shall consider the problem of interaction of particles. Taking this in mind, it is convenient to label the field potentials in the following way: the scalar potential produced by particle a at a position of particle b is denoted as $\varphi_a(\mathbf{r}_b)$.

The solutions of the steady-state equations (7.4) for the electromagnetic field scalar and vector potentials are given by

$$\varphi_a(\mathbf{r}_b)=\frac{q_a}{m_ac^2}\int\frac{\bar{\Psi}_a\left(E_a-q_a\varphi_b(\mathbf{r}_a)\right)\Psi_a}{r_{ba}}dV_a-i\mu_a\int\frac{\bar{\Psi}_a\boldsymbol{\alpha}_a\mathbf{r}_{ba}\Psi_a}{r_{ba}^3}dV_a, \quad (9.3)$$

$$\mathbf{A}_a(\mathbf{r}_b)=\frac{q_a}{m_ac}\int\frac{1}{r_{ba}}\left[\frac{i\hbar}{2}\left(\nabla\bar{\Psi}_a\cdot\Psi_a-\bar{\Psi}_a\nabla\Psi_a\right)-\frac{q_a}{c}\bar{\Psi}_a\mathbf{A}_b(\mathbf{r}_a)\Psi_a\right]dV_a+ \\ +\mu_a\int\frac{\bar{\Psi}_a\left[\boldsymbol{\Sigma}_a\mathbf{r}_{ba}\right]\Psi_a}{r_{ba}^3}dV_a, \quad (9.4)$$

where $\mathbf{r}_{ba}=\mathbf{r}_b-\mathbf{r}_a$, and the field potentials $\mathbf{A}_b(\mathbf{r}_a)$ and $\varphi_b(\mathbf{r}_a)$ are produced by the charges external with respect to the considered particle. The obtained equations have the very simple physical interpretation. The first term in the equation (9.3) has the form of the scalar potential produced by the space charge with the charge density

$$\rho_e(\mathbf{r}_a)=\frac{q_a}{m_ac^2}\bar{\Psi}_a\left(E_a-q_a\varphi_b(\mathbf{r}_a)\right)\Psi_a.$$

It is seen that the above equation coincides with the first term in the equation (7.7). The second term in the equation (9.3) has the form of the scalar potential produced by the space charge with the electric polarization vector (or dipole moment density) defined by the following equation

$$\mathbf{P}=-i\mu_0\bar{\Psi}\boldsymbol{\alpha}\Psi. \quad (9.5)$$

The first term in the equation (9.4) is the vector potential originated from the orbital motion of the particle. The second term in the equation (9.4) has the form of the vector potential produced by the space charge with the magnetic polarization vector (or magnetization) defined by the following equation

$$\mathbf{M} = \mu_0 \bar{\Psi} \mathbf{\Sigma} \Psi. \tag{9.6}$$

The equations (9.3), (9.4) provide the following equations for the strength of the electric and magnetic fields

$$\mathbf{E}_a(\mathbf{r}_b) = \int \frac{\mathbf{r}_{ba}\rho_e^{(a)}(\mathbf{r}_a)}{r_{ba}^3} dV_a + \int \frac{\left(3\mathbf{r}_{ba}(\mathbf{P}_a\mathbf{r}_{ba}) - \mathbf{P}_a r_{ba}^2\right)}{r_{ba}^5} dV_a, \tag{9.7}$$

$$\begin{aligned}\mathbf{H}_a(\mathbf{r}_b) = \int & \frac{\left(3\mathbf{r}_{ba}(\mathbf{M}_a\mathbf{r}_{ba}) - \mathbf{M}_a r_{ba}^2\right)}{r_{ba}^5} dV_a - \\ & - \frac{q_a}{m_a c} \int \frac{1}{r_{ba}^3} \left\{ \frac{i\hbar}{2} \left([\mathbf{r}_{ba}\nabla_a] \bar{\Psi}_a \cdot \Psi_a - \bar{\Psi}_a \cdot [\mathbf{r}_{ba}\nabla_a] \Psi_a \right) - \right. \\ & \left. - \frac{q_a}{c} \bar{\Psi}_a \left[\mathbf{r}_{ba}\mathbf{A}_b(\mathbf{r}_a)\right] \Psi_a \right\} dV_a, \end{aligned} \tag{9.8}$$

The physical meaning of the last equations is quite obvious, and these equations do not require the further discussion.

It is seen from the equation (9.5) that, in standard representation of the Dirac matrices, the electric polarization vector $\mathbf{P}$ is non-zero only in those states of the particle when both spinors of the bispinor wave function are non-zero. The equation (9.7) demonstrates explicitly the significant difference between the Dirac equation and equation (7.3). Indeed, according to equation (9.7), the particle of zero charge $q_0 = 0$ and non-zero magnetic moment $\mu_0 \neq 0$ can produce the electric field due to presence of non-zero electric polarization vector.

Along with the electric (9.5) and magnetic (9.6) polarization vectors we can introduce the electric $\mathbf{d}$ and magnetic $\mathbf{m}$ dipole moments:

$$\mathbf{d} = \int \mathbf{P}\, dV = -i\mu_0 \int \bar{\Psi}\boldsymbol{\alpha}\Psi\, dV, \quad \mathbf{m} = \int \mathbf{M}\, dV = \mu_0 \int \bar{\Psi}\mathbf{\Sigma}\Psi\, dV. \tag{9.9}$$

It is seen from the definition of $\mathbf{d}$ and $\mathbf{m}$, that the non-zero value of the electric and magnetic polarization vectors does not necessarily result in the non-zero value of the dipole and magnetic moments.

9.2 Symmetric form of the filed equations

The equations (9.3), (9.4) can be easily generalized for the case of the transient electromagnetic field. However, the interpretation of the equations (7.3)–(7.5) becomes more obvious if we write down the equations for the electric and magnetic fields, i.e. the set of the Maxwell equations. It is seen from the equations (9.1), (9.2) that the charge and current densities consist of the two terms. The first terms in these equations are proportional to the charge q_0 and do not include the spin operators:

$$\rho_e(\mathbf{r},t) = \frac{q_0}{m_0 c}\left[-\frac{i\hbar}{2c}\left(\frac{\partial\bar{\Psi}}{\partial t}\Psi - \bar{\Psi}\frac{\partial\Psi}{\partial t}\right) - \frac{q_0}{c}\bar{\Psi}\varphi\Psi\right], \tag{9.10}$$

$$\mathbf{j}_e(\mathbf{r},t) = \frac{q_0}{m_0}\left[\frac{i\hbar}{2}\left(\nabla\bar{\Psi}\cdot\Psi - \bar{\Psi}\cdot\nabla\Psi\right) - \frac{q_0}{c}\bar{\Psi}\mathbf{A}\Psi\right]. \tag{9.11}$$

There are a number of reasons to assume, that these parts of the charge density and current density are associated with the electric charge density and electric current density. Firstly, these terms are proportional to the electric charge q_0, while the rest terms are proportional to magneton μ_0. Secondly, the rest terms of the equations (9.1), (9.2) are proportional to the derivatives of the bilinear combinations of the wave functions, hence, they will not contribute to the integral charge and current in the steady-state case. Thirdly, these terms do not depend on the spin operators, while the rest terms depend on them. Fourthly, the charge and current densities defined by the equations (9.10) and (9.11), respectively, obey the continuity equation

$$\frac{\partial\rho_e}{\partial t} + \mathrm{div}\,\mathbf{j}_e = 0.$$

With the help of the standard definition of the electric and magnetic fields

$$\mathbf{E} = -\frac{1}{c}\frac{\partial\mathbf{A}}{\partial t} - \nabla\varphi, \quad \mathbf{B} = \mathrm{curl}\,\mathbf{A}, \tag{9.12}$$

we can rewrite the equations (7.4) in the following form

$$\mathrm{curl}\,(\mathbf{B} - 4\pi\mathbf{M}) = \frac{1}{c}\frac{\partial}{\partial t}(\mathbf{E} + 4\pi\mathbf{P}) + \frac{4\pi}{c}\mathbf{j}_e, \tag{9.13}$$

$$\mathrm{div}\,(\mathbf{E} + 4\pi\mathbf{P}) = 4\pi\rho_e, \tag{9.14}$$

where the vectors **P** and **M** are defined by the equations (9.5) and (9.6), respectively.

To form the Maxwell set of equations, the equations (9.13), (9.14) are supplied by the following two equations

$$\mathrm{div}\,\mathbf{B} = 0, \tag{9.15}$$

$$\mathrm{curl}\,\mathbf{E} = -\frac{1}{c}\frac{\partial\mathbf{B}}{\partial t}, \tag{9.16}$$

which follow directly from the definitions given by (9.12).

In the previous subsection we have shown that in the case of static fields the vectors **P** and **M** play the role of the electric and magnetic polarization vectors. It is seen from the equations (9.13), (9.14) that this interpretation is still true in transient case, too. Therefore we can introduce the electric displacement **D** and magnetic field strength **H** vectors

$$\mathbf{D} = \mathbf{E} + 4\pi\mathbf{P}, \quad \mathbf{H} = \mathbf{B} - 4\pi\mathbf{M}, \tag{9.17}$$

then the set of equations (9.13)–(9.16) takes the following form

$$\begin{aligned} \text{div } \mathbf{D} &= 4\pi\rho_e, \\ \text{div } \mathbf{H} &= 4\pi\rho_m, \\ \text{curl } \mathbf{H} &= \frac{1}{c}\frac{\partial \mathbf{D}}{\partial t} + \frac{4\pi}{c}\mathbf{j}_e, \\ \text{curl } \mathbf{D} &= -\frac{1}{c}\frac{\partial \mathbf{H}}{\partial t} - \frac{4\pi}{c}\mathbf{j}_m, \end{aligned} \tag{9.18}$$

where

$$\begin{aligned} \rho_m(\mathbf{r},t) &= -\mu_0 \text{ div } \left(\bar{\Psi}\boldsymbol{\Sigma}\Psi\right), \\ \mathbf{j}_m(\mathbf{r},t) &= \mu_0\left[\frac{\partial}{\partial t}\left(\bar{\Psi}\boldsymbol{\Sigma}\Psi\right) - ic \text{ curl } \left(\bar{\Psi}\boldsymbol{\alpha}\Psi\right)\right]. \end{aligned} \tag{9.19}$$

Thus, one can see that the use of definitions (9.17), which are similar to the definitions of the classical electrodynamics, results in the symmetric form of the Maxwell equations, where ρ_e and $\mathbf{j}_e$ play the role of the electric charge density and electric current density, and ρ_m and $\mathbf{j}_m$ play the role of the magnetic charge density and magnetic current density. It should be noted that the magnetic charge and current densities obey the continuity equation as well

$$\frac{\partial \rho_m}{\partial t} + \text{div } \mathbf{j}_m = 0. \tag{9.20}$$

In classical electrodynamics the following equations are used for the induced charge and current densities

$$\rho_e(\mathbf{r},t) = -\text{ div } \mathbf{P}_e, \quad \mathbf{j}_e(\mathbf{r},t) = \frac{\partial \mathbf{P}_e}{\partial t} + \text{curl } \mathbf{M}_e,$$

where the vectors $\mathbf{P}_e$ and $\mathbf{M}_e$ are associated with the internal medium fields (the applied external fields break the uniform charge distribution of the initially disordered macroscopic medium). There is a close analogy between these equations and equations (9.19), which can be written in the following form

$$\rho_m(\mathbf{r},t) = -\text{ div } \mathbf{M}, \quad \mathbf{j}_m(\mathbf{r},t) = \frac{\partial \mathbf{M}}{\partial t} + \text{curl } \mathbf{P}.$$

There is some difference between the electric and magnetic charge density, as it is seen from the equation (9.19). The magnetic charge density is identically equal to zero for the free particle, because the square modulus of the free-particle wave function does not depend on the coordinate. However, in the presence of the magnetic field the particle wave function changes, and the magnetic charge density becomes non-zero (see section 8.3). However, the magnetic charge, i.e. the spatial integral over the magnetic charge density $\int \rho_m(\mathbf{r}, t)\, dV$, is still identically equal to zero.

As already mentioned, when particle moves in the static magnetic field we can always assume that one of the spinors of the bispinor wave function is equal to zero. This is due to the diagonal form of the matrix $\boldsymbol{\Sigma}$ in the standard representation. In this case, according to definition (9.5), the electric polarization vector $\mathbf{P}$ is identically equal to zero. However, in the presence of the electric field, both spinors of the wave function become non-zero. We shall see later that it is the non-zero electric polarization vector that results in the appearance of the Lamb shift.

In the section 7.3 we have shown that the operator $\boldsymbol{\Sigma}$ is the generator of the three-dimensional rotation transformation, as a result the operator $\boldsymbol{\Sigma}$ is associated with the intrinsic angular momentum of a particle. The equation (9.6) establishes the linear relation between the magnetic polarization vector and spin. On the other hand, the operator $\boldsymbol{\alpha}$ is the generator of the four-dimensional rotation transformation, hence, the electric polarization vector is non-zero in a such motion of a particle, when not only the direction but the magnitude of the particle velocity is changed. The scalar wave function of the Klein–Gordon–Fock equation is invariant with respect to the three- and four-dimensional rotations, as a result the electric and magnetic polarization vectors are equal to zero in any state of the KGF particle. Therefore, we can assume that the operators of the electric and magnetic polarization vectors of an arbitrary spin particle are the generators of the three- and four-dimensional rotations of the appropriate equations.

9.3 Lamb shift

In this section we shall show how the Lamb shift can be interpreted. The solutions obtained in the section 8.2 enable us to write the explicit equations for the electric $\mathbf{P}$ and magnetic $\mathbf{M}$ polarization vectors for electron interacting with the Coulomb field. For the convenience purposes, let us label the wave functions (8.48) in the following way

$$\Psi^{(\pm)} = \Psi_{njm}^{(j=l\pm 1/2)}(\mathbf{r}) = C_{njm} \begin{pmatrix} u_{jlm} \\ i^{\pm 1} \zeta \sigma_r u_{jlm} \end{pmatrix} F(-n, \nu_{\pm}, r), \qquad (9.21)$$

where u_{jlm} is one of the spinors (8.36) and we have used the relation $\Omega^{(2)}_{jlm} = -i\sigma_r \Omega^{(1)}_{jlm}$. By substituting the equation (9.21) into the equation (9.5), we get

$$\mathbf{P}^{(\pm)} = -i\mu_0 \bar{\Psi}^{(\pm)} \boldsymbol{\alpha} \Psi^{(\pm)} = \pm \mathbf{e}_r \frac{2|\mu_0|\,\zeta}{1-\zeta^2} \bar{\Psi}^{(\pm)} \Psi^{(\pm)}. \tag{9.22}$$

Thus, the term in the Hamiltonian of the equation (8.35), describing the interaction of the electric polarization vector with the intra-atomic field $\mathbf{E}_a = \mathbf{e}_r \left(e/r^2\right)$, takes the form

$$\Delta E^{(j=l\pm 1/2)}_{PE} = -\int \mathbf{P}^{(\pm)} \mathbf{E}_a \, dV = \pm \frac{2|\mu_0 e|\,\zeta}{1-\zeta^2} \int \bar{\Psi}^{(\pm)} \Psi^{(\pm)} dr \tag{9.23}$$

Hence, at a given j, the energy is higher for the state with the smaller value of orbital angular momentum l. Particularly, the $nS_{1/2}$ level lies above the $nP_{1/2}$ level.

The equation for the magnetic polarization vector $\mathbf{M}$ is

$$\mathbf{M}^{(\pm)} = -|\mu_0| \left[\mathbf{e}_r \left(1-\zeta^2\right) u^+_{jlm} \sigma_r u_{jlm} + \mathbf{e}_\theta \left(1+\zeta^2\right) u^+_{jlm} \sigma_\theta u_{jlm}\right] \times \\ \times C^2_{njl} F^2(-n, \nu_\pm, r)\,. \tag{9.24}$$

For example, the magnetic moments at the $1S_{1/2}$ and $2P_{1/2}$ states are

$$\mathbf{M}\left(1S_{1/2}, m = \pm 1/2\right) = \\ = \mp \frac{|\mu_0|}{4\pi} \left[\mathbf{e}_r \cos\theta - \mathbf{e}_\theta \sin\theta - \zeta^2 (\mathbf{e}_r \cos\theta + \mathbf{e}_\theta \sin\theta)\right] C^2_{1,1/2,0} F^2(-1, \nu_+, r),$$

$$\mathbf{M}\left(2P_{1/2}, m = \pm 1/2\right) = \\ = \mp \frac{|\mu_0|}{4\pi} \left[\mathbf{e}_r \cos\theta + \mathbf{e}_\theta \sin\theta - \zeta^2 (\mathbf{e}_r \cos\theta - \mathbf{e}_\theta \sin\theta)\right] C^2_{2,1/2,1} F^2(-2, \nu_-, r).$$

It is seen, that the z projection of the magnetic polarization vector at the $1S_{1/2}$ state, along with the permanent component, has the additional

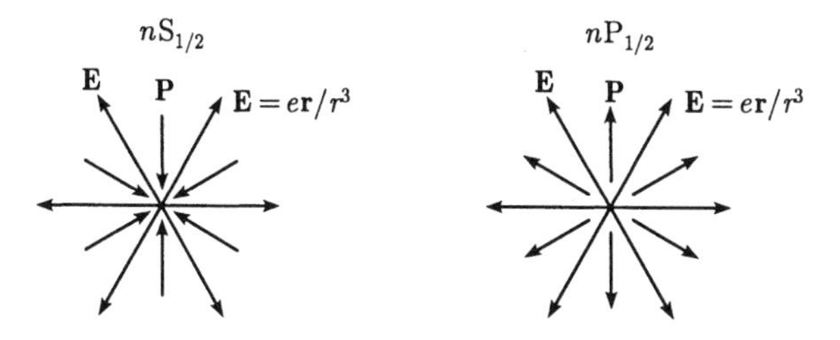

Figure 9.1. The mutual orientation of the intra-atomic electric field $\mathbf{E} = e\mathbf{r}/r^3$ and the electric polarization vector $\mathbf{P}$ for the $nS_{1/2}$ and $nP_{1/2}$ states

component proportional to α^2, which varies as $\cos(2\theta)$ with the angle θ. In the $2P_{1/2}$ state the main component varies with the angle θ and the additional small component is permanent. The interaction of the magnetic polarization vector with the intra-atomic magnetic field results in the well known Zeeman structure of atomic levels.

Thus, the splitting of the $nS_{1/2}$ and $nP_{1/2}$ states of electron in the Coulomb field is completely due to the interaction of the electric polarization vector with the intra-atomic electric field. Fig. 9.1 shows schematically the mutual orientation of the intra-atomic electric field $\mathbf{E} = e\mathbf{r}/r^3$ and the electric polarization vector $\mathbf{P}$ for the $nS_{1/2}$ and $nP_{1/2}$ states.

9.4 Neutron interaction with the static electric field

As we have discussed in section 1.3, the interest to the experiments on search of the electric dipole moment of elementary particles is enhanced significantly in the last time. The most of the experimental researches are devoted to the study of mechanism based on the violation of the CP invariance.

The analysis given above has shown that, on one hand, the equation (7.3) is the CPT invariant, and, on the other hand, this equation predicts the appearance of the induced electric polarization vector for the particle interacting with the electric field. The definition of the electric polarization vector given by the equation (9.5) does not violate the P invariance, because the upper and lower spinors of the bispinor wave function have the opposite parity. The vector $\mathbf{P}$, defined by equation (9.5), does not violate the T invariance, because the operator $\boldsymbol{\alpha}$ is the generator of the Lorentz transformation.

Hence, it can be anticipated that the interaction of the neutron with the electric field will be drastically different in the case when we accept the definition (9.5) and in the case when we accept the definition given in the section 1.3. One can see that there are a number of motivations to study the interaction of neutron with the electric and magnetic fields, because the results of this study can play the significant role both for the general theory of the spin-1/2 particles and atomic spectroscopy, especially for the theory of the Lamb shift.

9.4.1 Neutron reflection by the static electric field

Let us consider the problem on the neutron interaction with the static electric field of the following spatial profile $\mathbf{E} = \mathbf{e}E(z)$, where $\mathbf{e}$ is the arbitrary unit vector. This problem is of the general theoretical interest, because, on one hand, it enables us to reveal the mechanism of the neutron scattering by the electric field. On the other hand, this is an example of the exactly integrable scattering problem.

As we have repeatedly mentioned, in the case of the free particle or particle moving in the static magnetic field, the bispinor wave function is really spinor wave function, because if one of the spinors is equal zero in the initial state, it remains zero at any other stages of the particle evolution. The wave function normalization condition enables us to define the particle and antiparticle states on the basis of sign of the charge: $\int \rho\, dV = q_0$ for particle states and $\int \rho\, dV = -q_0$ for the antiparticle states. As far as the equations for upper and lower spinors of the bispinor wave function are independent in the absence of the electric field, then the particle interaction with the static magnetic field will not result in the appearance of antiparticles, and vise versa.

The situation is drastically changed when particle interacts with the electric field. In this case

$$\frac{1}{q_0}\int \rho(\mathbf{r})\, dV = \frac{1}{m_0 c^2}\int \varphi^{+}(\mathbf{r})\,(E - U(\mathbf{r}))\,\varphi(\mathbf{r})\, dV - \\ - \frac{1}{m_0 c^2}\int \chi^{+}(\mathbf{r})\,(E - U(\mathbf{r}))\,\chi(\mathbf{r})\, dV. \quad (9.25)$$

It is seen from the equation (7.3) that the ratio between the upper and lower spinors is varied with the variation of $U(r)$. Therefore, to preserve the charge conservation law (9.25) the ratio between the particle and antiparticle currents should be different at different spatial points. Thus, it can be assumed that the analysis of the problem on the neutron scattering by the spatial inhomogeneous static electric field can give us the further insight into the inerpretation of the particle and antiparticle solutions.

In the case of the neutral particle interacting with the electrostatic field $\mathbf{E}(\mathbf{r})$, the equation (7.3) takes the form

$$\left[\Delta + \kappa^2\right]\Psi(\mathbf{r}) = i\frac{2\mu_0 m_0}{\hbar^2}\left(\boldsymbol{\alpha}\mathbf{E}\right)\Psi(\mathbf{r}), \quad (9.26)$$

where

$$\kappa^2 = \frac{E^2 - m_0^2 c^4}{\hbar^2 c^2}. \quad (9.27)$$

Let the spatial profile of the static electric field be

$$\mathbf{E}(z) = \frac{\mathbf{E}_0}{\cosh^2(\beta z)}. \quad (9.28)$$

The Hamiltonian of the equation (9.26) commutes with the operator $\Sigma_E = (\mathbf{n}_E \boldsymbol{\Sigma})$, where $\mathbf{n}_E = \mathbf{E}_0/E_0$ is the unit vector. Hence, we can take the wave function in the following form

$$\Psi(\mathbf{r}) = \begin{pmatrix} \varphi f(\mathbf{r}) \\ i\chi g(\mathbf{r}) \end{pmatrix},$$

where the spinor φ is the eigenfunction of the equation

$$(\mathbf{n}_E \boldsymbol{\sigma})\, u_\sigma = \sigma u_\sigma,$$

and $\chi = (\mathbf{n}_E \boldsymbol{\sigma})\, \varphi$.

The Hamiltonian of the equation (9.26) commutes also with the operator of the momentum projection on the plane perpendicular to the z axis, therefore we can assume $\Psi(z)$.

By introducing the new variable

$$\eta = \frac{1}{1 + \exp(2\beta z)},$$

and the new unknown functions

$$f(z) = p(z)\cosh^{-ik}(\beta z), \quad g(z) = q(z)\cosh^{-ik}(\beta z),$$

we can transform the equations for the functions $f(z)$ and $g(z)$ to the hypergeometric type equations

$$\begin{aligned} \eta(1-\eta)\, p'' + (1+ik)(1-2\eta)\, p' - ik(1+ik)\, p &= -aq, \\ \eta(1-\eta)\, q'' + (1+ik)(1-2\eta)\, q' - ik(1+ik)\, q &= ap, \end{aligned} \tag{9.29}$$

where

$$k^2 = \frac{\kappa^2}{\beta^2}, \quad a = \frac{2\mu_0 m_0 E_0}{\hbar^2 \beta^2}. \tag{9.30}$$

The solutions of the obtained set of equations can be easily found with the help of the solution of the following equation

$$\eta(1-\eta)\, Q'' + (1+ik)(1-2\eta)\, Q' - ik(1+ik)\, Q + bQ = 0. \tag{9.31}$$

We require, that the solution of the one-dimensional scattering problem should have the following asymptotical form

$$f(z) = \begin{cases} t \cdot \exp(i\kappa z), & z \to \infty, \\ \exp(i\kappa z) + r \cdot \exp(-i\kappa z), & z \to -\infty. \end{cases} \tag{9.32}$$

The solution of the equation (9.31), having the required asymptotical form, is

$$Q(\eta) = \eta^{-ik} F\left(1 + \frac{1}{2}\left(\sqrt{1-4b} - 1\right), -\frac{1}{2}\left(\sqrt{1-4b} - 1\right), 1 - ik, \eta\right), \tag{9.33}$$

where $F(\alpha, \beta, \gamma, z)$ is the hypergeometric function. Therefore, the unknown functions $p(\eta)$ and $q(\eta)$ can be taken as

$$p(\eta) = AQ(\eta), \quad q(\eta) = BQ(\eta),$$

where A and B are the constants. The unknown parameter b of the function (9.33) is determined by the condition of the existence of the non-trivial solutions of the algebraic set of equations for the coefficients A and B. With the help of this condition we obtain the following two possible values of b:

$$b_{1,2} = \pm ia.$$

Thus, the general positive energy solution of the equation (9.26), satisfying the boundary conditions (9.32), is

$$\Psi(z) = A_1 \begin{pmatrix} \varphi \\ \chi \end{pmatrix} G(s, k, z) + A_2 \begin{pmatrix} \varphi \\ -\chi \end{pmatrix} G(s^*, k, z), \tag{9.34}$$

where $A_{1,2}$ are the constants,

$$s = \frac{1}{2}\left(\sqrt{1 + i4a} - 1\right), \tag{9.35}$$

and the function $G(s, k, z)$ is defined by

$$G(s, k, z) = F\left(1 + s, -s, 1 - ik, \frac{1}{1 + \exp(2\beta z)}\right) \exp(i\kappa z). \tag{9.36}$$

The function $G(s, k, z)$ has the following asymptotical form. At $z \to \infty$, the asymptotical form follows directly from the equation (9.36):

$$G(s, k, z)|_{z \to \infty} = \exp(i\kappa z). \tag{9.37}$$

To find the asymptotical form at $z \to \infty$, it is convenient to use the following transformation of the hypergeometric function

$$F(a, b, c, z) = \frac{\Gamma(c)\,\Gamma(c - a - b)}{\Gamma(c - a)\,\Gamma(c - b)} F(a, b, a + b + 1 - c, 1 - z) +$$
$$+ \frac{\Gamma(c)\,\Gamma(a + b - c)}{\Gamma(a)\,\Gamma(b)} (1 - z)^{c-a-b} F(c - a, c - b, c + 1 - a - b, 1 - z).$$

With the help of the last equation we get

$$G(s, k, z)|_{z \to -\infty} =$$
$$= \frac{\Gamma(1 - ik)\,\Gamma(-ik)}{\Gamma(-ik - s)\,\Gamma(1 - ik + s)} \exp(i\kappa z) + \frac{\Gamma(1 - ik)\,\Gamma(ik)}{\Gamma(1 + s)\,\Gamma(-s)} \exp(-i\kappa z). \tag{9.38}$$

9.4.2 Symmetry properties of wave function

As we have mentioned above, it is convenient to assume that the basis spinors are the eigenfunctions of the equation

$$(\mathbf{n}_E \boldsymbol{\sigma})\, w^{(\sigma=\pm 1)} = (\pm 1)\, w^{(\sigma=\pm 1)}.$$

Hence, the bispinors, appearing in the solution (9.34), are

$$u_1^{(\sigma)} = \begin{pmatrix} \varphi^{(\sigma)} \\ \chi^{(\sigma)} \end{pmatrix} = \begin{pmatrix} w^{(\sigma)} \\ \sigma w^{(\sigma)} \end{pmatrix}, \quad u_2^{(\sigma)} = \begin{pmatrix} \varphi^{(\sigma)} \\ -\chi^{(\sigma)} \end{pmatrix} = \begin{pmatrix} w^{(\sigma)} \\ -\sigma w^{(\sigma)} \end{pmatrix}. \quad (9.39)$$

The charge conjugation transformation is defined by the equation (7.91)

$$\Psi_C(\mathbf{r}, t) = \hat{C}\Psi(\mathbf{r}, t) = -i\lambda_C \gamma_2 \Psi^*(\mathbf{r}, t).$$

The solutions, charge conjugated to the two items of the general solution (9.34),

$$\Psi_{E,\sigma}^{(1,2)}(\mathbf{r}, t) = \Psi_\sigma^{(1,2)}(z) \exp\left(-i\frac{Et}{\hbar}\right),$$

are

$$\begin{aligned} \hat{C}\Psi_{E,\sigma=\pm 1}^{(1)} &= -i\lambda_C \begin{pmatrix} w^{(\sigma'=\mp 1)} \\ \sigma' w^{(\sigma'=\mp 1)} \end{pmatrix} G^*(s, k, z) \exp\left(i\frac{Et}{\hbar}\right), \\ \hat{C}\Psi_{E,\sigma=\pm 1}^{(2)} &= i\lambda_C \begin{pmatrix} w^{(\sigma'=\mp 1)} \\ -\sigma' w^{(\sigma'=\mp 1)} \end{pmatrix} G^*(s^*, k, z) \exp\left(i\frac{Et}{\hbar}\right). \end{aligned} \quad (9.40)$$

It is seen that, at the charge conjugation, the spinor $u_1^{(\sigma)}$ is transformed into the spinor $u_1^{(\sigma'=-\sigma)}$. The same occurs in the case of the free particle.

Let us assume that the coefficients $A_{1,2}$ in the equation (9.34) are equal to $A_{1,2} = C_1 \pm C_2$, then the equation (9.34) takes the form

$$\Psi_\sigma(z) = C_1 \begin{pmatrix} w^{(\sigma)} G_+(s, k, z) \\ \sigma w^{(\sigma)} G_-(s, k, z) \end{pmatrix} + C_2 \begin{pmatrix} w^{(\sigma)} G_-(s, k, z) \\ \sigma w^{(\sigma)} G_+(s, k, z) \end{pmatrix}, \quad (9.41)$$

where

$$G_\pm(s, k, z) = G(s, k, z) \pm G(s^*, k, z). \quad (9.42)$$

At the three-dimensional space inversion transformation ($P_3 f(\mathbf{r}) = f(-\mathbf{r})$) the polar vector $\mathbf{E}$ changes its direction. It means that, at the three-dimensional space inversion transformation, the parameter a, defined by the equation (9.30), changes its sign. In its turn, it means that the three-dimensional space inversion results in the following replacement $s \to s^*$. The latter follows from the definition of the parameter s (see (9.35)). Hence, in accordance with the definition (9.42), we get

$$P_3 G_\pm (s,k,z) = \pm G_\pm (s,k,-z),$$

as a result the solutions, at the coefficients C_1 and C_2 of the equation (9.41), are transformed under the three-dimensional space inversion with the opposite signs:

$$\gamma_4 P_3 \begin{pmatrix} w^{(\sigma)} G_\pm (s,k,z) \\ \sigma w^{(\sigma)} G_\mp (s,k,z) \end{pmatrix} = \pm i \begin{pmatrix} w^{(-\sigma)} G_\pm (s,k,-z) \\ \sigma w^{(-\sigma)} G_\mp (s,k,-z) \end{pmatrix}. \tag{9.43}$$

The plane wave, associated with the incident particle, is a superposition of the even and odd states (with respect to the space inversion). However, as far as the Hamiltonian of the equation (9.26) commutes with the relativistic parity operator, the equation (9.43) shows explicitly that the general solution (9.41) provides the possibility to choose appropriately the parity of the incident particle state.

9.4.3 Reflection and transmission coefficients

Let the wave function of the incident particle be

$$\Psi_0(z) = \sum_{\sigma=\pm1} \alpha_\sigma \begin{pmatrix} w^{(\sigma)} \\ 0 \end{pmatrix} \exp(i\kappa z), \tag{9.44}$$

where $|\alpha_{+1}|^2 + |\alpha_{-1}|^2 = 1$. Then, the wave functions of the reflected Ψ_r and transmitted Ψ_t particles are

$$\begin{aligned} \Psi_r(z) &= \sum_\sigma \alpha_\sigma \begin{pmatrix} w^{(\sigma)} r_1 \exp(-i\kappa z) \\ \sigma w^{(\sigma)} r_2 \exp(-i\kappa z) \end{pmatrix}, \\ \Psi_t(z) &= \sum_\sigma \alpha_\sigma \begin{pmatrix} w^{(\sigma)} t_1 \exp(i\kappa z) \\ \sigma w^{(\sigma)} t_2 \exp(i\kappa z) \end{pmatrix}. \end{aligned} \tag{9.45}$$

To find the coefficients r_i and t_i in the equations (9.45), we should use the asymptotical form of the function $G(s,k,z)$ at $z \to \pm\infty$. It is convenient to transform initially the function $G(s,k,z)$ to asymptotical form given by the equation (9.32). Thus, according to the equations (9.37) and (9.38), we have for the coefficients r and t of the asymptotical form of the function $G(s,k,z)$ the following equations

$$\begin{aligned} r(k,s) &= \frac{\Gamma(1+s-ik)\,\Gamma(-s-ik)\,\Gamma(ik)}{\Gamma(1+s)\,\Gamma(-s)\,\Gamma(-ik)}, \\ t(k,s) &= \frac{\Gamma(1-ik+s)\,\Gamma(-ik-s)}{\Gamma(1-ik)\,\Gamma(-ik)}, \end{aligned} \tag{9.46}$$

and, similarly, for the function $G(s^*, k, z)$:

$$r(k, s^*) = \frac{\Gamma(1+s^*-ik)\,\Gamma(-s^*-ik)\,\Gamma(ik)}{\Gamma(1+s^*)\,\Gamma(-s^*)\,\Gamma(-ik)},$$

$$t(k, s^*) = \frac{\Gamma(1-ik+s^*)\,\Gamma(-ik-s^*)}{\Gamma(1-ik)\,\Gamma(-ik)}.$$

Hence, the asymptotical forms of the functions $G_{\pm}(s, k, z)$ are:

$$G_+(s, k, z) = 2\exp(ikz) + (r(k, s) + r(k, s^*))\exp(-ikz),$$

$$G_-(s, k, z) = (r(k, s) - r(k, s^*))\exp(-ikz).$$

It can be easily seen from the last equations, that, to satisfy the initial state (9.44), we should assume $C_2 = 0$ in the general solution (9.41). Hence, the coefficients $r_{1,2}$ and $t_{1,2}$ in the equation (9.45) are given by

$$\begin{aligned} r_1 &= \frac{r(k, s) + r(k, s^*)}{2}, \quad r_2 = \frac{r(k, s) - r(k, s^*)}{2}, \\ t_1 &= \frac{t(k, s) + t(k, s^*)}{2}, \quad t_2 = \frac{t(k, s) - t(k, s^*)}{2}. \end{aligned} \tag{9.47}$$

As we have discussed in the previous chapter, the equations for the energy reflection R and transmission T coefficients follow from the continuity equation

$$\int\limits_{z\to\infty} \mathbf{j}\, d\mathbf{S} + \int\limits_{z\to-\infty} \mathbf{j}\, d\mathbf{S} = 0.$$

The equations for these coefficients were given in the subsection 8.4.1. Using the wave functions (9.44) and (9.45), we get

$$R = \frac{\left(\nabla\bar{\Psi}_r \cdot \Psi_r - \bar{\Psi}_r \cdot \nabla\Psi_r\right)\mathbf{e}_{-z}}{\left(\nabla\bar{\Psi}_0 \cdot \Psi_0 - \bar{\Psi}_0 \cdot \nabla\Psi_0\right)\mathbf{e}_z} = |r_1|^2 - |r_2|^2,$$

$$T = \frac{\left(\nabla\bar{\Psi}_t \cdot \Psi_t - \bar{\Psi}_t \cdot \nabla\Psi_t\right)\mathbf{e}_z}{\left(\nabla\bar{\Psi}_0 \cdot \Psi_0 - \bar{\Psi}_0 \cdot \nabla\Psi_0\right)\mathbf{e}_z} = |t_1|^2 - |t_2|^2.$$

If the equations (9.46) and (9.47) are applied to the last equations, the reflection R and transmission T coefficients become

$$\begin{aligned} R = |r_1|^2 - |r_2|^2 &= \frac{1}{2}\left[\frac{\sin^2(\pi s)}{\sin^2(\pi s) + \sinh^2(\pi k)} + \frac{\sin^2(\pi s^*)}{\sin^2(\pi s^*) + \sinh^2(\pi k)}\right], \\ T = |t_1|^2 - |t_2|^2 &= \frac{1}{2}\left[\frac{\sinh^2(\pi k)}{\sin^2(\pi s) + \sinh^2(\pi k)} + \frac{\sinh^2(\pi k)}{\sin^2(\pi s^*) + \sinh^2(\pi k)}\right]. \end{aligned} \tag{9.48}$$

It is seen, that

$$R + T = 1. \tag{9.49}$$

The obtained equations enable us to make a number of the general conclusions on the dependency of the reflection and transmission coefficients on the incident particle state:

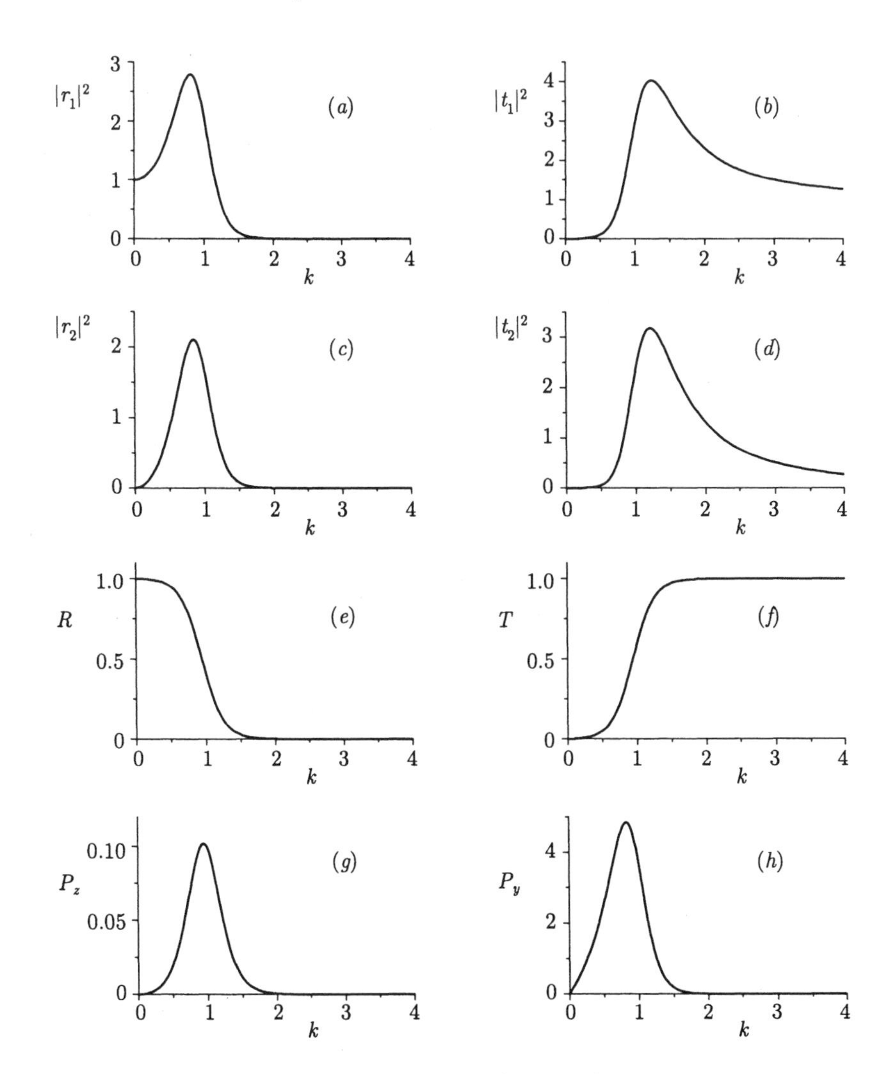

Figure 9.2. Energy spectra of the reflection $|r_1|^2$ (a), $|r_2|^2$ (c), R (e) and transmission $|t_1|^2$ (b), $|t_2|^2$ (d), T (f) coefficients, and electric polarization vector projections P_z (g), P_y (h) at $a = 2$. The parameters are defined by the following equations: k by (9.27) and (9.30); $r_{1,2}$ and $t_{1,2}$ by (9.47); R and T by (9.48); $P_{z,y}$ by (9.54)

(1) The equations (9.48) do not depend on the coefficients α_σ. Hence, the energy reflection and transmission coefficients do not depend on the polarization state of the incident particle.

(2) It follows from the equations (9.48) that the boundary of the reflection region is determined by the condition $k = k_0 = \mathrm{Im}\,(s)$. Hence, for k_0, we have

$$k_0 = s'' = \frac{1}{2}\sqrt{\frac{\sqrt{1+16a^2}-1}{2}}. \qquad (9.50)$$

(3) It follows from the last equation, that the boundary energy of reflection is defined by

$$\Delta E = E - m_0 c^2 \approx \frac{1}{4}\left(\sqrt{\left(\frac{\hbar^2\beta^2}{4m_0}\right)^2 + (2\mu_0 E_0)^2} - \frac{\hbar^2\beta^2}{4m_0}\right). \qquad (9.51)$$

Thus, the boundary energy of reflection depends on the ratio of the energy of the electric dipole interaction $\mu_0 E_0$ to the energy $\hbar^2\beta^2/(2m_0)$ that determines the change in kinetic energy for the particle scattered by a potential barrier of the spatial width $d = \beta^{-1}$.

The Fig. 9.2 shows the coefficients $|r_1|^2$ (a), $|t_1|^2$ (b), $|r_2|^2$ (c), $|t_2|^2$ (d), R (e), T (f) as functions of the wavenumber k for $a = 2$. In this case, the boundary value k_0 is $k_0 = 0.94$. It is seen from the graphs that the drastic increase in the coefficients $|r_i|^2$ and $|t_i|^2$, and the sharp drop in the reflection coefficient R occur when the magnitude of the wave vector k approaches to its boundary value $k = k_0$. It should be noted that the further increase in the value of the parameter a results in more pronounced increase in the coefficients $|r_i|^2$ and $|t_i|^2$ at $k = k_0$, and in more sharp drop of the reflection coefficient R.

9.4.4 Electric and magnetic polarization vectors of neutron scattered by electric field

One can see from the equation (9.51), that, in the case of the wide potential barrier ($\beta \to 0$), the boundary energy of reflection is the product $\mu_0 E_0$. According to the definition of the electric polarization vector $\mathbf{P}$, this product is the energy of the electric dipole interaction $-\int \mathbf{P}\mathbf{E}_0\, dV = -\mathbf{d}\mathbf{E}_0$ of a particle with the uniform electric field. However, it should be noted, that, in the case of the spatially inhomogeneous electric field, the interaction, described by $-\int \mathbf{P}\mathbf{E}\, dV$, is not purely electric dipole interaction. Indeed, if, for example, the function $\mathbf{P}(\mathbf{r})$ has the maximum at some spatial point $\mathbf{r} = \mathbf{r}_0$, then the expansion of the integral $-\int \mathbf{P}\mathbf{E}\, dV$ around the point $\mathbf{r} = \mathbf{r}_0$ includes the all space derivatives of the electric field strength. Thus, it should be more precisely to call the above interaction by the interaction of the induced polarization vector of a particle with the electric field.

Let us study the evolution of the electric and magnetic polarization vectors,

$$\mathbf{P} = -i\mu\bar{\Psi}\boldsymbol{\alpha}\Psi, \quad \mathbf{M} = \mu\bar{\Psi}\boldsymbol{\Sigma}\Psi,$$

in the process of the neutron scattering.

Using the wave function (9.44), we get for the electric and magnetic polarization vectors of the incident particle the following expressions:

$$\mathbf{P}_0 = 0, \tag{9.52}$$

$$\mathbf{M}_0 = \mathbf{e}_z\left(|\alpha_+|^2 - |\alpha_-|^2\right) + \mathbf{e}_x(\alpha_+^*\alpha_- + \alpha_+\alpha_-^*) - \mathbf{e}_y i(\alpha_+^*\alpha_- - \alpha_+\alpha_-^*). \tag{9.53}$$

Using the wave function (9.45), we get for the electric and magnetic polarization vectors of the reflected particle

$$\mathbf{P}_r = \mathbf{e}_z(r_2^* r_1 - r_1^* r_2) + \mathbf{e}_x i(\alpha_+^*\alpha_- - \alpha_+\alpha_-^*)(r_1^* r_2 + r_1 r_2^*) + \\ + \mathbf{e}_y(\alpha_+^*\alpha_- + \alpha_+\alpha_-^*)(r_1^* r_2 + r_1 r_2^*), \tag{9.54}$$

$$\mathbf{M}_r = \mathbf{e}_z\left(|\alpha_+|^2 - |\alpha_-|^2\right)R + \mathbf{e}_x(\alpha_+^*\alpha_- + \alpha_+\alpha_-^*)\left(|r_1|^2 + |r_2|^2\right) - \\ - \mathbf{e}_y i(\alpha_+^*\alpha_- - \alpha_+\alpha_-^*)\left(|r_1|^2 + |r_2|^2\right). \tag{9.55}$$

For shortness, we have omitted in the equations (9.52)–(9.55) the magneton μ_0.

The obtained equations enable us to make a number of the general conclusions concerning the electric and magnetic polarization vectors of the reflected particle:

(1) Independently of the polarization state of the incident particle, the state of the reflected particle is always characterized by the non-zero electric polarization vector directed along the direction of scattering electric field. Notice, that, as already mentioned above, the space inversion transformation results in the following substitution $r_1 \to r_1$ and $r_2 \to -r_2$. Thus, at the space inversion transformation, the electric polarization vector $\mathbf{P}$ changes its sign, and the magnetic polarization vector $\mathbf{M}$ remains invariable.

(2) The projection of the magnetic polarization vector on the direction of the scattering electric field is equal to the product of the magnetization vector of the incident wave and reflection coefficient: $M_{rz} = M_{0z}R$.

(3) If the incident particle is polarized along the direction of the scattering electric field (for example $\alpha_+ = 1$, $\alpha_- = 0$), then the projections of the reflected wave electric $\mathbf{P}_r$ and magnetic $\mathbf{M}_r$ polarization vectors on the plane perpendicular to the direction of the electric field is equal to zero.

(4) If the incident particle is polarized in the plane perpendicular to the direction of the scattering electric field (for example $\alpha_+ = \alpha_- = 1/\sqrt{2}$, hence $\mathbf{M}_0 = \mathbf{e}_x$), then, along with the non-zero longitudinal projection of the electric polarization vector P_z, there are the non-zero mutually perpendicular transversal projections of the vectors $\mathbf{P}_r$ and $\mathbf{M}_r$:

$$P_y = (r_1^* r_2 + r_1 r_2^*), \quad M_x = \left(|r_1|^2 + |r_2|^2\right).$$

The Fig. 9.2 shows the components P_z (curve (g)) and P_y (curve (f)) as functions of the wavenumber k at $a = 2$.

It should be mentioned in conclusion of this subsection, that the electric $\mathbf{P}_t$ and magnetic $\mathbf{M}_t$ polarization vectors of the transmitted wave are determined by the equations (9.54), (9.55), where the coefficients r_i should be replaced by the coefficients t_i.

9.4.5 Bound states of neutron and antineutron in the electric field

The equations (9.48) can be applied to almost all region of variation of the incoming parameters. There is only one exception, when the parameter s' is the positive integer Indeed, at

$$s' = n \tag{9.56}$$

the equations (9.48) take the form:

$$\begin{aligned} R &= \frac{\sinh^2(\pi s'')}{\sinh(\pi(k - s''))\sinh(\pi(k + s''))}, \\ T &= -\frac{\sinh^2(\pi k)}{\sinh(\pi(k - s''))\sinh(\pi(k + s''))}. \end{aligned} \tag{9.57}$$

It is seen from the equations (9.57) that the sum of the reflection and transmission coefficients is still equal to unity, $R + T = 1$. At the same time, the signs of the coefficients R and T are opposite. It can be easily understood that the negative values of the reflection and transmission coefficients correspond to the appearance of the antiparticles in the reflected or transmitted wave.

At

$$k = k_0 = s'' \tag{9.58}$$

the equations (9.57) have the singularity. We can also see this singularity in the asymptotical form of the function $G(s^*, k, z)$ at $z \to -\infty$:

$$\begin{aligned} G(s^*, k, z)|_{z\to-\infty} = &\frac{\Gamma(1 - ik)\Gamma(-ik)}{\Gamma(1 - ik + s^*)\Gamma(-ik - s^*)}\exp(i\kappa z) + \\ &+ \frac{\Gamma(1 - ik)\Gamma(ik)}{\Gamma(1 + s^*)\Gamma(-s^*)}\exp(-i\kappa z). \end{aligned} \tag{9.59}$$

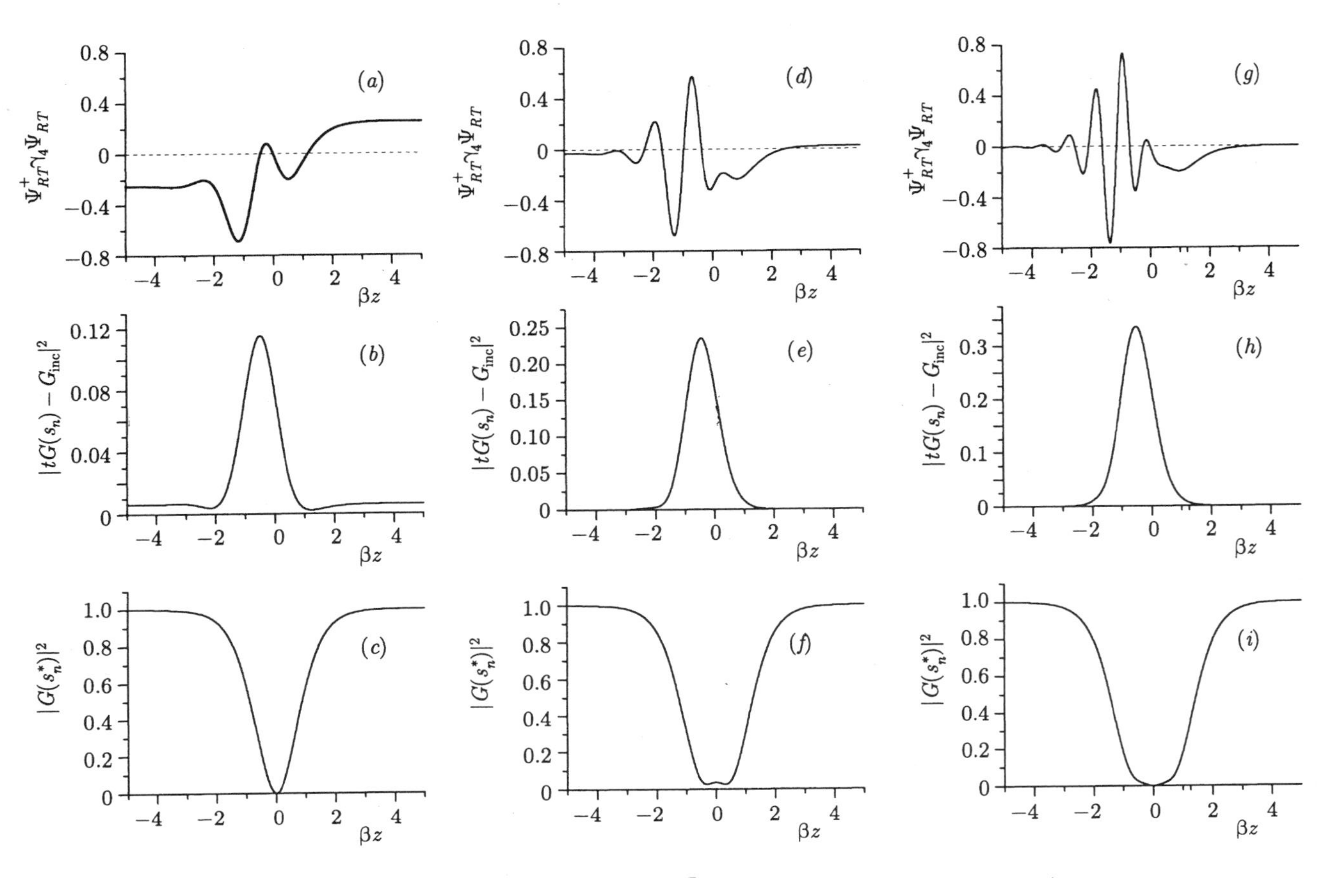

Figure 9.3. Spatial profiles of the functions $\bar{\Psi}_{RT}\Psi_{RT}$ (*a*, *d*, *g*), $|t(k,s_n)G(s_n,k,z) -$ $- G_{\rm inc}(z)|^2$ (*b*, *e*, *h*), and $|G(s_n^*,k,z)|^2$ (*c*, *f*, *i*) for $n = 1$ (*a*, *b*, *c*), 2 (*d*, *e*, *f*), 3 (*g*, *h*, *i*)

It can be easily seen, that if the conditions (9.56) and (9.58) hold, then the first item in the equation (9.59) turns to zero. To illustrate what happens in this case, Fig. 9.3 shows the spatial profile of the function $|G(s^*, k, z)|^2$ at $n = 1$ (c), 2 (f), 3 (i) and $k = s''$. It is seen that, in this case, the above function describes the state consisting of the equal number of neutrons and antineutrons uniformly distributed in whole space with the exception of region of the non-zero electric field. The second row in the Fig. 9.3 shows the profile of the function $|t(k,s)\,G(s,k,z) - G_{\text{inc}}(z)|^2$ for the same values of parameters, this function is the difference between the function (9.36), normalized to the unit current of incident particles, and the wave function of the incident particles $G_{\text{inc}}(z) = (1 - \tanh(\beta z/2))\exp(i\kappa z)/2$. Thus, the wave, corresponding to the solution $G(s,k,z)$, has the maximal amplitude in the region of the non-zero electric field. The solution

$$\Psi(z) = \begin{pmatrix} u\,(t(k,s)\,G(s,k,z) + G(s^*,k,z)) \\ v\,(t(k,s)\,G(s,k,z) - G(s^*,k,z)) \end{pmatrix} \tag{9.60}$$

includes the incident wave of the following type

$$\Psi_0(z) = \begin{pmatrix} u \\ v \end{pmatrix} \exp(i\kappa z) = \begin{pmatrix} u \\ 0 \end{pmatrix} \exp(i\kappa z) + \begin{pmatrix} 0 \\ v \end{pmatrix} \exp(i\kappa z), \tag{9.61}$$

i.e. the incident wave is the coherent superposition of the "pure" neutron and antineutron states. The upper row in fig. 9.3 shows the spatial profile of the function $\bar{\Psi}_{RT}(z)\,\Psi_{RT}(z)$, where

$$\Psi_{RT}(z) = \Psi(z) - \Psi_0(z)\,\frac{1 - \tanh(\beta z)}{2}.$$

It is seen that, in the case when the conditions (9.56) and (9.58) hold, the coherent superposition (9.61) has the probability to decay into the neutron and antineutron propagating in the opposite directions. However, with the increase of n, the energy of the incident wave (9.61) is mainly spent to excite the density oscillations of the neutron-antineutron pairs in the region of the non-zero electric field. It is seen from the lower row of graphics, that in the absence of the incident wave the density of the neutron-antineutron pairs was minimum in the region of the non-zero electric field. Thus, in the specific case of $s' = n$ and $k = k_0 = s''$, the scattering ceases to be elastic in the general sense, because the scattering of the neutron results in the appearance of antineutron in the reflected wave.

It should be noted that the linear independent solutions $G(s,k,z)$ or $G(s^*,k,z)$ can separately produce the coherent superposition of the type (9.61) at any s' and $k \neq s''$. However, these states are not scattered by the electric field, because, in this case, we have $\bar{\Psi}(z)\Psi(z) = 0$ at any z.

Notice also that the wave function (9.60) is applied only to the special case when $s' = n$ and $s'' = k$. The case of $s' = n$ and $k \neq s''$ is not required the special care, we can still use the equations obtained in the previous subsections. As we have mentioned above the specificity of this case is in the fact that the reflection and transmission coefficients are not restricted now by the conditions $0 \leq R \leq 1$ and $0 \leq T \leq 1$. If $|R| > 1$ and $|T| > 1$ it means that the total number of scattered particle exceeds the number of incident particles therefore one can say that scattering is inelastic in this case. However — and it is imperative — the energy conservation law still holds because the sum of the reflection and transmission coefficients is identically equal to unity, $R + T = 1$, as it follows from the equations (9.57). In complete analogy with the case considered in subsection 9.4.3 the amplitudes of waves increase significantly when k is approached to $k_0 = s''$. The above given discussion enables us now to explain this increase by the excitation of neutron-antineutron pairs accumulated in the region of the non-zero electric field.

It should be noted finally, that the phenomena, considered here, are in close similarity with the phenomena occurring under the electromagnetic waves propagation in the spatial inhomogeneous media, such, for example, as the excitation of the plasma oscillations and waves.

9.5 Neutron motion in superposition of electric and magnetic fields

Let us consider the neutron motion in the superposition of the static electric and magnetic fields. In this case the equation (7.3) takes the form

$$\left(\Delta + \kappa^2\right)\Psi(\mathbf{r}) = \left[-\frac{2m_0\mu_0}{\hbar^2}(\mathbf{\Sigma B}) + i\frac{2m_0\mu_0}{\hbar^2}(\boldsymbol{\alpha}\mathbf{E})\right]\Psi(\mathbf{r}). \tag{9.62}$$

We take the wave function in the general form

$$\Psi(\mathbf{r}) = \begin{pmatrix} uf(\mathbf{r}) \\ wg(\mathbf{r}) \end{pmatrix}, \tag{9.63}$$

where u is the arbitrary spinor, satisfying the normalization condition $u^+u = 1$, and $w = \sigma_n u$, here $\sigma_n = (\mathbf{n}\boldsymbol{\sigma})$, and $\mathbf{n}$ is the arbitrary unit vector. By substituting the equation (9.63) into the equation (9.62), we get

$$\begin{aligned} \left(\Delta + \kappa^2\right)uf(\mathbf{r}) - a(\mathbf{r})\sigma_{\mathrm{B}}uf(\mathbf{r}) &= -ib(\mathbf{r})\sigma_E\sigma_n ug(\mathbf{r}), \\ \left(\Delta + \kappa^2\right)wg(\mathbf{r}) - a(\mathbf{r})\sigma_{\mathrm{B}}wg(\mathbf{r}) &= -ib(\mathbf{r})\sigma_E\sigma_n \mathfrak{w}f(\mathbf{r}), \end{aligned} \tag{9.64}$$

where

$$\kappa^2 = \frac{E^2 - m_0^2 c^4}{\hbar^2 c^2}, \quad a(\mathbf{r}) = \frac{2m_0|\mu_0|}{\hbar^2} B(\mathbf{r}), \quad b(\mathbf{r}) = \frac{2m_0|\mu_0|}{\hbar^2} E(\mathbf{r}),$$

$\sigma_{\mathrm{B}} = (\mathbf{n}_{\mathrm{B}}\boldsymbol{\sigma})$, $\sigma_E = (\mathbf{n}_E\boldsymbol{\sigma})$, and $\mathbf{n}_{B,E}$ are the unit vectors of the directions of the electric and magnetic fields. In derivation of the equation (9.64) we have used the following formula

$$(\mathbf{A}\boldsymbol{\sigma})(\mathbf{B}\boldsymbol{\sigma}) = (\mathbf{AB}) + i\boldsymbol{\sigma}[\mathbf{AB}],$$

where $\mathbf{A}$ and $\mathbf{B}$ are the arbitrary vectors. This formula yields also the following equalities: $\sigma_n\sigma_n = 1$, and

$$(\mathbf{n}_E\boldsymbol{\sigma})(\mathbf{n}\boldsymbol{\sigma}) = (\mathbf{n}_E\mathbf{n}) + i\boldsymbol{\sigma}[\mathbf{n}_E\mathbf{n}].$$

In the case when the electric field is parallel or antiparallel to the magnetic field, i.e. $\mathbf{n}_E = \pm\mathbf{n}_{\mathrm{B}}$, it is convenient to assume $\mathbf{n} = \mathbf{n}_{\mathrm{B}}$, then we get

$$\sigma_E\sigma_n = \sigma_E\sigma_{\mathrm{B}} = \pm 1.$$

Thus, in the case of the parallel or antiparallel electric and magnetic fields, the set of equations (9.64) takes the form

$$\begin{aligned} \left(\Delta + \kappa^2 - a(\mathbf{r})\sigma_{\mathrm{B}}\right) uf(\mathbf{r}) &= \mp i b(\mathbf{r})\, ug(\mathbf{r}), \\ \left(\Delta + \kappa^2 - a(\mathbf{r})\sigma_{\mathrm{B}}\right) wg(\mathbf{r}) &= \mp i b(\mathbf{r})\, wf(\mathbf{r}). \end{aligned} \tag{9.65}$$

Assume the spinor u is the eigenfunction of the equation

$$\sigma_{\mathrm{B}} u_\sigma = \sigma u_\sigma,$$

then the spinor $w = \sigma_{\mathrm{B}} u$ is defined by

$$w_\sigma = \sigma_{\mathrm{B}} u_\sigma = \sigma u_\sigma.$$

In the case of the perpendicular electric and magnetic fields, i.e. $\mathbf{n}_E \perp \mathbf{n}_{\mathrm{B}}$, it is convenient to take the vector $\mathbf{n}$ in the following form

$$\mathbf{n} = [\mathbf{n}_{\mathrm{B}}\mathbf{n}_E].$$

In this case we have

$$\sigma_E\sigma_n = i\sigma_{\mathrm{B}},$$

and the set of equations (9.64) takes the form

$$\begin{aligned} \left(\Delta + \kappa^2 - a(\mathbf{r})\sigma_{\mathrm{B}}\right) uf(\mathbf{r}) &= b(\mathbf{r})\,\sigma_{\mathrm{B}} ug(\mathbf{r}), \\ \left(\Delta + \kappa^2 - a(\mathbf{r})\sigma_{\mathrm{B}}\right) wg(\mathbf{r}) &= b(\mathbf{r})\,\sigma_{\mathrm{B}} wf(\mathbf{r}), \end{aligned} \tag{9.66}$$

where

$$w_\sigma = \sigma_n u_\sigma = i\sigma\sigma_E u_\sigma.$$

9.5.1 Parallel fields

As already mentioned, it is more realistic to assume that the electric and magnetic fields are non-zero in half-space than in whole space. Therefore, the electric and magnetic fields are

$$\begin{Bmatrix} \mathbf{B}(z) \\ \mathbf{E}(z) \end{Bmatrix} = \begin{Bmatrix} \mathbf{B}_0 \\ \mathbf{E}_0 \end{Bmatrix} \frac{1}{1+\exp(-\beta z)} \tag{9.67}$$

and the equations (9.65) at $\mathbf{n}_E = \mathbf{n}_\mathrm{B}$ become

$$\begin{aligned} \left(\frac{d^2}{dz^2} + k^2 - \frac{a\sigma}{1+\exp(-z)}\right) f_\sigma(z) &= -i\frac{b}{1+\exp(-z)} g_\sigma(z), \\ \left(\frac{d^2}{dz^2} + k^2 - \frac{a\sigma}{1+\exp(-z)}\right) g_\sigma(z) &= -i\frac{b}{1+\exp(-z)} f_\sigma(z), \end{aligned} \tag{9.68}$$

where we have used the dimensionless coordinate $z' = \beta z$ (in the last equation the primes have been omitted), and

$$k = \frac{\kappa}{\beta}, \quad a = \frac{2m_0|\mu_0|\,B_0}{\hbar^2\beta^2}, \quad b = \frac{2m_0|\mu_0|\,E_0}{\hbar^2\beta^2}.$$

By introducing the new variable

$$\xi = -\exp(-z),$$

the equations (9.68) can be transformed to the following form

$$\begin{aligned} \xi^2(1-\xi)\, f_\sigma'' + \xi(1-\xi)\, f_\sigma' + \kappa^2(1-\xi)\, f_\sigma - a\sigma f_\sigma &= -ibg_\sigma, \\ \xi^2(1-\xi)\, g_\sigma'' + \xi(1-\xi)\, g_\sigma' + \kappa^2(1-\xi)\, g_\sigma - a\sigma g_\sigma &= -ibf_\sigma. \end{aligned} \tag{9.69}$$

The general solution of the equation (9.69) is

$$\begin{aligned} \Psi(z) = A_1 \begin{pmatrix} u \\ -w \end{pmatrix} G_1(\nu_1, z) + A_2 \begin{pmatrix} u \\ w \end{pmatrix} G_1(\nu_2, z) + \\ + B_1 \begin{pmatrix} u \\ -w \end{pmatrix} G_2(\nu_1, z) + B_2 \begin{pmatrix} u \\ w \end{pmatrix} G_2(\nu_2, z) \end{aligned} \tag{9.70}$$

where

$$\begin{aligned} G_1(\nu, z) &= t_1 \exp(i\nu z)\, F(-ik - i\nu, ik - i\nu, 1 - i2\nu, -\exp(-z)), \\ G_2(\nu, z) &= t_2 \exp(-i\nu z)\, F(-ik + i\nu, ik + i\nu, 1 + i2\nu, -\exp(-z)). \end{aligned}$$

and

$$\nu_1 = \sqrt{k^2 - a\sigma - ib}, \quad \nu_2 = \sqrt{k^2 - a\sigma + ib}. \tag{9.71}$$

It follows from the equations (9.71) that the function $\Psi(z)$ is finite at $z \to \infty$ only in the case when the following conditions hold

$$A_1 = 0, \quad B_2 = 0.$$

By taking into account this conditions, we can write the general solution in the following form

$$\Psi(z) = C_1 \begin{pmatrix} u\left(G_1(\nu, z) + G_2(\nu^*, z)\right) \\ w\left(G_1(\nu, z) - G_2(\nu^*, z)\right) \end{pmatrix} + C_2 \begin{pmatrix} u\left(G_1(\nu, z) - G_2(\nu^*, z)\right) \\ w\left(G_1(\nu, z) + G_2(\nu^*, z)\right) \end{pmatrix}, \tag{9.72}$$

where, in accordance with the equations (9.71), we have introduced the following notations: $\nu = \nu_2$, $\nu^* = \nu_1$.

The functions $G_{1,2}$ have the following asymptotical form

$$G_1(\nu, z) = \begin{cases} \exp(ikz) + r_1 \exp(-ikz), & z \to -\infty, \\ t_1 \exp(i\nu z), & z \to \infty, \end{cases}$$

$$G_2(\nu^*, z) = \begin{cases} \exp(ikz) + r_2 \exp(-ikz), & z \to -\infty, \\ t_2 \exp(-i\nu^* z), & z \to \infty, \end{cases}$$

where

$$r_1 = \frac{\Gamma(-ik - i\nu)\Gamma(1 - ik - i\nu)\Gamma(2ik)}{\Gamma(ik - i\nu)\Gamma(1 + ik - i\nu)\Gamma(-2ik)}, \quad t_1 = \frac{\Gamma(-ik - i\nu)\Gamma(1 - ik - i\nu)}{\Gamma(1 - i2\nu)\Gamma(-2ik)},$$

$$r_2 = \frac{\Gamma(-ik + i\nu^*)\Gamma(1 - ik + i\nu^*)\Gamma(2ik)}{\Gamma(ik + i\nu^*)\Gamma(1 + ik + i\nu^*)\Gamma(-2ik)}, \quad t_2 = \frac{\Gamma(-ik + i\nu^*)\Gamma(1 - ik + i\nu^*)}{\Gamma(1 + i2\nu^*)\Gamma(-2ik)}. \tag{9.73}$$

In complete analogy with the equation (9.41), the first term in the solution (9.72) corresponds to the incident particle, and the second term corresponds to the incident antiparticle. Therefore, the wave function of the incident neutron is

$$\Psi_0(z) = \begin{pmatrix} u \\ 0 \end{pmatrix} \exp(ikz).$$

In accordance with the equation (9.72), the wave function of the reflected neutron is defined by

$$\Psi_r(z) = \frac{1}{2} \begin{pmatrix} u\left(r_1 + r_2\right) \\ w\left(r_1 - r_2\right) \end{pmatrix} \exp(-ikz).$$

The definition of the energy reflection coefficient was given in subsection 9.4.3. By using the equations (9.73) for the energy reflection coefficient we obtain

$$R = \frac{1}{4}\left[(r_1 + r_2)^* (r_1 + r_2) - (r_1 - r_2)^* (r_1 - r_2)\right] \equiv 1.$$

Thus, there is the total reflection of the neutron incident on the semi-infinite parallel electric and magnetic fields. Notice, that the total reflection is completely due to the electric field. Indeed, we have seen in the previous chapter that the neutron of the energy, above the boundary energy of reflection, penetrates entirely into the magnetic field.

The characteristic length of the neutron penetration into the electric field is determined by the imaginary part of ν. At the condition $k^2 \gg b$ (i.e. $E^2 - m_0^2c^4 \gg eE_0\hbar c = e\varphi_0\hbar c/d$, where φ_0 is the voltage between the condenser disks placed at the distance d one from another, and we have assumed, for simplicity, that the neutron magneton is equal to the nuclear magneton), we obtain the following formula for the penetration depth

$$l = d\frac{\sqrt{E^2 - m_0^2c^4}}{U_0}, \tag{9.74}$$

where $U_0 = e\varphi_0$. At the energy of the incident neutron about a few MeV and for the reasonable value of the voltage, the penetration depth is about 100 m.

In the subsection 8.4.1 the spacial frequency of the spin precession in the uniform magnetic field has been calculated. It follows from the equation (9.71) that the spacial frequency of the spin precession in the parallel electric and magnetic fields is determined by the following expression

$$\Delta\nu = \mathrm{Re}\big(\sqrt{k^2 - a + ib}\big) - \mathrm{Re}\big(\sqrt{k^2 + a + ib}\big).$$

At $k^2 > b$ we get

$$\Delta\nu \approx \Delta\nu_0 + \frac{b^2}{8}\left(\frac{1}{\left(k^2 - a\right)^{3/2}} - \frac{1}{\left(k^2 + a\right)^{3/2}}\right) \tag{9.75}$$

where $\Delta\nu_0 = \sqrt{k^2 - a} - \sqrt{k^2 + a}$ is the spacial frequency of the spin precession in the uniform magnetic field. Thus, it is seen that the correction to the spin precession frequency ν_0 due to the presence of the non-zero parallel electric field is about

$$\Delta\nu - \Delta\nu_0 \approx \left(\frac{|\mu_0|}{E - m_0c^2}\right)^3 E_0^2 B_0 k.$$

By taking into account the equation for ν_0:

$$\Delta\nu_0 \approx \frac{|\mu_0|\, B_0}{E - m_0c^2}k,$$

for the relative correction we get the following equation

$$\frac{\Delta\nu - \Delta\nu_0}{\Delta\nu_0} \approx \left(\frac{|\mu_0|\, E_0}{E - m_0 c^2}\right)^2 .$$

Thus, the above analysis enables us to make the following very important conclusions. Firstly, in contrast to the equations of the section 1.3, the obtained correction to the spin precession frequency due to the non-zero parallel electric field is proportional to the square of the applied electric field, but not to the filed strength as it is prescribed by the mechanism based on the violation of CP invariance. Secondly, as well as the correction is proportional to the square of the applied electric field, then the reversion of the direction of the applied electric field does not result in the change of the spin precession frequency.

9.5.2 Crossed fields

In the case of the crossed electric and magnetic fields the equations for the radial wave functions (9.63) take the form of (9.66). By substituting the equation (9.67) for the electric and magnetic fields into the equation (9.66) and introducing the new variable $\xi = -\exp(-z)$, we finally get

$$\begin{aligned} &\xi^2(1-\xi)\, f_\sigma'' + \xi\,(1-\xi)\, f_\sigma' + \kappa^2(1-\xi)\, f_\sigma - a\sigma f_\sigma = b\sigma g_\sigma, \\ &\xi^2(1-\xi)\, g_\sigma'' + \xi\,(1-\xi)\, g_\sigma' + \kappa^2(1-\xi)\, g_\sigma + a\sigma g_\sigma = -b\sigma f_\sigma, \end{aligned} \tag{9.76}$$

where we have taken into account the following relationships

$$\sigma_{\mathrm{B}} w_\sigma = \sigma \sigma_{\mathrm{B}} \sigma_n \sigma_{\mathrm{B}} u_\sigma = -\sigma w_\sigma .$$

The general solution of the equations (9.76) is

$$\begin{aligned} \Psi(z) = A_1 \begin{pmatrix} u \\ -\varsigma w \end{pmatrix} G_1(\nu_1, z) + B_1 \begin{pmatrix} -\varsigma u \\ w \end{pmatrix} G_1(\nu_2, z) + \\ + A_2 \begin{pmatrix} u \\ -\varsigma w \end{pmatrix} G_2(\nu_1, z) + B_2 \begin{pmatrix} -\varsigma u \\ w \end{pmatrix} G_2(\nu_2, z), \end{aligned}$$

where the functions $G_{1,2}(z)$ have been defined in the previous subsection, and we have introduced the following notations

$$\begin{aligned} \varsigma &= \frac{b}{\sqrt{a^2 - b^2} + a}, \\ \nu_1 = \sqrt{k^2 - \sigma\sqrt{a^2 - b^2}}, &\quad \nu_2 = \sqrt{k^2 + \sigma\sqrt{a^2 - b^2}}. \end{aligned} \tag{9.77}$$

It can be easily seen that in the case $a < b$ (i.e. $B_0 < E_0$) the coefficients $\nu_{1,2}$ are the complex numbers. Hence, the equations (9.76) can be treated in a way similar to that used in the previous subsection. There is only one important difference: it follows from the above equations that the spin precession frequency is equal to zero.

Therefore, let us study the case of $a > b$. To satisfy the boundary conditions at $z \to \infty$ we should assume

$$A_2 = 0, \quad B_2 = 0.$$

It should be noted, that the case of the crossed fields differs qualitatively from the all above considered configurations. This difference is due to the commutation relation for the Hamiltonian and operator of spin projection on the direction of the magnetic field. In previous cases this projection of spin was the conservative variable. As a result the spinors u_σ were the eigenfunctions of the conservative spin projection. In the case of the crossed fields the spin projection operator $(\mathbf{n}_\mathrm{B}\mathbf{\Sigma})$ does not commute with the Hamiltonian of the equation (9.62), therefore the general solution of the equations (9.62) will be always the superposition of the spinors u_σ

$$\Psi(z) = \sum_{\sigma=\pm 1} \left[A_\sigma \begin{pmatrix} u_\sigma \\ -\varsigma w_\sigma \end{pmatrix} G_1(\nu_1^{(\sigma)}, z) + B_\sigma \begin{pmatrix} -\varsigma u_\sigma \\ w_\sigma \end{pmatrix} G_1(\nu_2^{(\sigma)}, z) \right], \tag{9.78}$$

where, in accordance with the equations (9.77), we have

$$\begin{aligned} \nu_1^{(+)} &= \nu_2^{(-)} = \sqrt{k^2 - \sqrt{a^2 - b^2}}, \\ \nu_1^{(-)} &= \nu_2^{(+)} = \sqrt{k^2 + \sqrt{a^2 - b^2}}. \end{aligned} \tag{9.79}$$

As far as the parameter $\varsigma < 1$ at $a > b$, hence, the first term in the equation (9.78) corresponds to the particle solution, because

$$\bar{\Psi}_1^{(\sigma)} \Psi_1^{(\sigma)} = \left| A_\sigma G_1(\nu_1^{(\sigma)}, z) \right|^2 (1 - \varsigma^2) > 0;$$

and the second term corresponds to the antiparticle solution, because

$$\bar{\Psi}_2^{(\sigma)} \Psi_2^{(\sigma)} = \left| B_\sigma G_1(\nu_2^{(\sigma)}, z) \right|^2 (\varsigma^2 - 1) < 0.$$

The asymptotical form of the function $G_1(\nu, z)$ was given in the previous subsection, with the help of this equation we get again the equations (9.73) for the reflection $r_{1,2}$ and transmission $t_{1,2}$ coefficients.

The principle difference of the wave function (9.78) from the solutions of other one-dimensional scattering problems is in the fact, that the incident particle with the given projection on the direction of the magnetic field corresponds the wave function of the following form

$$\Psi_0(z) = \frac{1}{\sqrt{1-\varsigma^2}} \begin{pmatrix} u_\sigma \\ -\varsigma w_\sigma \end{pmatrix} \exp(ikz).$$

It is seen that, in contrast to the previous cases, both upper and lower spinors of the bispinor wave function are non-zero. Of course, we can always choose the wave function in the form corresponding to the "pure" particle state. Indeed, by assuming $B_\sigma = \varsigma A_\sigma$ in the equation (9.78), we get

$$\Psi_0'(z) = \begin{pmatrix} u_\sigma \\ 0 \end{pmatrix} \exp(ikz).$$

Hence, the wave function Ψ_0 can be interpreted as a function describing the superposition of the 'pure' particle and antiparticle states.

The spacial frequency of the spin precession is defined by the expression

$$\Delta\nu = \sqrt{k^2 + \sqrt{a^2 - b^2}} - \sqrt{k^2 - \sqrt{a^2 - b^2}}.$$

If the energy of the incident neutron exceeds significantly the boundary energy of reflection, we get the following formula for the precession frequency

$$\Delta\nu = \frac{\sqrt{a^2 - b^2}}{k}.$$

It is seen that the precession frequency is quadratically depend on the applied electric field strength, therefore the reversion in the direction of the electric field will not result in the change of the precession frequency.

9.6 Geonium atom

The problem on the geonium atom has been treated in the previous chapter. However, we have not taken into account the presence of the electric field of the Penning trap. In the light of the previous discussion it looks useful to account it. The electrostatic potential of the Penning trap is described by the following equation

$$\varphi(\mathbf{r}) = \varphi_0 \frac{z^2 - \rho^2/2}{d^2}, \tag{9.80}$$

where the z axis is directed along the direction of the trap magnetic field. In this case the equation (7.3) takes the following form

$$\left[\Delta+\kappa^2+i\frac{|e|\,B_0}{\hbar c}\frac{\partial}{\partial\varphi}-\left(\frac{eB_0}{2\hbar c}\right)^2\rho^2-\frac{2EU(\rho,z)}{\hbar^2c^2}+\frac{U^2(\rho,z)}{\hbar^2c^2}\right]\Psi=$$
$$=\frac{\mu_0}{\mu_{\mathrm{B}}}\left(\frac{|e|\,B_0}{\hbar c}\Sigma_z+i\frac{U_0}{\hbar cd^2}\left(2z\alpha_z-\rho\alpha_\rho\right)\right)\Psi, \quad (9.81)$$

where $U_0=e\varphi_0>0$, and μ_0 is magnitude of the electron magneton.

Similar to the case of the neutron motion in the crossed electric and magnetic fields, the account of the electric field projection E_ρ results in the loss of conservation of the spin projection on the direction of the magnetic field. The non-zero projection E_ρ results in the magnetron motion of electron in the trap. In this case, the electron orbit takes the epicycle form. However, the electric field strength of the Penning trap is much smaller than the magnetic field strength, therefore we can neglect this projection of the electric field in the zero-order approximation. Assume additionally, that $U_0\ll m_0c^2$, then the equation (9.81) is simplified significantly. Due to the symmetry of the problem, the wave function of the equation (9.81) is

$$\Psi(\mathbf{r})=\begin{pmatrix} u_\sigma f_\sigma(\rho,z)\\ u_\sigma g_\sigma(\rho,z)\end{pmatrix}\exp(im\varphi). \quad (9.82)$$

where the spinors u_σ are the eigenfunctions of the equation $\sigma_z u_\sigma=\sigma u_\sigma$. By substituting the wave function (9.82) into the equation (9.81), for the radial wave functions we get the following equations

$$\left[\frac{\partial^2}{\partial\rho^2}+\frac{1}{\rho}\frac{\partial}{\partial\rho}-\frac{m^2}{\rho^2}-\nu^2\rho^2+\frac{\partial^2}{\partial z^2}-\frac{2EU_0}{\hbar^2c^2}\frac{z^2}{d^2}+\right.$$
$$\left.+\kappa^2-2\nu\left(m+\frac{\mu_0}{\mu_{\mathrm{B}}}\sigma\right)\right]f=i\frac{\mu_0}{\mu_{\mathrm{B}}}\frac{2U_0}{\hbar cd^2}z\sigma g,$$
$$\left[\frac{\partial^2}{\partial\rho^2}+\frac{1}{\rho}\frac{\partial}{\partial\rho}-\frac{m^2}{\rho^2}-\nu^2\rho^2+\frac{\partial^2}{\partial z^2}-\frac{2EU_0}{\hbar^2c^2}\frac{z^2}{d^2}+\right.$$
$$\left.+\kappa^2-2\nu\left(m+\frac{\mu_0}{\mu_{\mathrm{B}}}\sigma\right)\right]g=i\frac{\mu_0}{\mu_{\mathrm{B}}}\frac{2U_0}{\hbar cd^2}z\sigma f, \quad (9.83)$$

where $\nu=|e|B_0/(2\hbar c)$. The general solution of the equations (9.83) is

$$\Psi_\sigma(\mathbf{r})=C_1\begin{pmatrix} u_\sigma\left(G_\sigma(\rho,z)+G_\sigma^*(\rho,z)\right)\\ u_\sigma\left(G_\sigma(\rho,z)-G_\sigma^*(\rho,z)\right)\end{pmatrix}+$$
$$+C_2\begin{pmatrix} u_\sigma\left(G_\sigma(\rho,z)-G_\sigma^*(\rho,z)\right)\\ u_\sigma\left(G_\sigma(\rho,z)+G_\sigma^*(\rho,z)\right)\end{pmatrix}, \quad (9.84)$$

where $C_{1,2}$ are the normalization constants. The function $G_\sigma(\rho, z)$, satisfying the required boundary conditions, is

$$G_\sigma(\rho, z) = \left(\nu\rho^2\right)^{m/2} \exp\left(-\frac{\nu\rho^2}{2} - \frac{1}{2}\sqrt{\frac{2EU_0 d^2}{\hbar^2 c^2}}\frac{z^2}{d^2} - i\sigma\frac{\mu_0}{\mu_\mathrm{B}}\sqrt{\frac{U_0}{2E}}\frac{z}{d}\right)\times$$
$$\times L_{n_\rho}^{(m)}\left(\nu\rho^2\right) H_{n_z}\left(\left(\frac{2EU_0 d^2}{\hbar^2 c^2}\right)^{1/4}\frac{z}{d} + i\sigma\frac{\mu_0}{\mu_\mathrm{B}}\left(\frac{\hbar^2 U_0}{8m_0^3 c^4 d^2}\right)^{1/4}\right). \quad (9.85)$$

where $L_n^{(m)}(z)$ is the generalized Laguerre polynomial, $H_n(z)$ is the Hermite polynomial, n_ρ and n_z are the non-negative integers. The energy spectrum depends on the quantum numbers n_ρ, m, σ, n_z, and to find the explicit expression for it we should solve the following equation

$$\frac{E^2 - m_0^2 c^4}{\hbar^2 c^2} = \frac{|e|\, B_0}{\hbar c}\left(m + \frac{\mu_0}{\mu_\mathrm{B}}\sigma\right) + \frac{2\,|e|\, B_0}{\hbar c}\left(n_\rho + \frac{1+m}{2}\right) +$$
$$+ \frac{2\sqrt{2EU_0}}{\hbar c d}\left(n_z + \frac{1}{2}\right) + \left(\frac{\mu_0}{\mu_\mathrm{B}}\right)^2 \frac{U_0}{2Ed^2}. \quad (9.86)$$

By assuming, that $E - m_0 c^2 \ll m_0 c^2$, we can write the approximate solution of the last equation in the following form

$$E = \left[m_0^2 c^4 + 2m_0 c^2\left(\hbar\omega_H\left(n_\rho + m + \frac{1}{2}\right) + \mu_0 B_0 \sigma +\right.\right.$$
$$\left.\left. + \sqrt{\frac{2\hbar^2 U_0}{m_0 d^2}}\left(n_z + \frac{1}{2}\right)\right) + \left(\frac{\mu_0}{\mu_\mathrm{B}}\right)^2 \frac{\hbar^2 U_0}{2m_0 d^2}\right]^{1/2}. \quad (9.87)$$

It is seen that the approximate solution of the equation (9.86) includes the two additional terms in comparison with the spectrum obtained in the section 8.3. However, the exact solution of the equation (9.86) will include the infinite series of the additional terms. The accuracy of the measurements of the fundamental constants in the experiments with the Penning trap is so high that it could be necessary to account this difference.

The wave function (9.84), (9.85) differs also from the wave function obtained in section 8.3. The particle solution is given by the first term in the equation (9.84). Its upper and lower spinors are the real and imaginary parts of the function $G(\rho, z)$, respectively. It is seen from the equation (9.85) that at $U_0 \ll m_0 c^2$ the imaginary part of the function $G(\rho, z)$ is small, therefore the magnitude of the upper spinor is always greater than the magnitude of the lower spinor. Nevertheless the difference between the wave function (9.84) and the wave function of electron interacting with the uniform magnetic field results in the essential difference in the magnitude of the observable variables. Firstly, the non-zero value of the lower spinor results in the non-zero value of the electric polarization vector. Indeed, according to the definition of the electric polarization vector, we get

$$\mathbf{P} = -i\mu_0 \bar{\Psi}\boldsymbol{\alpha}\Psi = -i\mathbf{e}_z \mu_0 \sigma \left(\varphi_\sigma^+ \chi_\sigma - \varphi_\sigma \chi_\sigma^+\right),$$

hence,

$$|\mathbf{P}| \sim \mathrm{Re}\,(G)\,\mathrm{Im}\,(G)\,.$$

Secondly, the magnetic polarization vector is also changed

$$\mathbf{M} = \mu_0 \bar{\Psi}\boldsymbol{\Sigma}\Psi = \mathbf{e}_z \mu_0 \sigma \left(\varphi_\sigma^+ \varphi_\sigma - \chi_\sigma^+ \chi_\sigma\right),$$

hence,

$$|\mathbf{M}| \sim (\mathrm{Re}\,(G))^2 - (\mathrm{Im}\,(G))^2\,.$$

It is this difference in the magnitudes of the electric and magnetic polarization vectors, that results in the difference of the spectrum (9.87) from the spectrum of electron in the uniform magnetic field.

9.7 Hyperfine structure of hydrogenic spectra: comparison with the experimental data

In the previous chapter we have shown that the energy spectrum of electron in the Coulomb field includes the splitting of the states with the same value of the total angular momentum j and different values of the orbital angular momentum l. This is the principle difference between the Dirac equation and equation (7.3). However, we have already discussed that the problem on the electron motion in the Coulomb field is not equivalent to the hydrogen atom problem, because the latter problem is the two-particle problem. The analysis of the different approximations of the two-particle problem, given in the previous chapters, has shown that the shift of the states with $l = 0$ exceeds the shift of the states with $l > 0$, and the magnitude of the shift decreases with the increase of l. Hence, the experimentally measured frequencies of transitions between the initial $l_i = 0$ and final $l_f > 0$ states will be most strongly differ from those calculated for electron in the Coulomb field. On the other hand, the frequencies of transitions between the levels $l_i > 0$ and $l_f > 0$ may be much closer to the spectrum of the one-particle problem.

The analysis of the two-particle problem for the equation (7.3) will be given in the next chapter. Here, we compare the spectra obtained from the solution of the Dirac equation and equation (7.3) for the problem on the electron motion in the Coulomb field. Let us denote

$\Delta E = E_{nj}^{(l)} - E_{nj}$, where $E_{nj}^{(l)}$ is defined by the equation (8.49) and E_{nj} is defined by the equation (6.132). It is convenient to express the shift ΔE in terms of the power series in the fine structure constant. The shifts in the energy of $nS_{1/2}$, $nP_{1/2}$, and $nP_{3/2}$ states are given by

$$\frac{\Delta E\left(nS_{1/2}\right)}{m_0c^2} = \frac{\gamma_e^2 - 1}{2n^3}(Z\alpha)^4 + \\ + \frac{\gamma_e^2 - 1}{8n^5}\left[\left(7-\gamma_e^2\right)n^2 + 3\left(3-\gamma_e^2\right)n - 6\right](Z\alpha)^6 + \ldots,$$

$$\frac{\Delta E\left(nP_{1/2}\right)}{m_0c^2} = -\frac{\gamma_e^2 - 1}{6n^3}(Z\alpha)^4 - \\ - \frac{\gamma_e^2 - 1}{216n^5}\left[\left(19+11\gamma_e^2\right)n^2 + 9\left(5+\gamma_e^2\right)n - 54\right](Z\alpha)^6 + \ldots, \tag{9.88}$$

$$\frac{\Delta E\left(nP_{3/2}\right)}{m_0c^2} = \frac{\gamma_e^2 - 1}{12n^3}(Z\alpha)^4 + \\ + \frac{\gamma_e^2 - 1}{1728n^5}\left[\left(37+5\gamma_e^2\right)n^2 + 18\left(7-\gamma_e^2\right)n - 216\right](Z\alpha)^6 + \ldots,$$

where $n = n_r + 1$ is the principle quantum number for the S states, and $n = n_r + 2$ is the principle quantum number for the P states.

It follows from the equations (9.88) that the first non-vanishing correction in the expansions is proportional to $\alpha^4\left(\gamma_e^2 - 1\right)$. By taking into account the relation $m_0c^2\alpha^2 = 2\,\mathrm{Ry}$, we can see that the first non-vanishing correction is proportional to $Z^4\alpha^2/n^3$, this is the typical dependency of the hyperfine shifts on the nucleus charge Z, principle quantum number n, and fine structure constant α. The $nS_{1/2}$ and $nP_{3/2}$ levels move up and the $nP_{1/2}$ level moves down with respect to their position in the frames of the Dirac theory. This is in qualitative agreement with the experimental spectra. Notice, that in the next chapter, we shall see that the account for the reduced electron mass results in the upwards shift of all these levels.

For the Lamb shift we obtain

$$\Delta E_L\left(nS_{1/2} - nP_{1/2}\right) = \frac{3\left(\gamma_e^2 - 1\right)}{4n^3}m_0c^2Z^4\alpha^4.$$

Accounting that the experimentally measured value of the electron magneton is satisfactory approximated by the formula $\gamma_e = \mu_e/\mu_\mathrm{B} \approx$ $\approx (1 + \alpha/2\pi)$, we get

$$\Delta E_L = \frac{3m_0c^2Z^4\alpha^5}{2\pi n^3} + \ldots. \tag{9.89}$$

The level shift due to the lowest order radiative corrections is given by [9]:

$$\Delta E\,(l,j) = \frac{mc^2(Z\alpha)^4\,\alpha}{\pi n^3}\Big\{A_{40} + A_{41}\ln{(Z\alpha)^{-2}} + A_{50}Z\alpha +$$
$$+ + (Z\alpha)^2\Big[A_{62}\ln^2(Z\alpha)^{-2} + A_{61}\ln{(Z\alpha)^{-2}}\Big] + \ldots\Big\}, \quad (9.90)$$

where the coefficients A_{nm} of the expansion in series on the powers of $(Z\alpha)^n \ln^m (Z\alpha)^{-2}$ are the sums of the electron self-energy, vacuum polarization, and anomalous magnetic moment contributions.

It is seen that the equation (9.89) differs from the equation (9.90) only in the value of the coefficients.

Fig. 9.4, *a* shows the Lamb shift, calculated with the help of equation (8.49), for the $2s_{1/2} - 2p_{1/2}$ (curve *1*) and $3p_{3/2} - 3d_{3/2}$ (curve *2*) levels, and the product $\Delta E n^3$ (curve *3*) as functions of the principle

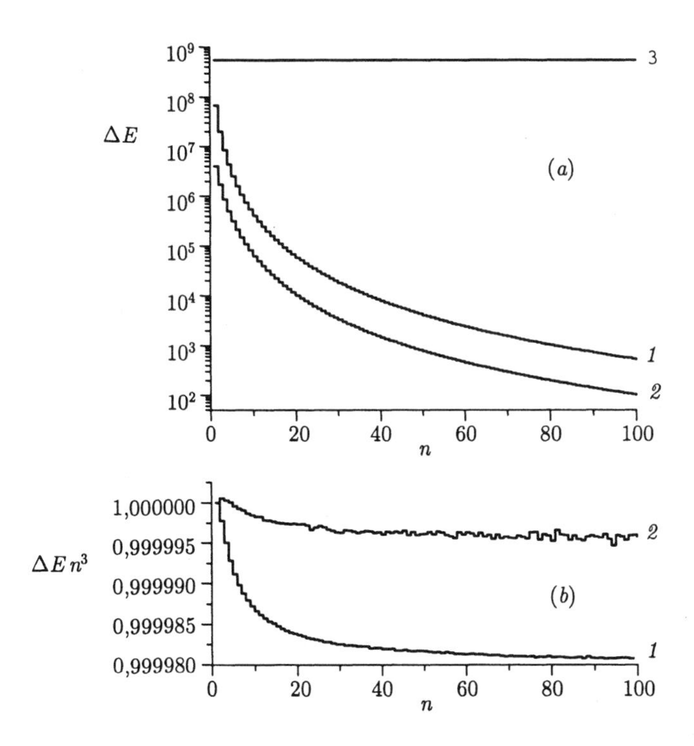

Figure 9.4. (*a*) The Lamb shift, calculated with the help of equation (8.49), for the $2s_{1/2} - 2p_{1/2}$ (curve *1*) and $3p_{3/2} - 3d_{3/2}$ (curve *2*) levels, and the product $\Delta E n^3$ (curve *3*) as functions of the principle quantum number n. (*b*) The product $\Delta E n^3$ as function of n

quantum number n. In Fig. 9.4, b the product $\Delta E n^3$ as function of n is shown in the magnified scale. It is seen that the leading term, in lowest order in $Z\alpha$, in the expansions (9.88) makes the main contribution, because the product $\Delta E n^3$ differs from unity on the value about $4 \cdot 10^{-6}$ for the shift of the $3p_{3/2} - 3d_{3/2}$ levels and about $2 \cdot 10^{-5}$ for the shift of the $2s_{1/2} - 2p_{1/2}$ levels.

The agreement of the experimental and calculated data can be illustrated with the help the results of the precision measurements [10] of the frequencies of the $8d_{5/2} - 8d_{3/2}$ and $12d_{5/2} - 12d_{3/2}$ transitions in hydrogen and deuterium.

To simplify the theoretical calculations, it is convenient to rewrite the equation (8.49) in the following form

$$\Delta E_{nj}^{(l=j\mp 1/2)} = m_0 c^2 - E_{nj}^{(l=j\mp 1/2)} =$$

$$= \frac{m_0 c^2 Z^2 \alpha^2}{(n+\nu_\pm)^2} \frac{1}{\sqrt{1+\frac{Z^2\alpha^2}{(n+\nu_\pm)^2}}\left(1+\sqrt{1+\frac{Z^2\alpha^2}{(n+\nu_\pm)^2}}\right)}.$$

By taking into account that

$$m_e c^2 \alpha^2 = 4\pi\hbar R_\infty c,$$

for the position of the level in the frequency units we get

$$\nu_{nl}^{(l=j\mp 1/2)} = \frac{2R_\infty c Z^2}{(n+\nu_\pm)^2} \frac{m_n}{m_n+m_e} \frac{1}{\sqrt{1+\frac{Z^2\alpha^2}{(n+\nu_\pm)^2}}\left(1+\sqrt{1+\frac{Z^2\alpha^2}{(n+\nu_\pm)^2}}\right)}, \tag{9.91}$$

where we have accounted the correction associated with the finite nucleus mass m_n. This correction results in the replacement of the electron mass m_e by the reduced mass $m_r = m_e m_n/(m_e + m_n)$. The correction due to the finite nucleus mass has been calculated in the Chapter 3 for the case of the Schrödinger equation. The structure of the equation (8.35) is similar to the structure of the Schrödinger equation, therefore, it could be anticipated, that this specific correction will be the same. The detailed analysis of the corrections associated with nucleus motion will be given in the next chapter.

Table 9.1 and Table 9.2 give the results of the experimental measurements, made by de Beauvoir, et al. [10], for the frequencies of the $8(12)D_J$–$2S_{1/2}$ transitions in hydrogen and deuterium. These measurements enables us to determine the frequencies of the $8(12)D_{5/2}$–$8(12)D_{3/2}$ transitions. These figures are shown in bold face. The

Table 9.1. The frequency of $8D_J - 2S_{1/2}$ transitions in hydrogen and deuterium [10] and comparison of experimental and theoretical data for $8D_{5/2} - 8D_{3/2}$ transitions

	Experiment [10]	
	Hydrogen, MHz	*Deuterium*, MHz
ν_0	770649000	770859000
$\nu(8D_{3/2} - 2S_{1/2}) - \nu_0$	504.4500(83)	195.7018(63)
$\nu(8D_{5/2} - 2S_{1/2}) - \nu_0$	561.5842(64)	252.8495(59)
$\nu(8D_{5/2} - 8D_{3/2})$	**57.1342**	**57.1477**
	Theory	
$\nu(8D_{5/2} - 8D_{3/2})$	**57.1293**	**57.14487**

Table 9.2. The frequency of $12D_J - 2S_{1/2}$ transitions in hydrogen and deuterium [10] and comparison of experimental and theoretical data for $12D_{5/2} - 12D_{3/2}$ transitions

	Experiment [10]	
	Hydrogen, MHz	*Deuterium*, MHz
ν_0	799191000	799409000
$\nu(12D_{3/2} - 2S_{1/2}) - \nu_0$	710.4727(93)	168.0380(86)
$\nu(12D_{5/2} - 2S_{1/2}) - \nu_0$	727.4037(70)	184.9668(68)
$\nu(12D_{5/2} - 12D_{3/2})$	**16.931**	**16.9288**
	Theory	
$\nu(12D_{5/2} - 12D_{3/2})$	**16.9272**	**16.9318**

theoretical values of the transition frequencies are calculated with the help of the equation (9.91), where we have used the following values of the fundamental constants [9]: $R_\infty c = 3.289841960368 \cdot 10^{15}$ Hz, $\alpha = 0.0072973525333285885$, $m_n/m_e = 1836.152667$, $\gamma_e = 1.0011596521884$. The results of the theoretical calculations are shown in the lower rows of the tables.

It is seen that there is the difference between the experimentally measured and theoretically calculated (with the help of the formula (9.91)) frequencies of transitions. However, the difference is about a few kHz, which is smaller than the uncertainty of the measurements.

Chapter 10

HYDROGEN ATOM

10.1 Action principle

The action is the additive function, therefore the action of an ensemble of particles is the sum of the actions (7.1) for the individual particles, where we should take into account that the field, that acts on each individual particle, is produced by all other particles of ensemble

$$S=\frac{1}{8\pi}\int\left[\left(\frac{1}{c}\frac{\partial\mathbf{A}}{\partial t}+\nabla\varphi\right)^{2}-(\mathrm{curl}\;\mathbf{A})^{2}\right]dV\,dt-$$
$$-\sum_{\substack{a,b\\(a\neq b)}}\frac{1}{2m_a}\int\left[\left(\frac{\hbar}{c}\frac{\partial\bar{\Psi}_a}{\partial t}-iq_a\varphi_b\left(\mathbf{r}_a\right)\bar{\Psi}_a\right)\left(-\frac{\hbar}{c}\frac{\partial\Psi_a}{\partial t}-iq_a\varphi_b\left(\mathbf{r}_a\right)\Psi_a\right)+\right.$$
$$+\left(i\hbar\nabla\bar{\Psi}_a-\frac{q_a}{c}\mathbf{A}_b\left(\mathbf{r}_a\right)\bar{\Psi}_a\right)\left(-i\hbar\nabla\Psi_a-\frac{q_a}{c}\mathbf{A}_b\left(\mathbf{r}_a\right)\Psi_a\right)+$$
$$\left.+m_a^2c^2\bar{\Psi}_a\Psi_a\right]dV_a\,dt+$$
$$+\sum_{\substack{a,b\\(a\neq b)}}\mu_a\int\bar{\Psi}_a\left[i\boldsymbol{\alpha}_a\left(\frac{1}{c}\frac{\partial\mathbf{A}_b\left(\mathbf{r}_a\right)}{\partial t}+\nabla\varphi_b\left(\mathbf{r}_a\right)\right)+\right.$$
$$\left.+\boldsymbol{\Sigma}_a\,\mathrm{curl}\;\mathbf{A}_b\left(\mathbf{r}_a\right)\right]\Psi_a\,dV_a\,dt \quad (10.1)$$

The variation of action (10.1) with respect to $\bar{\Psi}_a$ results in the equation, which is similar to the wave equation for an individual particle interacting with the external electromagnetic field:

$$\sum_{b(\neq a)}\left\{\frac{1}{c^2}\left(i\hbar\frac{\partial}{\partial t}-q_a\varphi_b\left(\mathbf{r}_a\right)\right)^{2}-\left[\left(\mathbf{p}_a-\frac{q_a}{c}\mathbf{A}_b\left(\mathbf{r}_a\right)\right)^{2}+\right.\right.$$
$$\left.\left.+m_a^2c^2+2m_a\mu_a\left(i\boldsymbol{\alpha}_a\mathbf{E}_b\left(\mathbf{r}_a\right)-\boldsymbol{\Sigma}_a\mathbf{B}_b\left(\mathbf{r}_a\right)\right)\right]\right\}\Psi_a=0. \quad (10.2)$$

The variation of action with respect to the field potentials results in the equations for the electromagnetic field potentials, in which the current density four vector is the sum of the current density four vectors of the individual particles:

$$\Delta \mathbf{A} - \frac{1}{c^2}\frac{\partial^2 \mathbf{A}}{\partial t^2} = -\frac{4\pi}{c}\sum_a \mathbf{j}_a(\mathbf{r}, t), \tag{10.3}$$

$$\Delta \varphi - \frac{1}{c^2}\frac{\partial^2 \varphi}{\partial t^2} = -4\pi \sum_a \rho_a(\mathbf{r}, t). \tag{10.4}$$

We have used the Lorentz gauge in derivation of the field equations (10.3), (10.4):

$$\frac{1}{c}\frac{\partial \varphi}{\partial t} + \operatorname{div} \mathbf{A} = 0. \tag{10.5}$$

The components of the current density four vector, appearing in the equations (10.3), (10.4), are defined by the following equations

$$\mathbf{j}_a(\mathbf{r}, t) = \frac{q_a}{m_a}\left[\frac{i\hbar}{2}\left(\nabla\bar{\Psi}_a \cdot \Psi_a - \bar{\Psi}_a \nabla \Psi_a\right) - \frac{q_a}{c}\sum_{b(\neq a)} \bar{\Psi}_a \mathbf{A}_b(\mathbf{r}, t)\Psi_a\right] + \\ + c\mu_a \operatorname{curl}\left(\bar{\Psi}_a \boldsymbol{\Sigma}_a \Psi_a\right) - i\mu_a \frac{\partial}{\partial t}\left(\bar{\Psi}_a \boldsymbol{\alpha}_a \Psi_a\right), \tag{10.6}$$

$$\rho_a(\mathbf{r}, t) = \frac{q_a}{m_a c}\left[-\frac{i\hbar}{2c}\left(\frac{\partial \bar{\Psi}_a}{\partial t}\Psi_a - \bar{\Psi}_a \frac{\partial \Psi_a}{\partial t}\right) - \frac{q_a}{c}\sum_{b(\neq a)} \bar{\Psi}_a \varphi_b(\mathbf{r}, t)\Psi_a\right] + \\ + i\mu_a \nabla\left(\bar{\Psi}_a \boldsymbol{\alpha}_a \Psi_a\right). \tag{10.7}$$

The generalized momenta canonically conjugate to the fields $\mathbf{A}$, $\bar{\Psi}$ and Ψ are given by the variational derivatives of the Lagrangian function L with respect to $\dot{\mathbf{A}}$, $\dot{\bar{\Psi}}$ and $\dot{\Psi}$, which we can read off from (10.1) as

$$\begin{aligned} \mathbf{\Pi} &= \frac{\partial L}{\partial \dot{\mathbf{A}}} = \frac{1}{4\pi c}\left(\frac{1}{c}\frac{\partial \mathbf{A}}{\partial t} + \nabla \varphi\right) + i\sum_a \frac{\mu_a}{c}\bar{\Psi}_a \boldsymbol{\alpha}_a \Psi_a, \\ \bar{\Pi}_a &= \frac{\partial L}{\partial \dot{\Psi}_a} = \frac{\hbar}{2m_a c}\left(\frac{\hbar}{c}\frac{\partial \bar{\Psi}_a}{\partial t} - i\frac{q_a}{c}\varphi \bar{\Psi}_a\right), \\ \Pi_a &= \frac{\partial L}{\partial \dot{\bar{\Psi}}_a} = \frac{\hbar}{2m_a c}\left(\frac{\hbar}{c}\frac{\partial \Psi_a}{\partial t} + i\frac{q_a}{c}\varphi \Psi_a\right). \end{aligned} \tag{10.8}$$

The Hamiltonian function is given by the sum of all canonical momenta, times the time-derivatives of the corresponding fields, minus the Lagrangian function

$$H = \mathbf{\Pi}\dot{\mathbf{A}} + \sum_{a} (\bar{\Pi}_a \dot{\Psi}_a + \dot{\bar{\Psi}}_a \Pi_a) - L = \frac{1}{8\pi}\left[\frac{1}{c^2}\left(\frac{\partial \mathbf{A}}{\partial t}\right)^2 - (\nabla\varphi)^2 + (\mathrm{curl}\ \mathbf{A})^2\right] +$$

$$+ \sum_{\substack{a,b \\ (a\neq b)}} \frac{1}{2m_a}\left[\frac{\hbar^2}{c^2}\frac{\partial \bar{\Psi}_a}{\partial t}\frac{\partial \Psi_a}{\partial t} - \frac{q_a^2}{c^2}\varphi_b^2 \bar{\Psi}_a \Psi_a + \right.$$

$$\left. + \left(i\hbar\nabla\bar{\Psi}_a - \frac{q_a}{c}\mathbf{A}_b \bar{\Psi}_a\right)\left(-i\hbar\nabla\Psi_a - \frac{q_a}{c}\mathbf{A}_b\Psi_a\right) + m_a^2 c^2 \bar{\Psi}_a \Psi_a\right] -$$

$$- \sum_{\substack{a,b \\ (a\neq b)}} \mu_a \left(i\bar{\Psi}_a \boldsymbol{\alpha}_a \nabla\varphi_b \Psi_a + \bar{\Psi}_a \mathbf{\Sigma}_a \mathbf{B}_b \Psi_a\right). \quad (10.9)$$

Thus, the energy of ensemble of particles coupled by the electromagnetic field is given by

$$E = \frac{1}{8\pi}\int \left(\mathbf{E}^2 + \mathbf{B}^2\right) dV +$$

$$+ \sum_{\substack{a,b \\ (a\neq b)}} \frac{1}{2m_a c^2}\int \left(-i\hbar\frac{\partial \bar{\Psi}_a}{\partial t} - q_a\varphi_b(\mathbf{r}_a)\bar{\Psi}_a\right)\left(i\hbar\frac{\partial \Psi_a}{\partial t} - q_a\varphi_b(\mathbf{r}_a)\Psi_a\right) dV_a +$$

$$+ \sum_{\substack{a,b \\ (a\neq b)}} \frac{1}{2m_a}\int \bar{\Psi}_a \left[\left(-i\hbar\nabla - \frac{q_a}{c}\mathbf{A}_b(\mathbf{r}_a)\right)^2 + m_a^2 c^2\right] \Psi_a \, dV_a -$$

$$- \sum_{\substack{a,b \\ (a\neq b)}} \mu_a \int \bar{\Psi}_a \mathbf{\Sigma}_a \mathbf{B}_b(\mathbf{r}_a) \Psi_a \, dV_a. \quad (10.10)$$

To derive the equation (10.10) we have used the following vectorial equalities

$$(\nabla\varphi)^2 = -\varphi\Delta\varphi + \mathrm{div}\,(\varphi\nabla\varphi), \quad (10.11)$$

$$\frac{1}{c}\,\mathrm{div}\left(\varphi\frac{\partial \mathbf{A}}{\partial t}\right) = \frac{1}{c}\frac{\partial \mathbf{A}}{\partial t}\nabla\varphi + \frac{1}{c}\varphi\frac{\partial}{\partial t}\,\mathrm{div}\ \mathbf{A} = \frac{1}{c}\frac{\partial \mathbf{A}}{\partial t}\nabla\varphi - \frac{1}{c^2}\varphi\frac{\partial^2\varphi}{\partial t^2}. \quad (10.12)$$

It is seen that the equation (10.10) has the structure similar to that of equation (4.51) for the non-relativistic spin-1/2 particle. The difference between these equations is in the additional terms in (10.10):

$$\int \left(\frac{2m_a c^2}{\hbar^2}\bar{\Pi}_a \Pi_a + \frac{m_a c^2}{2}\bar{\Psi}_a \Psi_a\right) dV.$$

The appearance of the generalized momentum Π_a in the equation (10.10) is quite natural, because the action (10.1) is the quadratic form of the time derivative operator. In the steady-state case these additional terms are

$$\frac{1}{2m_a c^2}\int \bar{\Psi}_a \left[(E_a - q_a\varphi)^2 + m_a^2 c^4\right] \Psi_a \, dV_a.$$

As the result, the equation (10.10) takes the relativistic invariant form, in contrast to the equation (4.51).

10.2 Steady-state case

The ensemble of the primary interest for us is the hydrogen atom, therefore, we assume that the ensemble consists of the two particles. In steady-state case the wave functions of particles are

$$\Psi_a(\mathbf{r}, t) = \Psi_a(\mathbf{r}) \exp\left(-i\frac{E_a t}{\hbar}\right).$$

According to the variational principle the energy functional has extremum at the eigenstates of the particles of ensemble. Therefore, the main goal of our treatment is to find the extremal values of the energy functional (10.10). The energy of the ensemble is the sum of the kinetic energy of particles, the energy of the electromagnetic field produced by them, and the energy of their interaction. As we have discussed in the previous chapters, it is convenient to exclude the electromagnetic field variables from the energy functional, and then vary it with respect to the wave functions of particles. The field variables are excluded with the help of the solutions of the equations for the electromagnetic field potentials (10.3) and (10.4). The field energy is the sum of energies of the electromagnetic fields produced by each particle, and the energy of their interaction, which depends on the mutual position of particles. Notice, that the energy of the field produced by a particle is accounted in its rest energy, therefore for the field energy in the stead-state case we get

$$\begin{aligned} E_f &= \frac{1}{8\pi}\int \left(\mathbf{E}^2 + \mathbf{B}^2\right) dV = \frac{1}{8\pi}\int \left((\nabla\varphi)^2 + (\operatorname{curl}\mathbf{A})^2\right) dV = \\ &= \frac{1}{2}\int \left[\varphi_n(\mathbf{r}_e)\,\rho_e(\mathbf{r}_e) + \frac{1}{c}\mathbf{A}_n(\mathbf{r}_e)\,\mathbf{j}_e(\mathbf{r}_e)\right] dV_e + \\ &+ \frac{1}{2}\int \left[\varphi_e(\mathbf{r}_n)\,\rho_n(\mathbf{r}_n) + \frac{1}{c}\mathbf{A}_e(\mathbf{r}_n)\,\mathbf{j}_n(\mathbf{r}_n)\right] dV_n, \quad (10.13) \end{aligned}$$

where $\varphi_b(\mathbf{r}_a)$ and $\mathbf{A}_b(\mathbf{r}_a)$ are the field potentials produced by the particle $b = (n, e)$ at the position of particle $a = (e, n)$. To derive the equation (10.13) we have used the equation (10.11) and the following formula

$$\operatorname{curl}\mathbf{A} \cdot \operatorname{curl}\mathbf{A} = \operatorname{div}\left[\mathbf{A}\operatorname{curl}\mathbf{A}\right] + \mathbf{A}\operatorname{curl}\operatorname{curl}\mathbf{A}.$$

By substituting the equations (10.6), (10.7) into the equation (10.13) we get for E_f the following equation

$$
\begin{aligned}
E_f = \frac{q_e}{2m_e c^2} \int \bar{\Psi}_e \left(E_e - q_e \varphi_n (\mathbf{r}_e)\right) \varphi_n (\mathbf{r}_e) \Psi_e \, dV_e + \\
+ \frac{q_n}{2m_n c^2} \int \bar{\Psi}_n \left(E_n - q_n \varphi_e (\mathbf{r}_n)\right) \varphi_e (\mathbf{r}_n) \Psi_n \, dV_n - \\
- \frac{i\mu_e}{2} \int \bar{\Psi}_e \boldsymbol{\alpha}_e \nabla_e \varphi_n (\mathbf{r}_e) \Psi_e \, dV_e - \frac{i\mu_n}{2} \int \bar{\Psi}_n \boldsymbol{\alpha}_n \nabla_n \varphi_e (\mathbf{r}_n) \Psi_n \, dV_n - \\
- \frac{q_e}{2m_e c} \int \bar{\Psi}_e \mathbf{A}_n \mathbf{p}_e \Psi_e \, dV_e + \frac{q_n}{2m_n c} \int \bar{\Psi}_n \mathbf{A}_e \mathbf{p}_n \Psi_n \, dV_n - \\
- \frac{q_e^2}{2m_e c^2} \int \bar{\Psi}_e \mathbf{A}_n^2 \Psi_e \, dV_e - \frac{q_n^2}{2m_n c^2} \int \bar{\Psi}_n \mathbf{A}_e^2 \Psi_n dV_n + \\
+ \frac{\mu_e}{2} \int \bar{\Psi}_e \boldsymbol{\Sigma}_e \mathbf{B}_n (\mathbf{r}_e) \Psi_e \, dV_e + \frac{\mu_n}{2} \int \bar{\Psi}_n \boldsymbol{\Sigma}_n \mathbf{B}_e (\mathbf{r}_n) \Psi_n \, dV_n. \quad (10.14)
\end{aligned}
$$

Thus, the total energy of the atom is

$$
\begin{aligned}
E = \frac{E_e}{2m_e c^2} \int \bar{\Psi}_e \left(E_e - q_e \varphi_n\right) \Psi_e \, dV_e + \frac{E_n}{2m_n c^2} \int \bar{\Psi}_n \left(E_n - q_n \varphi_e\right) \Psi_n \, dV_n + \\
+ \frac{1}{2m_e} \int \bar{\Psi}_e \left(\mathbf{p}_e^2 + m_e^2 c^2\right) \Psi_e \, dV_e + \frac{1}{2m_n} \int \bar{\Psi}_n \left(\mathbf{p}_n^2 + m_n^2 c^2\right) \Psi_n \, dV_n - \\
- \frac{q_e}{2m_e c} \int \bar{\Psi}_e \mathbf{A}_n \mathbf{p}_e \Psi_e \, dV_e - \frac{q_n}{2m_n c} \int \bar{\Psi}_n \mathbf{A}_e \mathbf{p}_n \Psi_n \, dV_n - \\
- \frac{i\mu_e}{2} \int \bar{\Psi}_e \boldsymbol{\alpha}_e \nabla_e \varphi_n \Psi_e \, dV_e - \frac{i\mu_n}{2} \int \bar{\Psi}_n \boldsymbol{\alpha}_n \nabla_n \varphi_e \Psi_n \, dV_n - \\
- \frac{\mu_e}{2} \int \bar{\Psi}_e \boldsymbol{\Sigma}_e \mathbf{B}_n \Psi_e \, dV_e - \frac{\mu_n}{2} \int \bar{\Psi}_n \boldsymbol{\Sigma}_n \mathbf{B}_e \Psi_n \, dV_n. \quad (10.15)
\end{aligned}
$$

The total energy of the atom is the sum of the energies of the electron and nucleus $E = E_e + E_n$, therefore we can rewrite the equation (10.15) in the following form

$$
\begin{aligned}
\int \bar{\Psi}_e \left\{ \frac{1}{2m_e} \left(\mathbf{p}_e^2 + \frac{m_e^2 c^4 - E_e^2}{c^2} \right) + \frac{E_e}{2m_e c^2} q_e \varphi_n (\mathbf{r}_e) - \right. \\
\left. - \frac{q_e}{2m_e c} \mathbf{A}_n (\mathbf{r}_e) \mathbf{p}_e - \frac{i\mu_e}{2} \boldsymbol{\alpha}_e \nabla_e \varphi_n (\mathbf{r}_e) - \frac{\mu_e}{2} \boldsymbol{\Sigma}_e \mathbf{B}_n (\mathbf{r}_e) \right\} \Psi_e \, dV_e + \\
+ \int \bar{\Psi}_n \left\{ \frac{1}{2m_n} \left(\mathbf{p}_n^2 + \frac{m_n^2 c^4 - E_n^2}{c^2} \right) + \frac{E_n}{2m_n c^2} q_n \varphi_e (\mathbf{r}_n) - \right. \\
\left. - \frac{q_n}{2m_n c} \mathbf{A}_e (\mathbf{r}_n) \mathbf{p}_n - \frac{i\mu_n}{2} \boldsymbol{\alpha}_n \nabla_n \varphi_e (\mathbf{r}_n) - \frac{\mu_n}{2} \boldsymbol{\Sigma}_n \mathbf{B}_e (\mathbf{r}_n) \right\} \Psi_n \, dV_n = 0. \\
(10.16)
\end{aligned}
$$

By comparing the last equation with the equation (4.55), we can see that the equation (10.16) has the relativistic invariant form and include additional terms describing the interaction of the electron and nucleus electric polarization vectors with the electric field produced by the particle of the opposite charge

$$-\frac{1}{2}\left[\int \mathbf{P}_e \mathbf{E}_n\left(\mathbf{r}_e\right) dV_e + \int \mathbf{P}_n \mathbf{E}_e\left(\mathbf{r}_n\right) dV_n\right].$$

The variation of the equation (10.16) with respect to the functions $\bar{\Psi}_e$ and $\bar{\Psi}_n$ results in the set of the equations for the wave functions Ψ_e and Ψ_n. The solution of these equations together with the equations for the field potentials enable us to determine the energy eigenvalues $E_{e,n}^{(i)}$ and eigenfunctions $\Psi_{e,n}^{(i)}$. For example, in the limit of free particles $|\mathbf{r}_e - \mathbf{r}_n| \to \infty$, we get $E_a = \sqrt{\hbar^2 k_a^2 c^2 + m_a^2 c^4}$. However, in the case of interacting particles we must take into account that the field potentials are the functionals of the wave functions of particles. These functionals are defined by the equations (9.3) and (9.4). As already mentioned, it is more convenient to exclude the field potentials from the functional (10.16) and then vary the obtained functional over the product of the wave functions of the electron and nucleus. By substituting the equations (9.3) and (9.4) into the equation (10.16), we get

$$\begin{aligned}
\int \bar{\Psi}_e \bar{\Psi}_n \Bigg\{ & -\frac{E_n \hbar^2}{2 m_e m_n c^2}\Delta_e - \frac{E_e \hbar^2}{2 m_e m_n c^2}\Delta_n + \frac{m_e^2 c^4 - E_e^2}{2 m_e c^2}\frac{E_n}{m_n c^2} + \\
& + \frac{m_n^2 c^4 - E_n^2}{2 m_n c^2}\frac{E_e}{m_e c^2} + \frac{E_e E_n}{m_e m_n c^4}\frac{q_e q_n}{r} - \frac{q_e^2 q_n^2}{2 m_e m_n c^4}\left(\frac{E_e^2}{m_e c^2} + \frac{E_n^2}{m_n c^2}\right)\frac{1}{r^2} + \\
& + i\mu_e q_n \frac{E_n}{m_n c^2}\frac{\boldsymbol{\alpha}_e \mathbf{r}_{en}}{r^3} - i\mu_n q_e \frac{E_e}{m_e c^2}\frac{\boldsymbol{\alpha}_n \mathbf{r}_{en}}{r^3} - \frac{q_e q_n}{2c^2}\frac{1}{r}\left(\mathbf{v}_n \frac{\mathbf{p}_e}{m_e} + \mathbf{v}_e \frac{\mathbf{p}_n}{m_n}\right) - \\
& - \frac{q_e \mu_n \hbar}{m_e c}\frac{\boldsymbol{\Sigma}_n \mathbf{l}_e}{r^3} - \frac{q_n \mu_e \hbar}{m_n c}\frac{\boldsymbol{\Sigma}_e \mathbf{l}_n}{r^3} + \mu_e \mu_n \frac{3(\boldsymbol{\alpha}_e \mathbf{r}_{en})(\boldsymbol{\alpha}_n \mathbf{r}_{en}) - \boldsymbol{\alpha}_e \boldsymbol{\alpha}_n r^2}{r^5} - \\
& - \mu_e \mu_n \frac{3(\boldsymbol{\Sigma}_e \mathbf{r}_{en})(\boldsymbol{\Sigma}_n \mathbf{r}_{en}) - \boldsymbol{\Sigma}_e \boldsymbol{\Sigma}_n r^2}{r^5} + H_h \Bigg\} \Psi_e \Psi_n = 0, \quad (10.17)
\end{aligned}$$

where $r = |\mathbf{r}_{en}|$, and, similar to section 4.3, v_b is defined by

$$\mathbf{v}_b = \frac{1}{m_b}\left(\mathbf{p}_b - \frac{q_b}{c}\mathbf{A}_a\left(\mathbf{r}_b\right) - \frac{i\hbar\left|\mathbf{r}_a - \mathbf{r}_b\right|}{2}\nabla_b \frac{1}{\left|\mathbf{r}_a - \mathbf{r}_b\right|}\right), \qquad (10.18)$$

The variation of the equation (10.17) with respect to the function $\bar{\Psi}_e \bar{\Psi}_n$ results in the equation for the wave function $\Psi_e \Psi_n$, the Hamiltonian of which coincides with the expression in the braces of (10.17). Let us discuss the physical meaning of each term in the Hamiltonian.

(1) It is well known that the binding energy in hydrogen atom is much smaller than the rest energy of electron and proton. In its turn, $m_e c^2 \ll \ll m_n c^2$. Hence, $E_e \ll E_n$ and $E_n \approx m_n c^2$. By taking into account these inequalities, the first six terms of the equation (10.17), in the limit of the infinitely heavy nucleus, can be transformed to the following form

$$H_C = -\frac{E_n \hbar^2}{2m_e m_n c^2}\Delta_e - \frac{E_e \hbar^2}{2m_e m_n c^2}\Delta_n + \frac{m_e^2 c^4 - E_e^2}{2m_e c^2}\frac{E_n}{m_n c^2} + $$
$$+ \frac{m_n^2 c^4 - E_n^2}{2m_n c^2}\frac{E_e}{m_e c^2} + \frac{E_e E_n}{m_e m_n c^4}\frac{q_e q_n}{r} - \frac{q_e^2 q_n^2}{2m_e m_n c^4}\left(\frac{E_e^2}{m_e c^2} + \frac{E_n^2}{m_n c^2}\right)\frac{1}{r^2} \approx$$
$$\approx -\frac{\hbar^2}{2m_e c^2}\Delta_e + \frac{m_e^2 c^4 - E_e^2}{2m_e c^2} + \frac{E_e}{m_e c^2}\frac{q_e q_n}{r} - \frac{1}{2m_e c^2}\frac{q_e^2 q_n^2}{r^2} =$$
$$= \frac{1}{2m_e c^2}\left[p_e^2 c^2 + m_e^2 c^4 - \left(E_e - q_e \varphi_n\left(\mathbf{r}_e\right)\right)^2\right]. \quad (10.19)$$

Thus, the first six terms of the Hamiltonian are the kinetic energy of the electron and nucleus and the potential energy of their Coulomb interaction. In the limit of the infinitely heavy nucleus these terms coincide with the spin independent part of the Hamiltonian for electron moving in the Coulomb field.

(2) The next two terms

$$H_{PE} = i\mu_e q_n \frac{E_n}{m_n c^2}\frac{\boldsymbol{\alpha}_e \mathbf{r}_{en}}{r^3} - i\mu_n q_e \frac{E_e}{m_e c^2}\frac{\boldsymbol{\alpha}_n \mathbf{r}_{en}}{r^3} \quad (10.20)$$

describe the interaction of the electric polarization vector of a particle with the electric field produced by the particle of the opposite charge.

(3) In the section 4.3 we have already met the following term

$$-\frac{q_e q_n}{2c^2}\frac{1}{r}\left(\mathbf{v}_n \frac{\mathbf{p}_e}{m_e} + \mathbf{v}_e \frac{\mathbf{p}_n}{m_n}\right)$$

With the help of transformation

$$\mathbf{v}_n \mathbf{p}_e = \frac{1}{r_{en}^2}\left\{\left[\mathbf{r}_{en}\mathbf{v}_n\right]\left[\mathbf{r}_{en}\mathbf{p}_e\right] + \left(\mathbf{r}_{en}\mathbf{v}_n\right)\left(\mathbf{r}_{en}\mathbf{p}_e\right) + \frac{i\hbar}{m_n}\left(\mathbf{r}_{en}\mathbf{v}_e\right) + \frac{\hbar^2}{2m_n}\right\},$$
$$\mathbf{v}_e \mathbf{p}_n = \frac{1}{r_{en}^2}\left\{\left[\mathbf{r}_{en}\mathbf{v}_e\right]\left[\mathbf{r}_{en}\mathbf{p}_n\right] + \left(\mathbf{r}_{en}\mathbf{v}_e\right)\left(\mathbf{r}_{en}\mathbf{p}_n\right) - \frac{i\hbar}{m_e}\left(\mathbf{r}_{en}\mathbf{v}_n\right) + \frac{\hbar^2}{2m_e}\right\},$$

and gauge condition

$$\operatorname{div}_a \mathbf{A}_b\left(\mathbf{r}_a\right) = -\frac{q_b}{c}\int \frac{\bar{\Psi}_b \mathbf{r}_{ab}\mathbf{v}_b \Psi_b}{r_{ab}^3}\, dV_b = 0,$$

this term is transformed to the following form

$$H_{ll} = \frac{q_e q_n \hbar^2}{2m_e m_n c^2}\frac{\mathbf{l}_e \mathbf{l}_n + \mathbf{l}_n \mathbf{l}_e - 1}{r^3}, \quad (10.21)$$

where

$$\hbar \mathbf{l}_e = [\mathbf{r}_{en}\mathbf{p}_e], \quad \hbar \mathbf{l}_n = [\mathbf{r}_{ne}\mathbf{p}_n]. \tag{10.22}$$

The last item in the equation (10.21) is due to the fact that the operators $\mathbf{l}_e$ and $\mathbf{l}_n$ are the non-commutative operators (as we mentioned in the section 4.3). The Hamiltonian (10.21) describes the interaction of the orbital angular momenta of the particles.

(4) The next two terms

$$H_{ls} = -\frac{q_e \mu_n \hbar}{m_e c}\frac{\boldsymbol{\Sigma}_n \mathbf{l}_e}{r^3} - \frac{q_n \mu_e \hbar}{m_n c}\frac{\boldsymbol{\Sigma}_e \mathbf{l}_n}{r^3}. \tag{10.23}$$

describe the spin-orbital interaction.

(5) The term

$$H_{PP} = \mu_e \mu_n \frac{3(\boldsymbol{\alpha}_e \mathbf{r}_{en})(\boldsymbol{\alpha}_n \mathbf{r}_{en}) - \boldsymbol{\alpha}_e \boldsymbol{\alpha}_n r^2}{r^5} \tag{10.24}$$

and

$$H_{MM} = -\mu_e \mu_n \frac{3(\boldsymbol{\Sigma}_e \mathbf{r}_{en})(\boldsymbol{\Sigma}_n \mathbf{r}_{en}) - \boldsymbol{\Sigma}_e \boldsymbol{\Sigma}_n r^2}{r^5} \tag{10.25}$$

describe the interaction of the electric and magnetic polarization vectors of particles.

It is seen that the interactions, appearing in the Hamiltonian H_C and H_{PE}, depend on the distance as $1/r$ and $1/r^2$. The rest interactions depend on the distance as $1/r^3$. In the Chapter 4 we have shown that the corrections, contributed by the interactions proportional to $1/r^3$, are about $\mathrm{Ry} \cdot \alpha^2 m_e/m_n$.

We have not discussed the terms contributing the corrections of the highest order in α. In the equation (10.17) these terms are denoted as H_h. They are

$$\begin{aligned} H_h = &-\frac{q_n \varphi_e(\mathbf{r}_n)}{2m_e m_n c^4}\left(p_e^2 c^2 + m_e^2 c^4 - E_e^2\right) - \frac{q_e \varphi_n(\mathbf{r}_e)}{2m_e m_n c^4}\left(p_n^2 c^2 + m_n^2 c^4 - E_n^2\right) + \\ &+ \frac{E_e q_n^2}{2m_e m_n c^4}\frac{q_e}{r}\left(\varphi_e - \frac{E_e}{m_e c^2}\frac{q_e}{r}\right) - \frac{E_n q_e^2}{2m_e m_n c^4}\frac{q_n}{r}\left(\varphi_n - \frac{E_n}{m_n c^2}\frac{q_n}{r}\right) + \\ &+ i\mu_e \frac{\boldsymbol{\alpha}_e \mathbf{r}_{ne}}{r^3}\frac{q_n^2 \varphi_e(\mathbf{r}_n)}{2m_n c^2} + i\mu_n \frac{\boldsymbol{\alpha}_n \mathbf{r}_{en}}{r^3}\frac{q_e^2 \varphi_n(\mathbf{r}_e)}{2m_e c^2} - \\ &- \mu_e \frac{q_n^2}{2m_n c}\frac{\mathbf{A}_e [\mathbf{r}_{ne}\boldsymbol{\Sigma}_e]}{r^3} - \mu_n \frac{q_e^2}{2m_e c}\frac{\mathbf{A}_n [\mathbf{r}_{en}\boldsymbol{\Sigma}_n]}{r^3}. \end{aligned} \tag{10.26}$$

10.3 Integrals of motion

By varying the equation (10.17) with respect to $\bar{\Psi}_e(\mathbf{r}_e)\bar{\Psi}_n(\mathbf{r}_n)$ we get the following equation

$$(H_C + H_{PE} + H_{ll} + H_{ls} + H_{PP} + H_{MM})\Psi_e(\mathbf{r}_e)\Psi_n(\mathbf{r}_n) = 0, \tag{10.27}$$

where we have omitted H_h.

It is convenient to introduce the relative position radius vector $\mathbf{r}$ and the center-of-mass radius vector $\mathbf{R}$:

$$\mathbf{r} = \mathbf{r}_e - \mathbf{r}_n, \quad \mathbf{R} = \frac{E_e \mathbf{r}_e + E_n \mathbf{r}_n}{E_e + E_n}. \tag{10.28}$$

In this case, the orbital angular momentum operator is

$$\hbar \mathbf{L} = [\mathbf{r}_e \mathbf{p}_e] + [\mathbf{r}_n \mathbf{p}_n] = [\mathbf{r}\mathbf{p}] + [\mathbf{R}\mathbf{P}], \tag{10.29}$$

where

$$\mathbf{p} = -i\hbar \frac{\partial}{\partial \mathbf{r}}, \quad \mathbf{P} = -i\hbar \frac{\partial}{\partial \mathbf{R}}.$$

The total angular momentum operator

$$\mathbf{J} = \mathbf{L} + \mathbf{S} \tag{10.30}$$

is the sum of the orbital momentum operator and spin

$$\mathbf{S} = \frac{1}{2}\left(\boldsymbol{\Sigma}_e + \boldsymbol{\Sigma}_n\right). \tag{10.31}$$

By using the definitions (10.28), we obtain

$$\frac{E_n}{m_n c^2} \frac{p_e^2}{2m_e} + \frac{E_e}{m_e c^2} \frac{p_n^2}{2m_n} = -\frac{\hbar^2 E}{2m_e m_n c^2} \left(\Delta_r + \frac{E_e E_n}{E^2} \Delta_R\right). \tag{10.32}$$

In the section 4.3 we have shown that the total angular momentum operator commutes with part of the terms of the Hamiltonian of equation (10.27). This part is $H_C + H_{ll} + H_{ls} + H_{MM}$. It can be easily shown that the rest two terms of the Hamiltonian of equation (10.27) commute also with the operator $\mathbf{J}$. Indeed, it was shown above that the operator $[\mathbf{r}\mathbf{p}] + \frac{\hbar}{2}\boldsymbol{\Sigma}_a$ commutes with the product $\boldsymbol{\alpha}_a \mathbf{E}(\mathbf{r})$. Hence,

$$[\mathbf{J}, H_{PE}] = 0.$$

By taking into account the commutation relations for the operators $\boldsymbol{\Sigma}$ and $\boldsymbol{\alpha}$:

$$[\Sigma_i, \alpha_j] = 2ie_{ijk}\alpha_k,$$

we get

$$\left[\left(\boldsymbol{\Sigma}_e + \boldsymbol{\Sigma}_n\right)_i, \boldsymbol{\alpha}_e \boldsymbol{\alpha}_n\right] = 2ie_{ijk}\left(\boldsymbol{\alpha}_e\right)_k \left(\boldsymbol{\alpha}_n\right)_j + 2ie_{ijk}\left(\boldsymbol{\alpha}_e\right)_j \left(\boldsymbol{\alpha}_n\right)_k = 0.$$

Thus, the operators of the orbital momentum $\mathbf{L}$ and spin $\mathbf{S}$ do not separately commute with the Hamiltonian of the equation (10.27). The conservative variable is the total angular momentum:

$$[\mathbf{J}, (H_0 + H_{PE} + H_{ll} + H_{ls} + H_{PP} + H_{MM})] = 0. \tag{10.33}$$

The Hamiltonian of the equation (10.27) depends only on the radius vector $\mathbf{r} = \mathbf{r}_e - \mathbf{r}_n$, hence, the operator of the total momentum of atom, $\mathbf{P} = \mathbf{p}_e + \mathbf{p}_n$, commutes with the Hamiltonian

$$[\mathbf{P}, (H_0 + H_{PE} + H_{ll} + H_{ls} + H_{PP} + H_{MM})] = 0. \tag{10.34}$$

Hence, the atomic wave functions are the eigenfunctions of the atomic total momentum, total angular momentum, and its projection.

10.4 Angular dependency of hydrogen atom wave functions

In the case of the motionless atom the equation (10.27) takes the form

$$(H_C + H_{PE} + H_{HF} + H_{PP})\, \Psi_e \Psi_n = 0, \tag{10.35}$$

where

$$H_C = -\frac{\hbar^2 E}{2m_e m_n c^2}\Delta - \frac{E_e E_n}{m_e m_n c^4}\frac{Ze^2}{r} - \\ - \frac{1}{2m_e m_n c^4}\left(\frac{E_e^2}{m_e c^2} + \frac{E_n^2}{m_n c^2}\right)\frac{Z^2 e^4}{r^2} + \frac{m_e^2 c^4 E_n + m_n^2 c^4 E_e - E_e E_n E}{2m_e m_n c^4}, \tag{10.36}$$

$$H_{PE} = -i\frac{Ze^2\hbar}{2m_e m_n c^3}\left(E_n \gamma_e \frac{\boldsymbol{\alpha}_e \mathbf{r}}{r^3} - E_e \gamma_n \frac{\boldsymbol{\alpha}_n \mathbf{r}}{r^3}\right), \tag{10.37}$$

$$H_{HF} = -\frac{4\mu_B \mu_N}{r^3}\mathbf{l}^2 + \frac{2\mu_B \mu_N}{r^3}\left(\gamma_e \boldsymbol{\Sigma}_e + \gamma_n \boldsymbol{\Sigma}_n\right)\mathbf{l} + \\ + \frac{\mu_B \mu_N \gamma_e \gamma_n}{r^3}\left(3(\boldsymbol{\Sigma}_e \mathbf{e})(\boldsymbol{\Sigma}_n \mathbf{e}) - \boldsymbol{\Sigma}_e \boldsymbol{\Sigma}_n\right), \tag{10.38}$$

$$H_{PP} = -\frac{\mu_B \mu_N \gamma_e \gamma_n}{r^3}\left(3(\boldsymbol{\alpha}_e \mathbf{e})(\boldsymbol{\alpha}_n \mathbf{e}) - \boldsymbol{\alpha}_e \boldsymbol{\alpha}_n\right), \tag{10.39}$$

where we have used the definitions introduced in the section 4:

$$\mu_e = -\gamma_e \mu_B, \quad \mu_n = \gamma_n \mu_N, \tag{10.40}$$

$\gamma_{e(n)}$ is the gyromagnetic ratio of the electron (nucleus), the Bohr magneton μ_B and nuclear magneton μ_N are defined by the well known equations:

$$\mu_B = -\frac{q_e \hbar}{2m_e c} = \frac{|e|\, h}{2m_e c}, \quad \mu_N = \frac{q_n \hbar}{2m_n c} = \frac{Z\, |e|\, \hbar}{2m_n c}.$$

We have divided the Hamiltonian of the equation (10.35) into the four items. The Hamiltonian H_C is the operator of the kinetic energy and the potential energy of the Coulomb interaction. The Hamiltonian H_{PE} describes the interaction of the electric polarization vectors of particles with the electric field produced by the second particle. The Hamiltonian H_{HF} describes the hyperfine interactions appropriate to the non-relativistic spin-1/2 particle. These interactions were investigated in the Chapter 4. The Hamiltonian H_{PP} describes the interaction of the electric polarization vectors of particles. As already mentioned, the first two terms in the equation (10.35) include the interaction depending on the distance as $1/r$ and $1/r^2$, the second two terms include the interaction depending on distance as $1/r^3$. The Hamiltonian H_C does not depend on the spin operators. The Hamiltonian H_{HF} depends on the diagonal spin operator $\boldsymbol{\Sigma}_a = \begin{pmatrix} \boldsymbol{\sigma}_a & 0 \\ 0 & \boldsymbol{\sigma}_a \end{pmatrix}$. The Hamiltonians H_{PE} and H_{PP} depend on the antidiagonal operator $\boldsymbol{\alpha}_a = \begin{pmatrix} 0 & \boldsymbol{\sigma}_a \\ \boldsymbol{\sigma}_a & 0 \end{pmatrix}$. These specific features explain the convenience of dividing the Hamiltonian into the four groups.

The wave function of the equation (10.35) is the direct product of the bispinor wave functions

$$\Psi = \begin{pmatrix} w_1 \\ w_2 \\ w_3 \\ w_4 \end{pmatrix} = \Psi_e \Psi_n = \begin{pmatrix} \varphi_e \varphi_n \\ \chi_e \varphi_n \\ \varphi_e \chi_n \\ \chi_e \chi_n \end{pmatrix}. \tag{10.41}$$

The Hamiltonian of the equation (10.35) depends only on the radius vector $\mathbf{r}$, hence, in complete analogy with the equation (4.87), for the products of spinors we have

$$w_n(\mathbf{r}) = f_n(r)\, \Omega^{(0)}_{j,l=j,m}(\theta, \varphi) + \sum_{\sigma=-1}^{+1} g^{(n)}_{\sigma}(r)\, \Omega^{(1)}_{j,l=j-\sigma,m}(\theta, \varphi),$$

where the second order spinors $\Omega^{(s)}_{jlm}$ are defined by the equations (4.84)–(4.88).

The matrix of the space inversion transformation is $S_P = \lambda_P \gamma_4$ (see Chapter 7). Hence, the atomic wave function is transformed under space inversion in the following way

$$S^{(e)}_P S^{(n)}_P \begin{pmatrix} w_1 \\ w_2 \\ w_3 \\ w_4 \end{pmatrix} = \lambda^{(e)}_P \lambda^{(n)}_P \begin{pmatrix} w_1 \\ -w_2 \\ -w_3 \\ w_4 \end{pmatrix}.$$

Thus, by taking into account the parity properties of the spinors $\Omega^{(s)}_{jlm}$, the general solution of equation (9.33) has the following form

$$\begin{aligned} w_1(\mathbf{r}) &= f_1(r)\,\Omega^{(0)}_{j,l=j,m}(\theta,\varphi) + g^{(1)}_0(r)\,\Omega^{(1)}_{j,l=j,m}(\theta,\varphi), \\ w_2(\mathbf{r}) &= \sum_{\sigma=\pm1} g^{(2)}_\sigma(r)\,\Omega^{(1)}_{j,l=j-\sigma,m}(\theta,\varphi), \\ w_3(\mathbf{r}) &= \sum_{\sigma=\pm1} g^{(3)}_\sigma(r)\,\Omega^{(1)}_{j,l=j-\sigma,m}(\theta,\varphi), \\ w_4(\mathbf{r}) &= f_4(r)\,\Omega^{(0)}_{j,l=j,m}(\theta,\varphi) + g^{(4)}_0(r)\,\Omega^{(1)}_{j,l=j,m}(\theta,\varphi), \end{aligned} \tag{10.42}$$

In complete analogy with the discussion given in section 8.2, the second linear independent solution is

$$w_1' = w_2, \quad w_2' = w_1, \quad w_3' = w_4, \quad w_4' = w_3.$$

In the Chapter 4 the angular matrix elements were obtained for all items of the Hamiltonian H_{HF}. By using the equations (4.84) and (4.88), we can easily calculate the matrix elements of the Hamiltonian H_{PE}. The non-zero matrix elements are

$$\begin{aligned} \int \Omega^{(1)+}_{j,l=j-1,m}(\boldsymbol{\sigma}_{e,n}\mathbf{e})\,\Omega^{(1)}_{j,l=j,m}\sin\theta\,d\theta\,d\varphi &= -i\sqrt{\frac{j+1}{2j+1}}, \\ \int \Omega^{(1)+}_{j,l=j+1,m}(\boldsymbol{\sigma}_{e,n}\mathbf{e})\,\Omega^{(1)}_{j,l=j,m}\sin\theta\,d\theta\,d\varphi &= i\sqrt{\frac{j}{2j+1}}, \\ \int \Omega^{(1)+}_{j,l=j-1,m}(\boldsymbol{\sigma}_{e}\mathbf{e})\,\Omega^{(0)}_{j,l=j,m}\sin\theta\,d\theta\,d\varphi &= i\sqrt{\frac{j}{2j+1}}, \\ \int \Omega^{(1)+}_{j,l=j+1,m}(\boldsymbol{\sigma}_{e}\mathbf{e})\,\Omega^{(0)}_{j,l=j,m}\sin\theta\,d\theta\,d\varphi &= i\sqrt{\frac{j+1}{2j+1}}, \\ \int \Omega^{(1)+}_{j,l=j-1,m}(\boldsymbol{\sigma}_{n}\mathbf{e})\,\Omega^{(0)}_{j,l=j,m}\sin\theta\,d\theta\,d\varphi &= -i\sqrt{\frac{j}{2j+1}}, \\ \int \Omega^{(1)+}_{j,l=j+1,m}(\boldsymbol{\sigma}_{n}\mathbf{e})\,\Omega^{(0)}_{j,l=j,m}\sin\theta\,d\theta\,d\varphi &= -i\sqrt{\frac{j+1}{2j+1}}, \end{aligned} \tag{10.43}$$

The matrix elements (10.43) and matrix elements obtained in the Chapter 4 enable us to write down now the equations for the radial wave functions $f^{(n)}(r)$ and $g^{(n)}(r)$.

10.5 Equations for radial wave functions

It is convenient to introduce the operator L:

$$H = -\frac{\hbar^2 E}{2m_e m_n c^2} L.$$

Substituting the equations (10.42) into the equation (10.35) and using the angular matrix elements, we get the following equations for the radial wave functions

$$\begin{aligned} L_C w_1 + L_{PE}^{(e)} w_2 + L_{PE}^{(n)} w_3 &= -L_{HF} w_1 - L_{PP} w_4, \\ L_C w_2 + L_{PE}^{(e)} w_1 + L_{PE}^{(n)} w_4 &= -L_{HF} w_2 - L_{PP} w_3, \\ L_C w_3 + L_{PE}^{(e)} w_4 + L_{PE}^{(n)} w_1 &= -L_{HF} w_3 - L_{PP} w_2, \\ L_C w_4 + L_{PE}^{(e)} w_3 + L_{PE}^{(n)} w_2 &= -L_{HF} w_4 - L_{PP} w_1. \end{aligned} \tag{10.44}$$

The labeled operators L, appearing in the equation (10.44), are

$$L_C = \Delta + \frac{2E_e E_n}{E\hbar c}\frac{Z\alpha}{r} + \frac{Z^2\alpha^2}{r^2}\frac{1}{E}\left(\frac{E_e^2}{m_e c^2} + \frac{E_n^2}{m_n c^2}\right) - \\ - \frac{m_e^2 c^4 E_n + m_n^2 c^4 E_e - E_e E_n E}{E\hbar^2 c^2}, \tag{10.45}$$

$$L_{PE} = L_{PE}^{(e)} + L_{PE}^{(n)} = \frac{iZ\alpha}{r^2}\left(\frac{E_n}{E}\gamma_e \boldsymbol{\sigma}_e \mathbf{e} - \frac{E_e}{E}\gamma_n \boldsymbol{\sigma}_n \mathbf{e}\right), \tag{10.46}$$

$$L_{HF} = \frac{Z\alpha^2}{r^3}\frac{m_r c^2}{E a_{\mathrm{B}}}\left(2\mathbf{l}^2 - \gamma_e \boldsymbol{\sigma}_e \mathbf{l} - \gamma_n \boldsymbol{\sigma}_n \mathbf{l} - \frac{\gamma_e \gamma_n}{2}\left(3(\boldsymbol{\sigma}_e \mathbf{e})(\boldsymbol{\sigma}_n \mathbf{e}) - \boldsymbol{\sigma}_e \boldsymbol{\sigma}_n\right)\right), \tag{10.47}$$

$$L_{PP} = \frac{Z\alpha^2}{r^3}\frac{m_r c^2}{E a_{\mathrm{B}}}\frac{\gamma_e \gamma_n}{2}\left(3(\boldsymbol{\sigma}_e \mathbf{e})(\boldsymbol{\sigma}_n \mathbf{e}) - \boldsymbol{\sigma}_e \boldsymbol{\sigma}_n\right), \tag{10.48}$$

where $\boldsymbol{\sigma}_{a,b}$ are the Pauli matrices.

The energy parameters in the equations (10.44)–(10.48) are defined by the conventional equations: $E = E_e + E_n$ and $E_e = m_e c^2 - \Delta E_e$, $E_n = m_n c^2 - \Delta E_n$, where ΔE_a is the binding energy. In the case of the motionless atom, $\mathbf{p}_e + \mathbf{p}_n = 0$, we have $\Delta E_n / \Delta E_e \sim m_e/m_n \ll 1$. It can be easily shown, with the help of the last relationship, that

$$\begin{aligned} &\frac{E_e E_n}{E} \approx \frac{m_r}{m_e} E_e, \quad \frac{1}{E}\left(\frac{E_e^2}{m_e c^2} + \frac{E_n^2}{m_n c^2}\right) \approx 1, \\ &\frac{m_e^2 c^4 E_n + m_n^2 c^4 E_e + E_e E_n E}{E} \approx m_e^2 c^4 - E_e^2, \end{aligned} \tag{10.49}$$

where $m_r = m_e m_n / (m_e + m_n)$ is the reduced electron mass.

Let us start with the zero order approximation. In the zero order approximation we neglect the interactions depending on the distance as $1/r^3$. As we have seen in the previous chapters, in the zero order approximation, in principle, it is possible to find the exact solution of the equations. The equations (10.44) are the coupled set of equations for the eight radial functions $f_{1,4}$, $g_0^{(1,4)}$, $g_{\pm 1}^{(2,3)}$, $g_{\pm 1}^{(2,3)}$. We write down the equations for the radial functions of the spinors w_1 and w_2, because the remaining equations have the same structure. By substituting the equations (10.42) into the equations (10.44) and accounting for the approximations (10.49), we get

$$\left(\frac{d^2}{dr^2}+\frac{2}{r}\frac{d}{dr}+\frac{2E_e m_r}{\hbar c m_e}\frac{Z\alpha}{r}+\frac{Z^2\alpha^2-j\left(j+1\right)}{r^2}-\kappa^2\right)f_1+$$
$$+\gamma_e\frac{Z\alpha}{r^2}\frac{m_r}{m_e}\left(\sqrt{\frac{j+1}{2j+1}}g_{-1}^{(2)}+\sqrt{\frac{j}{2j+1}}g_{+1}^{(2)}\right)+$$
$$+\gamma_n\frac{Z\alpha}{r^2}\frac{m_r}{m_n}\left(\sqrt{\frac{j+1}{2j+1}}g_{-1}^{(3)}+\sqrt{\frac{j}{2j+1}}g_{+1}^{(3)}\right)=0, \quad (10.50)$$

$$\left(\frac{d^2}{dr^2}+\frac{2}{r}\frac{d}{dr}+\frac{2E_e m_r}{\hbar c m_e}\frac{Z\alpha}{r}+\frac{Z^2\alpha^2-j\left(j+1\right)}{r^2}-\kappa^2\right)g_0^{(1)}+$$
$$+\gamma_e\frac{Z\alpha}{r^2}\frac{m_r}{m_e}\left(\sqrt{\frac{j}{2j+1}}g_{-1}^{(2)}-\sqrt{\frac{j+1}{2j+1}}g_{+1}^{(2)}\right)-$$
$$-\gamma_n\frac{Z\alpha}{r^2}\frac{m_r}{m_n}\left(\sqrt{\frac{j}{2j+1}}g_{-1}^{(3)}-\sqrt{\frac{j+1}{2j+1}}g_{+1}^{(3)}\right)=0, \quad (10.51)$$

$$\left(\frac{d^2}{dr^2}+\frac{2}{r}\frac{d}{dr}+\frac{2E_e m_r}{\hbar c m_e}\frac{Z\alpha}{r}+\frac{Z^2\alpha^2-\left(j+1\right)\left(j+2\right)}{r^2}-\kappa^2\right)g_{-1}^{(2)}-$$
$$-\gamma_e\frac{Z\alpha}{r^2}\frac{m_r}{m_e}\left(\sqrt{\frac{j+1}{2j+1}}f_1+\sqrt{\frac{j}{2j+1}}g_0^{(1)}\right)-$$
$$-\gamma_n\frac{Z\alpha}{r^2}\frac{m_r}{m_n}\left(\sqrt{\frac{j+1}{2j+1}}f_4-\sqrt{\frac{j}{2j+1}}g_0^{(4)}\right)=0, \quad (10.52)$$

$$\left(\frac{d^2}{dr^2}+\frac{2}{r}\frac{d}{dr}+\frac{2E_e m_r}{\hbar c m_e}\frac{Z\alpha}{r}+\frac{Z^2\alpha^2-j\left(j-1\right)}{r^2}-\kappa^2\right)g_{+1}^{(2)}-$$
$$-\gamma_e\frac{Z\alpha}{r^2}\frac{m_r}{m_e}\left(\sqrt{\frac{j}{2j+1}}f_1-\sqrt{\frac{j+1}{2j+1}}g_0^{(1)}\right)-$$
$$-\gamma_n\frac{Z\alpha}{r^2}\frac{m_r}{m_n}\left(\sqrt{\frac{j}{2j+1}}f_4+\sqrt{\frac{j}{2j+1}}g_0^{(4)}\right)=0. \quad (10.53)$$

It becomes clear, if we compare the obtained equations with the equations (8.37), (8.38), that the solution of the above eight coupled equations should be taken in the form

$$f_i\left(r\right)=f_{0i}G\left(\nu,r\right), \quad g_\sigma^{(i)}\left(r\right)=g_{0\sigma}^{(i)}G\left(\nu,r\right), \quad (10.54)$$

where

$$G\left(\nu,r\right)=\exp\left(-\kappa r\right)r^{\nu-1}F\left(\nu-\frac{E_e m_r Z\alpha}{\hbar c m_e\kappa},2\nu,2\kappa r\right), \quad (10.55)$$

here $F(p, q, z)$ is the confluent hypergeometric function. The condition of the existence of the non-trivial solutions of the set of algebraic equations for the constants f_{0i}, $g_{0\sigma}^{(i)}$ provides the equation for the free parameter ν.

10.6 Perturbation theory

Up to now, we have taken into account the interactions, that decrease with the distance not faster than $1/r^2$. The contributions of the terms of the Hamiltonian, depending on the distance as $1/r^3$, are relatively small. Indeed, for the ratio of the mean value of the kinetic energy, $\langle K \rangle = \langle (\hbar^2/(2m_r))\, \Delta \rangle$, to the mean value of $\langle H_{HF} + H_{PP} \rangle$, we get

$$\frac{\langle H_{HF} + H_{PP} \rangle}{\langle K \rangle} \approx \frac{Ze^2}{Ea_{\mathrm{B}}} = Z\alpha^2 \frac{m_r}{m_n}.$$

Hence, to account the interactions, depending on the distance as $1/r^3$, we can use the perturbation theory.

The general principles of constructing of the perturbation theory series for the equation (7.3) are the same as for any other quantum mechanical equation. Let we know the solution of the eigenvalue problem

$$H_0(E_n)\, \Psi_n(\mathbf{r}) = 0, \tag{10.56}$$

where H_0 is the Hamiltonian of the equation (7.3) or the Hamiltonian $H_C + H_{PE}$ of the equation (10.35). We would like to calculate the approximated wave functions and energy eigenvalues of the equation

$$(H_0(E) + \delta H)\, \Psi(\mathbf{r}) = 0. \tag{10.57}$$

Let us express the wave function of the equation (10.57) in terms of the eigenfunctions of the equation (10.56)

$$\Psi(\mathbf{r}) = \sum_n c_n \Psi_n(\mathbf{r}). \tag{10.58}$$

By substituting the wave function (10.58) into the equation (10.57), we get

$$\sum_m c_m \frac{E_m - E}{m_0 c^2} \left(\frac{E + E_m}{2} - U(\mathbf{r}) \right) \Psi_m(\mathbf{r}) + \sum_m c_m \delta H \Psi_m(\mathbf{r}) = 0. \tag{10.59}$$

By varying the energy functional of the equation (10.57) in the space of the wave functions (10.58), we can, in principle, to get the solutions of the equation (10.57) at any arbitrary ratio of H_0 and δH. However, when the corrections are small, i.e. $|E - E_n| \ll E_n$, then, by applying the normalization condition (8.20), we get

$$(E_n - E)\, c_n + \sum_m \delta H_{nm} c_m = 0, \tag{10.60}$$

where $\delta H_{nm} = \int \bar{\Psi}_n \delta H \Psi_m \, dV$. If the diagonal elements of the Hamiltonian δH are only non-zero, $\delta H_{nm} = \delta H_{nn} \delta_{nm}$, then the corrected energy eigenvalue E'_n is determined by

$$E'_n = E_n + \delta H_{nn}. \tag{10.61}$$

If the non-diagonal elements of the Hamiltonian δH are non-zero, then, according to (10.60), we obtain the set of the coupled equations. For example, if the Hamiltonian δH couples the two neighboring levels, i.e. $\delta H_{n,n+1} \neq 0$, the energy distance between which is comparable with the mean value of δH, then the set of equations (10.60) is

$$\begin{aligned} (E_n + \delta H_{nn} - E)\, c_n + \delta H_{n,n+1} c_{n+1} &= 0, \\ (E_{n+1} + \delta H_{n+1,n+1} - E)\, c_{n+1} + \delta H_{n+1,n} c_n &= 0. \end{aligned} \tag{10.62}$$

The solutions of these equations are

$$E^{(1,2)} = \frac{E_n + E_{n+1} + \delta H_{nn} + \delta H_{n+1,n+1}}{2} \pm \\ \pm \sqrt{\left(\frac{E_n - E_{n+1} + \delta H_{nn} - \delta H_{n+1,n+1}}{2}\right)^2 + \delta H_{n+1,n} \delta H_{n,n+1}}. \tag{10.63}$$

If the corrections due to the Hamiltonian δH are small in comparison with the energy distance between the coupling levels, i.e. $|\delta H_{nn}| \ll \ll |E_n - E_{n+1}|$, then we can use the following approximated form of the solutions (10.63)

$$\begin{aligned} E^{(1)} &= E_n + \delta E_n \approx E_n + \delta H_{nn} + \frac{\delta H_{n+1,n} \delta H_{n,n+1}}{E_n - E_{n+1}}, \\ E^{(2)} &= E_{n+1} + \delta E_{n+1} \approx E_{n+1} + \delta H_{n+1,n+1} - \frac{\delta H_{n+1,n} \delta H_{n,n+1}}{E_n - E_{n+1}}. \end{aligned}$$

It is seen, that the corrections, due to the cross-interaction of the two levels, move the levels in the opposite sides. The state of the corrected energy $E^{(1)}$ is the superposition of the non-perturbed states n and $n+1$. According to the equations (10.62), for the amplitude of the impurity state $n+1$ we get

$$c_{n+1} = \frac{\delta H_{n+1,n}}{E^{(1)} - E_{n+1} - \delta H_{n+1,n+1}} c_n \approx \frac{\delta H_{n+1,n}}{E_n - E_{n+1}} c_n.$$

Thus, if the non-diagonal elements of the Hamiltonian δH are small in comparison with the energy distance between the coupling levels, then the amplitude of the impurity state is small.

Due to the cumbersome form of appropriate equations, it is impossible here to write down the corrections, associated with the Hamiltonian $H_{HF} + H_{PP}$. Nevertheless, the numerical calculations, based on the application of the above discussed algorithm, result in the hydrogenic spectra, which are in the reasonable good coincidence with the experimental data.

10.7 The case of $j = 0$

In the case of $j = 0$ the number of the equations (10.50)–(10.53) is halved, because instead of the four linear independent spinors $\Omega^{(s)}_{jlm}$ we have only two. In this case, the equations (10.42) become

$$w_1 = f_1\Omega^{(0)}_{000}, \quad w_2 = g^{(2)}_{-1}\Omega^{(1)}_{010}, \quad w_3 = g^{(3)}_{-1}\Omega^{(1)}_{010}, \quad w_4 = f_4\Omega^{(0)}_{000}. \tag{10.64}$$

The second linear independent solution is

$$w_1 = g^{(1)}_{-1}\Omega^{(1)}_{010}, \quad w_2 = f_2\Omega^{(0)}_{000}, \quad w_3 = f_3\Omega^{(0)}_{000}, \quad w_4 = g^{(4)}_{-1}\Omega^{(1)}_{010}. \tag{10.65}$$

If the wave function is given by the equation (10.64), then the parameter ν in the equation (10.55) is

$$\nu_i = \frac{1}{2}\left(1 + \sqrt{1 - 4\gamma_i}\right), \tag{10.66}$$

where

$$\gamma_1 = -1 + Z^2\alpha^2 + \sqrt{1 - Z^2\alpha^2\left(\gamma_e\frac{m_r}{m_e} + \gamma_n\frac{m_r}{m_n}\right)^2},$$

$$\gamma_2 = -1 + Z^2\alpha^2 + \sqrt{1 - Z^2\alpha^2\left(\gamma_e\frac{m_r}{m_e} - \gamma_n\frac{m_r}{m_n}\right)^2},$$

$$\gamma_3 = -1 + Z^2\alpha^2 - \sqrt{1 - Z^2\alpha^2\left(\gamma_e\frac{m_r}{m_e} + \gamma_n\frac{m_r}{m_n}\right)^2},$$

$$\gamma_4 = -1 + Z^2\alpha^2 - \sqrt{1 - Z^2\alpha^2\left(\gamma_e\frac{m_r}{m_e} - \gamma_n\frac{m_r}{m_n}\right)^2}.$$

The radial wave functions (10.54), (10.55) satisfy the boundary conditions at $r = 0$ and $r \to \infty$, if the following condition holds

$$\nu_i - \frac{E_e Z\alpha m_r}{\hbar c m_e \kappa} = -n_r,$$

where n_r is the non-negative integer, which is called by the radial quantum number. The last equation results in the following equation for the energy spectrum

$$E_{n,j=0}^{(i=1,\dots,4)} = \frac{m_e c^2 (n_r + \nu_i)}{\sqrt{(n_r + \nu_i)^2 + (Z\alpha m_r/m_e)^2}}. \tag{10.67}$$

Hence, the electron energy spectrum, $\Delta E = m_e c^2 - E_e$, is defined by

$$\Delta E_{n,j=0}^{(i=1,\dots,4)} = \\ = \frac{m_e c^2}{\sqrt{1 + \left(\frac{Z\alpha m_r}{(n_r + \nu_i) m_e}\right)^2} \left(1 + \sqrt{1 + \left(\frac{Z\alpha m_r}{(n_r + \nu_i) m_e}\right)^2}\right)} \left(\frac{Z\alpha m_r}{(n_r + \nu_i) m_e}\right)^2. \tag{10.68}$$

The eigenfunctions, related to the eigenvalues (10.67), are

$$\Psi_1(\mathbf{r}) = \begin{pmatrix} \Omega_{000}^{(0)} \\ -\varsigma_1 \Omega_{010}^{(1)} \\ -\varsigma_1 \Omega_{010}^{(1)} \\ \Omega_{000}^{(0)} \end{pmatrix} G(\nu_1, r), \quad \Psi_2(\mathbf{r}) = \begin{pmatrix} -\Omega_{000}^{(0)} \\ \varsigma_2 \Omega_{010}^{(1)} \\ -\varsigma_2 \Omega_{010}^{(1)} \\ \Omega_{000}^{(0)} \end{pmatrix} G(\nu_2, r), \tag{10.69}$$

$$\Psi_3(\mathbf{r}) = \begin{pmatrix} \varsigma_1 \Omega_{000}^{(0)} \\ -\Omega_{010}^{(1)} \\ -\Omega_{010}^{(1)} \\ \varsigma_1 \Omega_{000}^{(0)} \end{pmatrix} G(\nu_3, r), \quad \Psi_4(\mathbf{r}) = \begin{pmatrix} \varsigma_2 \Omega_{000}^{(0)} \\ -\Omega_{010}^{(1)} \\ \Omega_{010}^{(1)} \\ -\varsigma_2 \Omega_{000}^{(0)} \end{pmatrix} G(\nu_4, r), \tag{10.70}$$

where $G(\nu_i, r)$ is defined by the equation (10.55), and

$$\varsigma_1 = \frac{Z\alpha \left(\gamma_e \frac{m_r}{m_e} + \gamma_n \frac{m_r}{m_n}\right)}{1 + \sqrt{1 - Z^2\alpha^2 \left(\gamma_e \frac{m_r}{m_e} + \gamma_n \frac{m_r}{m_n}\right)^2}},$$

$$\varsigma_2 = \frac{Z\alpha \left(\gamma_e \frac{m_r}{m_e} - \gamma_n \frac{m_r}{m_n}\right)}{1 + \sqrt{1 - Z^2\alpha^2 \left(\gamma_e \frac{m_r}{m_e} - \gamma_n \frac{m_r}{m_n}\right)^2}}.$$

If the wave function is given by the equation (10.65), then we again obtain the equation (10.67) for the energy spectrum, and explicit form of the wave functions is

$$\Psi_5(\mathbf{r}) = \begin{pmatrix} \varsigma_1\Omega^{(1)}_{010} \\ -\Omega^{(0)}_{000} \\ -\Omega^{(0)}_{000} \\ \varsigma_1\Omega^{(1)}_{010} \end{pmatrix} G(\nu_1, r), \quad \Psi_6(\mathbf{r}) = \begin{pmatrix} \varsigma_2\Omega^{(1)}_{010} \\ -\Omega^{(0)}_{000} \\ \Omega^{(0)}_{000} \\ -\varsigma_2\Omega^{(1)}_{010} \end{pmatrix} G(\nu_2, r). \quad (10.71)$$

$$\Psi_7(\mathbf{r}) = \begin{pmatrix} \Omega^{(1)}_{010} \\ -\varsigma_1\Omega^{(0)}_{000} \\ -\varsigma_1\Omega^{(0)}_{000} \\ \Omega^{(1)}_{010} \end{pmatrix} G(\nu_3, r), \quad \Psi_8(\mathbf{r}) = \begin{pmatrix} -\Omega^{(1)}_{010} \\ \varsigma_2\Omega^{(0)}_{000} \\ -\varsigma_2\Omega^{(0)}_{000} \\ \Omega^{(1)}_{010} \end{pmatrix} G(\nu_4, r). \quad (10.72)$$

Similar to the solutions of the problem on the electron motion in the Coulomb field, the solutions (10.69)–(10.72) are twice degenerated with the respect to the pairs of the particles and antiparticles.

10.8 Internal parity

The Dirac adjoint function to the wave function (10.41) is $\bar{\Psi} = (w_1^+, -w_2^+, -w_3^+, w_4^+)$, then, by taking into account the inequalities $\varsigma_{1,2} \ll 1$, we get $\bar{\Psi}_{1,2,7,8}\Psi_{1,2,7,8} > 0$ and $\bar{\Psi}_{3,4,5,6}\Psi_{3,4,5,6} < 0$. In the case of the electron motion in the Coulomb field, the solutions, corresponding to the electron, have been chosen on the basis of the normalization condition, it reads for electron solutions as $\bar{\Psi}\Psi > 0$. The solutions of the two-particle problem (10.44) with $\bar{\Psi}\Psi > 0$ correspond to the pair of particles or antiparticles. If $\bar{\Psi}\Psi < 0$, then the two-particle wave function is the product of the one-particle wave functions, one of which is particle wave function and another is antiparticle ones. The transformation 'particle-antiparticle' is realized by the matrices $\gamma_5^{(e)}$ and $\gamma_5^{(n)}$. The action of these matrices on the two-particle wave function is defined by

$$\gamma_5^{(e)} \begin{pmatrix} w_1 \\ w_2 \\ w_3 \\ w_4 \end{pmatrix} = - \begin{pmatrix} w_2 \\ w_1 \\ w_4 \\ w_3 \end{pmatrix}, \quad \gamma_5^{(n)} \begin{pmatrix} w_1 \\ w_2 \\ w_3 \\ w_4 \end{pmatrix} = - \begin{pmatrix} w_3 \\ w_4 \\ w_1 \\ w_2 \end{pmatrix}.$$

It is seen, that there are the following relationships between the wave functions, corresponding to the same energy eigenvalue,

$$\gamma_5^{(e)}\Psi_1 = \Psi_5, \quad \gamma_5^{(n)}\Psi_1 = \Psi_5, \quad \gamma_5^{(e)}\Psi_3 = \Psi_7, \quad \gamma_5^{(n)}\Psi_3 = \Psi_7,$$

$$\gamma_5^{(e)}\Psi_2 = -\Psi_6, \quad \gamma_5^{(n)}\Psi_2 = \Psi_6, \quad \gamma_5^{(e)}\Psi_4 = -\Psi_8, \quad \gamma_5^{(n)}\Psi_4 = \Psi_8.$$

Thus, the matrices $\gamma_5^{(e)}$ and $\gamma_5^{(n)}$ transform the wave functions (10.69), (10.70) to the same final wave functions, however, the sign of the final wave functions is the same for the wave functions of the odd indexes, and sign is opposite for the wave functions of the even indexes.

Let us introduce the operator of the internal parity

$$\Gamma_5 = \gamma_5^{(e)}\gamma_5^{(n)}. \tag{10.73}$$

This operator transforms each particle wave function into the antiparticle wave function. The action of this operator on the wave function (10.41) is defined by

$$\gamma_5^{(e)}\gamma_5^{(n)} \begin{pmatrix} w_1 \\ w_2 \\ w_3 \\ w_4 \end{pmatrix} = \begin{pmatrix} w_4 \\ w_3 \\ w_2 \\ w_1 \end{pmatrix}.$$

By applying this operator to the wave functions (10.69)–(10.72), we get

$$\Gamma_5\Psi_{1,3,5,7} = \Psi_{1,3,5,7}, \quad \Gamma_5\Psi_{2,4,6,8} = -\Psi_{2,4,6,8}. \tag{10.74}$$

It should be noted, that the operators $\gamma_5^{(e)}$ and $\gamma_5^{(n)}$ commute separately with the Hamiltonian of the two-particle problem. The operator Γ_5 commutes with the two-particle Hamiltonian, too. Hence, the obtained symmetry properties remain invariable, even in the presence of the external electromagnetic field. This is an extremely important statement. Indeed, it is seen from the equations (10.66), (10.67), that the energy spectra of the two-particle systems, having the different internal symmetry, are different. However, the external electromagnetic field could not change the internal symmetry of the two-particle system, therefore these two different energy spectra correspond to the two different physical objects.

We have shown in the Chapter 7, that the matrix γ_5 is the matrix of the CPT-transformation, hence, the internal symmetry defines the parity with respect to the CPT-transformation. The difference in the internal structure of the two-particle system of the different internal parity can be illustrated by the following way. Let us determine the radial projections of the electric polarization vectors

$$\mathbf{P}_a^{(i)} = -i\mu_a\bar{\Psi}_i\boldsymbol{\alpha}\Psi_i$$

in the states, described by the wave functions (10.69) and (10.72). They are

$$d_e^{(1)} = \int \mathbf{e}\mathbf{P}_e^{(1)} dV = -4\mu_{\mathrm{B}}\gamma_e\varsigma_1, \quad d_n^{(1)} = \int \mathbf{e}\mathbf{P}_n^{(1)} dV = -4\mu_N\gamma_n\varsigma_1,$$
$$d_e^{(2)} = \int \mathbf{e}\mathbf{P}_e^{(2)} dV = -4\mu_{\mathrm{B}}\gamma_e\varsigma_2, \quad d_n^{(2)} = \int \mathbf{e}\mathbf{P}_n^{(2)} dV = 4\mu_N\gamma_n\varsigma_2,$$
$$d_e^{(3)} = \int \mathbf{e}\mathbf{P}_e^{(3)} dV = 4\mu_{\mathrm{B}}\gamma_e\varsigma_1, \quad d_n^{(3)} = \int \mathbf{e}\mathbf{P}_n^{(3)} dV = 4\mu_N\gamma_n\varsigma_1,$$
$$d_e^{(4)} = \int \mathbf{e}\mathbf{P}_e^{(4)} dV = 4\mu_{\mathrm{B}}\gamma_e\varsigma_2, \quad d_n^{(4)} = \int \mathbf{e}\mathbf{P}_n^{(4)} dV = -4\mu_N\gamma_n\varsigma_2,$$

where $\mathbf{e} = \mathbf{r}/r$. In the two-particle states of the even internal parity the directions of the electric polarization vectors of the constituent particles coincide, and in the states of the odd internal parity the directions of the electric polarization vectors of the constituent particles are opposite.

It is seen that the solutions (10.69) and (10.71) correspond to the nS states, and solutions (10.70) and (10.72) correspond to the nP states of the two-particle problem. In the nS states the vector $\mathbf{d}_e$ is directed oppositely to the direction of the intra-atomic field, in the nP states it is directed along the intra-atomic field. In the atomic systems of the even internal parity the vector $\mathbf{d}_n$ is parallel to the vector $\mathbf{d}_e$, in the atomic systems of the odd internal parity the vector $\mathbf{d}_n$ is antiparallel to the vector $\mathbf{d}_e$. In the two-particle systems, for which $|\mu_e| \gg |\mu_n|$ (like in hydrogen atom), the nS states move upward, and the nP states move downward. The difference in the value of shift of the nS states for the atomic systems of the different internal parity in given by

$$\frac{\Delta E(nS)}{m_e c^2} = \frac{(n_r + \nu_1)}{\sqrt{(n_r + \nu_1)^2 + (Z\alpha m_r/m_e)^2}} - \frac{(n_r + \nu_2)}{\sqrt{(n_r + \nu_2)^2 + (Z\alpha m_r/m_e)^2}} =$$
$$= \frac{2\gamma_e\gamma_n}{n^3}\frac{m_r}{M}(Z\alpha)^4 + \frac{2\gamma_e\gamma_n}{n^5}\frac{m_r}{M}\times$$
$$\times \left[4n^2 + 6n - 3 - \left(\left(\gamma_e\frac{m_r}{m_e}\right)^2 + \left(\gamma_n\frac{m_r}{m_n}\right)^2\right)n(n+3)\right](Z\alpha)^6 + \ldots, \tag{10.75}$$

for the nP states we have

$$\frac{\Delta E(nP)}{m_e c^2} = \frac{(n_r + \nu_3)}{\sqrt{(n_r + \nu_3)^2 + (Z\alpha m_r/m_e)^2}} - \frac{(n_r + \nu_4)}{\sqrt{(n_r + \nu_4)^2 + (Z\alpha m_r/m_e)^2}} =$$
$$= -\frac{2\gamma_e\gamma_n}{3n^3}\frac{m_r}{M}(Z\alpha)^4 - \frac{2\gamma_e\gamma_n}{27n^5}\frac{m_r}{M}\times$$
$$\times \left[4n^2 + 18n - 27 + \left(\left(\gamma_e\frac{m_r}{m_e}\right)^2 + \left(\gamma_n\frac{m_r}{m_n}\right)^2\right)n(11n+9)\right](Z\alpha)^6 + \ldots, \tag{10.76}$$

where $M = m_e + m_n$, $n = n_r + 1$ is the principle quantum number for the nS states, and $n = n_r + 2$ is the principle quantum number for the nP states. By taking into account, that $m_e c^2 \alpha^2 = 2\text{Ry}$, it can be easily seen, that the leading term of expansion, lowest order in α, is proportional to $\left(Z^4\alpha^2 m_r/(n^3 M)\right)$ Ry.

The position of the energy levels, defined by the equation (10.68), depends on the ratio m_e/m_n. Let us estimate the shifts of the levels (10.68) with respect to the Dirac's spectrum (6.133) in the limit of the infinitely heavy nucleus $m_n \to \infty$. For the even atomic systems we get

$$\frac{\Delta E_{\text{even}}(nS)}{m_e c^2} = \frac{Z^4\alpha^4}{2n^3}\left[\left(\gamma_e + \gamma_n \frac{m_e}{m_n}\right)^2 - 1\right] + \dots,$$

$$\frac{\Delta E_{\text{even}}(nP)}{m_e c^2} = -\frac{Z^4\alpha^4}{6n^3}\left[\left(\gamma_e + \gamma_n \frac{m_e}{m_n}\right)^2 - 1\right] + \dots$$

For the odd atomic systems we have

$$\frac{\Delta E_{\text{odd}}(nS)}{m_e c^2} = \frac{Z^4\alpha^4}{2n^3}\left[\left(\gamma_e - \gamma_n \frac{m_e}{m_n}\right)^2 - 1\right] + \dots,$$

$$\frac{\Delta E_{\text{odd}}(nP)}{m_e c^2} = -\frac{Z^4\alpha^4}{6n^3}\left[\left(\gamma_e - \gamma_n \frac{m_e}{m_n}\right)^2 - 1\right] + \dots$$

Taking in mind that the nucleus magnetic moment is equal to $\mu_n = \gamma_n \mu_N$, the ratio m_e/m_n was kept non-zero only in the terms proportional to $\mu_e \pm \mu_n$. Comparing the last equations with the equations (9.88), we can see that the account for the nucleus spin results, in the lowest order approximation, in the replacement of the electron magnetic moment by the summary magnetic moment $|\mu_e| + |\mu_n|$ for the even atomic systems, and by the difference magnetic moment $|\mu_e| - |\mu_n|$ for the odd atomic systems.

Thus, the analysis presented here has shown that the choice between the two wave functions, corresponding the same energy eigenvalue, is determined by the internal structure of the atomic system, whether it consists of the two particles or particle-antiparticle pair. The choice of the appropriate spectral series, among two alternative, is unambiguously prescribed by the internal parity of the atomic system.

References

[1] de Brogile L. *Comptes Rendus.* **177**, 5–7 (1923); *Nature* **112**, 540 (1923)

[2] Schrödinger E. *Ann. Phys.* **79**, 361, 489; **80**,437; **81**, 109 (1926)

[3] Pauli W. *Z. f. Phys.* **36**, 336 (1926)

[4] Dirac P.A.M. *Proc. Roy. Soc.* **A117**, 610 (1928)

[5] Bethe H.A. and Salpeter E.E. *Quantum mechanics of one- and two- electron atom.* Springer-Verlag, Berlin-Göttingen-Heidelberg, 1957.

[6] Sapirstein J.R. and Yennie D.R. In: *Quantum Electrodynamics*, ed. By T. Kinoshita, World Scientific, Singapure, 1990.

[7] Sobelman I.I. *Atomic spectra and radiative transitions.* Springer Series on Atoms and Plasmas, Vol. 12. Springer-Verlag, Berlin-Heidelberg-NY, 1991.

[8] Beyer H.F., Kluge H.-J., and Shevelko V.P. *X-Ray radiation of highly charged ions.* Springer Series on Atoms and Plasmas, Springer-Verlag, Berlin-Heidelberg-NY, 1997.

[9] Mohr P.J. and Taylor B.N. *Rev. Mod. Phys.* **72**, 351 (2000)

[10] de Beauvoir B., Schwob C., Acef O., et al. *Eur. Phys. J.* D **12**, 61 (2000)

[11] Eides M.I., Grotch H., and Shelyuto V.A. *Phys. Reports.* **342**, 63 (2001)

[12] Bauch A. and Telle H.R. *Rep. Prog. Phys.* **65**, 789 (2002)

[13] Kinoshita T. *Rep. Prog. Phys.* **59**, 1459 (1996)

[14] Flowers J.L., and Petley B.W. *Rep. Prog. Phys.* **64**, 1191 (2001)

[15] Dehmelt H. "Experiments with rest isolated subatomic particle", *Nobel lectures in Physics* — 1989, The Nobel Foundation (1990)

[16] Brown L.S. and Gabrielse G. *Rev. Mod. Phys.* **58**, 233 (1986)

[17] Van Dyck R.S., Jr., Schwinberg P.B., and Dehmelt H.G. *Phys. Rev. Lett.* **59**, 26–29 (1987)

[18] Van Dyck R.S., Jr., Schwinberg P.B., and Dehmelt H.G., in *The Electron*, edited by D. Hestenes and A. Weingartshofer (Kluwer Academic, Netherlands), pp. 239–293 (1991)

[19] Van Dyck R.S., Jr., Moore F.L., Farnham D.L., and Schwinberg P.B. *Bull. Am. Phys. Soc.* **31**, 244–244 (1986)

[20] Farnham D.L., Van Dyck R.S., Jr., and Schwinberg P.B. *Phys. Rev. Lett.* **75**, 3598–3601 (1995)

[21] Niering M., Holwarth R., Reichert J., Pokasov P., Udem Th., Weitz M., Hansch T.W., et al. *Phys. Rev. Lett.* **84**, 5496 (2000)

[22] Maas F.E., et al. *Phys. Lett.* A **187**, 247–254 (1994)

[23] Fee M.S., Chu S., Mills A.P., Jr., Chichester R.J., Zuckerman D.M., Shaw E.D., and Danzmann K., *Phys. Rev.* A **48**, 192–219, (1993)

[24] Lee T.D. and Yang C.N. *Phys. Rev.* **104**, 254 (1956)

[25] Landau L.D. *JETP.* **5**, 405 (1957)

[26] Christenson J.H., Cronin J.W., Fitch V.L., and Turlay R. *Phys. Rev. Lett.* **13**, 138 (1964)

[27] Khriplovich I.B., Lamoreaux S.K. *CP violation without strangeness.* Springer-Verlag, Berlin-Heidelberg-NY, 1977

[28] Commins E.D. *CP violation in atomic and nuclear physics.* In: *Advances in Atomic, Molecular and Optical physics*, v. 40, Academic Press, 1999.

[29] Purcell E.M. and Ramsey N.F. *Phys. Rev.* **78**, 807 (1950); Smith J.H., Purcell E.M., and Ramsey N.F. *Phys. Rev.* **108**, 120 (1957)

[30] Ramsey N.F. *Phys. Rep.* **43**, 410 (1978)

[31] "*The neutron electric dipole moment (EDM) experiment at psi*", http:\\ucn.web.psi.ch.

[32] Schwinger J. *Phys. Rev.* **73**, 407 (1948)

[33] Gerasimov S.B., Lebedev A.I., and Petrunkin V.A. *JETP.* **43**, 1872 (1962)

[34] Shull C.G. *Phys. Rev. Lett.* **10**, 297 (1963)

[35] Aleksandrov Yu.A. *JETP.* **33**, 294 (1957)

[36] Foldy L.L. *Phys. Rev. Lett.* **3**, 105 (1959)

[37] Thaler R.M. *Phys. Rev.* **114**, 827 (1959)

[38] Walt M. and Fossan D.B. *Phys. Rev.* **137**, B629 (1965)

[39] Elwyn A.J., Monahan J.E., Lane R.O., Langsdorf A., Jr., and Mooring F.P. *Phys. Rev.* **142**, 758 (1966)

[40] Schiff L.I. *Phys. Rev.* **132**, 2194 (1963)

[41] Sandars P.G.H. *Phys. Lett.* **14**, 194 (1965)

[42] Sandars P.G.H. *Phys. Lett.* **22**, 290 (1966)

[43] Commins E.D., Ross S.B., DeMille D., and Regan B.C. *Phys. Rev.* A **50**, 2960 (1994)

[44] Regan B.C., Commins E.D., Schmidt C.J., DeMille D. *Phys. Rev. Lett.* **88**, 071805-1 (2002)

[45] Jacobs J.P., Klipstein W.M., Lamoreaux S.K., Heckel B.R., and Fortson E.N. *Phys. Rev.* A **52**, 3521 (1995)

[46] Rosenberry M.A. and Chupp T.E. *Phys. Rev. Lett.* **86**, 22 (2001)

[47] Lande A., *Z. f. Phys.* **5**, 231 (1921); **15**, 189 (1923); **19**, 112 (1923)

[48] Compton A.H., *J. Frank. Inst.* **192**, 145 (1921)

[49] Uhlenbeck G.E., and Goudsmith S., *Naturwiss.* **13**, 953 (1925)

[50] Klein O. *Z. f. Phys.* **37**, 895 (1926)

[51] Fock V. *Z. f. Phys.* **38**, 242 (1926)

[52] Gordon W. *Z. f. Phys.* **40**, 117 (1926)

[53] Sommerfeld A. *Ann. Phys.* **51**, 1, 125 (1916)

[54] Andreev A.V. *JETP.* **116**, 793 (1999)

[55] Darwin C.G. *Proc. Roy. Soc.* **A118**, 654; **A120**, 621 (1928)

[56] Gordon W. *Z. f. Phys.* **48**, 11 (1928)

[57] Lamb W.E.,Jr. and Retherford R.C. *Phys. Rev.* **72**, 241 (1947)

[58] Andreev A.V. *Theory of spin-1/2 particles and hyperfine structure of the atomic spectra.* Fizmatlit, Moscow, 2003 [In Russian].

[59] Andreev A.V. *Las. Phys. Lett.* **1**, 69 (2004)

Index